Nucleic Acids and Molecular Biology

Volume 38

Series Editor

Allen W. Nicholson, Department of Biology, Temple University, Philadelphia, PA, USA

Since its inception in 1987, Nucleic Acids and Molecular Biology has emerged as a high-visibility series that brings to the scientific community in-depth analyses of cutting-edge topics in the field. Each volume offers a focused set of reviews, written by leading experts, that provides a comprehensive, multi-perspective analysis of the chosen topic. As such, Nucleic Acids and Molecular Biology serves as an enduring research resource for scientists, academicians, and students alike.

Akio Kanai • Allen W. Nicholson
Editors

Molecular Evolution of RNA Regulatory Enzymes and Their Systems

Springer

Editors
Akio Kanai
Institute for Advanced Biosciences
Keio University
Tsuruoka, Japan

Allen W. Nicholson
Department of Biology
Temple University
Philadelphia, PA, USA

ISSN 0933-1891 ISSN 1869-2486 (electronic)
Nucleic Acids and Molecular Biology
ISBN 978-3-032-03299-7 ISBN 978-3-032-03300-0 (eBook)
https://doi.org/10.1007/978-3-032-03300-0

This Springer imprint is published by the registered company Springer Nature Switzerland AG
The registered company address is: Gewerbestrasse 11, 6330 Cham, Switzerland

Preface

With the advent of high-throughput next-generation sequencing technologies, the genomic sequences of an ever-increasing range of organisms have been decoded, resulting in the accumulation of vast amounts of data on protein amino acid sequences and RNA nucleotide sequences. Genomic technologies and methods of sequence analysis continue to advance rapidly, and are now allowing large-scale comparative analyses—spanning thousands to tens of thousands of species—rather than being limited to traditional model organisms. As a result, researchers are now better equipped to identify lineages that have undergone substantial molecular evolution, particularly when investigating key evolutionary transitions of specific proteins or RNA molecules.

In the twenty-first century, biology at the RNA level has gained significant attention, driven by ongoing discoveries of numerous functional non-coding RNAs and gene regulatory systems, including CRISPR. Moreover, owing to these advances, high-resolution pictures have emerged of the evolutionary trajectories of RNA-related enzymes and RNA-binding proteins—key participants in the central dogma (DNA to RNA to protein).

Evolution occurs at multiple hierarchical levels, including the amino acid sequences of proteins, the organization of functional domains, higher-order structural configurations, and regulatory network architectures. The contributors to this volume were encouraged to explicitly consider the level(s) of evolution in their analyses of selected proteins or regulatory systems. Naturally, integrative discussions that span multiple levels were strongly encouraged and welcomed.

This volume comprises 11 reviews that focus on the molecular evolution of RNA-related enzymes, RNA-binding proteins, and RNA regulatory systems. The scope is not confined to newly discovered RNA regulatory factors or systems that have only recently emerged. Rather, the primary aim is to revisit and highlight RNA-associated proteins and regulatory mechanisms that play essential and fundamental roles in gene expression.

Dr. Akira Hirata examines the evolutionary history of RNA polymerase II from its archaeal origins; Drs. Sota Yagi and Shunsuke Tagami elucidate ancient protein folds found in RNA polymerase and ribosomal proteins; Dr. Naoyuki Kataoka discusses the regulation of exon recognition by serine-arginine-rich splicing factors and heterogeneous nuclear ribonucleoproteins in eukaryotes; Dr. George H. Jones explores the evolution of bacterial tRNA nucleotidyltransferases; Dr. Federica Briani analyzes the evolutionary origin and conservation of polyribonucleotide phosphorylase across bacteria and eukaryotes; Dr. Béatrice Clouet-d'Orval and coauthors present new insights into RNA processing and decay pathways in archaea, with a focus on the involvement of the archaeal-specific Ski2-like RNA helicases.

In addition, Dr. Anna Kurzynska-Kokorniak and coauthors provide a comparative investigation of RNase III enzymes and their accessory domains; Dr. Teppei Morita reviews conserved and diverse roles of bacterial small RNAs; Dr. Yangjie Chao and coauthors present an analysis of the molecular evolution of bacterial RNase E and its associated systems; Dr. Yoshizumi Ishino provides a comprehensive overview of the diversity and evolution of CRISPR–Cas systems; and Dr. Marcos de la Peña and coauthors analyze small self-cleaving ribozymes and their involvement in genome dynamics and gene regulation in eukaryotes.

Together, these contributions provide a multifaceted evolutionary perspective on fundamental RNA-related molecules and regulatory systems, encompassing amino acid sequences, domain architectures, tertiary structures, and complex regulatory frameworks. As editors, we are confident that this volume will serve as a valuable resource for advancing a systematic understanding of genetic regulation at the molecular level. We would be especially gratified if this book inspires readers to investigate their own proteins or RNA regulatory systems from a similar evolutionary perspective.

Tsuruoka, Japan — Akio Kanai
Philadelphia, PA, USA — Allen W. Nicholson

Contents

Part I Evolution of Transcription and Translation Systems

1 From Simplicity to Complexity: The Evolution of RNA Polymerase II from Its Archaeal Ancestor 3
Akira Hirata

2 Ancient Protein Folds in RNA Polymerase and Ribosomes 33
Sota Yagi and Shunsuke Tagami

Part II Evolution of RNA Processing, Modification, and Decay Mechanisms

3 Regulation of Exon Recognition by Serine-Arginine Rich Splicing Factors and Heterogeneous Ribonucleoproteins in Higher Eukaryotes 65
Naoyuki Kataoka

4 Evolution of Bacterial tRNA Nucleotidyltransferases 79
George H. Jones

5 Polyribonucleotide Phosphorylase: An Ancient Enzyme Conserved in Bacteria and Eukarya 97
Federica Briani

6 RNA Processing and Decay Pathways in Archaea: Emerging Insights on the Function and Evolution of the Archaeal Specific Ski2-Like RNA Helicases 123
Charles Cirio, Maël Ansart, Emmanuelle Schmitt, Marie Bouvier, and Béatrice Clouet-d'Orval

Part III Evolution of Small RNA-Mediated Regulations

7 RNase III Enzymes and Their Accessory Domains 161
Agnieszka Szczepanska, Klaudia Wojcik, and Anna Kurzynska-Kokorniak

8 Conserved and Diverse Functions of Bacterial Small RNAs 203
Teppei Morita

9 Molecular Basis and Evolution of Bacterial RNase E and Its Systems 225
Fang Liu, Philippe Icyishaka, Sarah L. Svensson, Masatoshi Miyakoshi, and Yanjie Chao

10 CRISPR–Cas Systems ~ Diversity and Evolution ~ 257
Yoshizumi Ishino

11 Small Self-Cleaving Ribozymes in Eukaryotes 283
Marcos de la Peña, Olga Rueda, and Amelia Cervera

Part I
Evolution of Transcription and Translation Systems

Chapter 1
From Simplicity to Complexity: The Evolution of RNA Polymerase II from Its Archaeal Ancestor

Akira Hirata

Abstract Archaeal RNA polymerase (RNAP) is widely recognized as the structural and functional ancestor of eukaryotic RNA polymerase II (Pol II). This chapter examines the evolutionary diversification of RNAP subunits and associated transcription factors, with emphasis on their roles in the emergence of complex regulatory systems. Key molecular components—Rpo3, Rpo4/7, TFE, Spt4/5, and TFS—are analyzed in terms of their conserved core structures and lineage-specific adaptations. Comparative structural studies reveal that, while the catalytic architecture of RNAP has been preserved, regulatory domains and additional functionalities have been incrementally acquired. Notably, the incorporation of an iron–sulfur (Fe–S) cluster in Rpo3 and the expansion of KOW domains in the elongation factor Spt5 are interpreted as critical steps toward coupling transcription with RNA processing. These findings support the view that increasing transcriptional complexity evolved through elaboration upon ancient molecular scaffolds. A conceptual framework is thus proposed to trace the evolutionary transition from archaeal transcription to the eukaryotic gene regulatory machinery.

Keywords RNA polymerase evolution · Archaeal transcription · Eukaryotic transcription · Transcription factors · Fe–S cluster · Domain architecture evolution · Structural biology

---

Supplementary Information The online version contains supplementary material available at https://doi.org/10.1007/978-3-032-03300-0_1.

A. Hirata (✉)
Department of Natural Science, Division of Science and Technology, Graduate School of Science and Technology for Innovation, Tokushima University, Tokushima, Japan
e-mail: ahirata@tokushima-u.ac.jp

A. Kanai, A. W. Nicholson (eds.), *Molecular Evolution of RNA Regulatory Enzymes and Their Systems*, Nucleic Acids and Molecular Biology 38,
https://doi.org/10.1007/978-3-032-03300-0_1

1.1 Introduction

Transcription is a fundamental biological process conserved across all domains of life. It governs the accurate conversion of genetic information into functional RNA molecules. While bacterial RNAP is a multi-subunit complex comprising five core subunits, it is architecturally simpler than the more intricate RNAPs found in archaea and eukaryotes. Archaeal RNAP and eukaryotic RNA polymerases—including RNA polymerase I (Pol I), RNA polymerase II (Pol II), and RNA polymerase III (Pol III)—share a highly similar multi-subunit organization, reflecting their evolutionary kinship. Among them, archaeal RNAP most closely resembles eukaryotic Pol II in both subunit composition and overall architecture.

1.2 Background

Seminal studies in the 1980s by Wolfgang Zillig and colleagues first uncovered the striking structural and functional similarities between archaeal RNAP and eukaryotic Pol II (Zillig 1981, 1989). These discoveries laid the foundation for the now widely accepted hypothesis that the eukaryotic transcriptional apparatus originated from an archaeal ancestor. Unlike the simpler bacterial system, both archaeal and eukaryotic RNAPs exhibit a complex multi-subunit organization built around a conserved catalytic core.

Subsequent structural and biochemical investigations have deepened our understanding of the similarities and distinctions between archaeal RNAP and Pol II (Hirata et al. 2008b; Korkhin et al. 2009; Werner and Grohmann 2011). While both systems feature a core of approximately 12 subunits, eukaryotic Pol II has evolved additional elements, such as the carboxy-terminal domain (CTD) of Rpb1. This CTD serves as a platform for coupling transcription with RNA processing events including capping, splicing, and polyadenylation (Eick and Geyer 2013). Moreover, transcription initiation in archaea requires only a minimal set of basal transcription factors—TATA-binding protein (TBP), transcription factor B (TFB), and transcription factor E (TFE). In contrast, eukaryotic transcription involves a more elaborate array of general transcription factors (GTFs) and coactivators, such as the Mediator complex (Geiger 2010; Malik and Roeder 2010).

The evolutionary expansion of transcriptional complexity is reflected not only in the acquisition of additional regulatory factors but also in the diversification of the RNAP subunits themselves. Subunits such as Rpo3 and Rpo7, and associated factors involved in transcription initiation and elongation—including Spt4/5, TFE, and TFS—have undergone substantial functional adaptation from their archaeal origins. Recent advances in cryo-electron microscopy (cryo-EM) have revolutionized our understanding of RNAP structure (Cramer 2021). High-resolution cryo-EM reconstructions of archaeal RNAP and eukaryotic Pol II have revealed detailed subunit organization, conformational states, and interaction networks that were previously inaccessible through crystallography alone (Bernecky et al. 2016, 2020). These insights provide a robust molecular framework for studying the evolutionary trajectory of transcription systems.

1.3 Objective of This Chapter

This chapter aims to discuss the structural and functional evolution of transcriptional machinery from archaeal RNAP to eukaryotic Pol II. Particular focus is placed on subunits Rpo3, Rpo7, and the elongation and initiation factors TFE, Spt4/5, and TFS. By integrating structural comparisons, biochemical insights, and recent cryo-EM findings, we explore how these components contributed to the emergence of complex gene regulation and multicellular life. Comprehensive reviews by Werner and colleagues have highlighted the evolutionary conservation and diversification of RNAP subunits across bacteria, archaea, and eukaryotes, supporting the notion of a common ancestry for transcriptional machinery (Werner and Grohmann 2011; Blombach and Werner 2022).

1.4 Overview of Archaeal RNAP and Transcription Mechanism

Archaeal RNAP is essential for the transcription of genetic information and closely resembles in structure and function eukaryotic Pol II. These observations support the idea that the core transcription machinery of eukaryotes evolved from an archaeal ancestor. The archaeal RNAP is a multi-subunit enzyme composed of approximately 12 subunits, organized around a highly conserved catalytic core. As summarized in Table 1.1, the subunit composition of archaeal RNAP closely parallels that of eukaryotic Pol II. High-resolution structural studies, including X-ray crystallography and cryo-electron microscopy (cryo-EM), have revealed that the catalytic core adopts a double-psi beta-barrel (DPBB) fold that is universally conserved across all domains of life (Mushegian and Koonin 2020; Koonin 2021). The overall architecture of archaeal RNAP and eukaryotic Pol II is shown in Fig. 1.1. Both enzymes are primarily composed of two major architectural modules: a central catalytic core and a peripheral stalk subcomplex. In archaea, the stalk is formed by the Rpo4 and Rpo7 subunits, while in eukaryotes, the analogous Rpb4 and Rpb7 subunits form the stalk (Armache et al. 2005). Notably, such a stalk structure is absent in bacterial RNAPs, underscoring a fundamental difference between archaeal/eukaryotic transcription systems and the bacterial machinery. The stalk subcomplexes are positioned on the surface of the core and contribute to the interaction with transcription factors and RNA during transcription. Detailed functional aspects of the stalk will be discussed in a later section. The transcription cycle mediated by

Table1.1 Subunit composition of archaeal RNA polymerase (RNAP) and eukaryotic RNA polymerase II (Pol II)

Archaea	Eukaryotic Pol II
Rpo1 (A)	Rpb1
(Rpo1´(A´)+1´´(A´´))*	
Rpo2 (B)**	Rpb2
Rpo3 (D)	Rpb3
Rpo11 (L)	Rpb11
Rpo6 (K)	Rpb6
Rpo5 (H)	Rpb5
Rpo7 (E)	Rpb7
Rpo4 (F)	Rpb4
Rpo10 (N)	Rpb10
Rpo12 (P)	Rpb12
Rpo8 (G)	Rpb8
Rpo13***	Rpb9

This table summarizes the subunit composition of archaeal RNAP and its correspondence to subunits of eukaryotic Pol II. Each subunit is color-coded as shown in the corresponding figures.
**In Archaea, two nomenclature systems for subunits are commonly used.*
**Most Rpo1 subunits are split into two separate polypeptides.*
***In methanogenic archaea, Rpo2 is divided into two distinct subunits.*
****Rpo13 is conserved in crenarchaeal RNAP. This subunit is absent from eukaryotic RNA polymerases.*

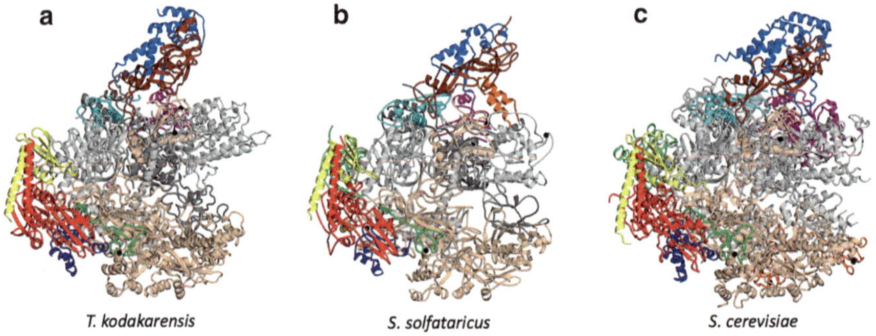

Fig. 1.1 Ribbon diagrams of X-ray crystal structures of RNA polymerases from *Thermococcus kodakarensis* (**a**, PDB: 4QIW), *Sulfolobus solfataricus* (**b**, PDB:3HKZ), and RNA polymerase II from *Saccharomyces cerevisiae* (**c**, PDB:1WCM). Each subunit is depicted in a distinct color consistent with Table 1.1: Rpo1, Rpo1′ and Rpb1 (gray), Rpo2 and Rpb2 (wheat), Rpo3 and Rpb3 (red), Rpo4 and Rpb4 (brown), Rpo5 and Rpb5 (purple), Rpo6 and Rpb6 (cyan), Rpo7 and Rpb7 (blue), Rpo8 and Rpb8 (green), Rpo13 and Rpb9 (orange), Rpo10 and Rpb10 (dark blue), Rpo11 and Rpb11 (yellow) and Rpo12 and Rpb12 (light green)

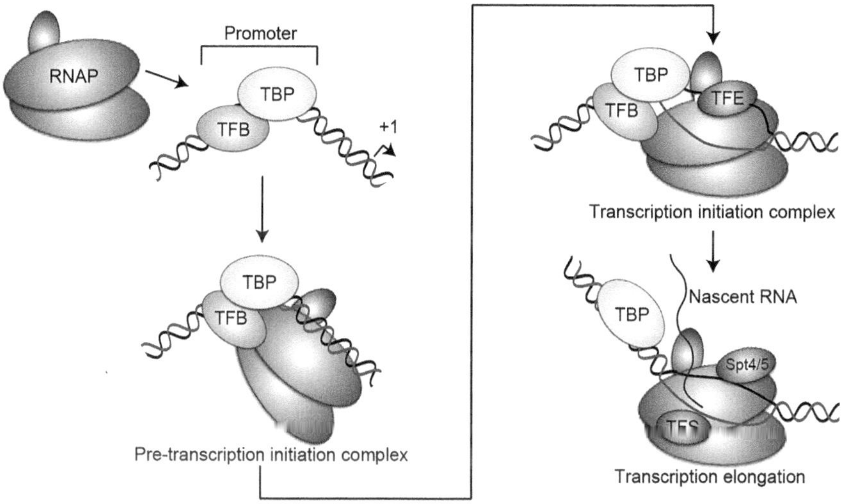

Fig. 1.2 Schematic representation of the promoter recognition and transcription elongation process by archaeal RNA polymerase (RNAP). The transcription start site is designated as +1

archaeal RNAP is depicted in Fig. 1.2. Transcription initiation begins with recognition of promoter elements by the TATA-binding protein (TBP) and transcription factor B (TFB), and then recruits RNAP (Hausner et al. 1996; Bell et al. 1999). The DNA duplex exhibits a pronounced bend at the TATA box, located approximately 20–30 base pairs upstream of the transcription start site (+1), induced by TBP binding. This conformational change facilitates the binding of TFB, which in turn promotes the recruitment of RNAP to the promoter region. Transcription factor E (TFE) helps with promoter melting and open complex stabilization (Ouhammouch et al. 2004; Grünberg et al. 2007; Werner and Grohmann 2011). Following initiation, RNAP transitions into the elongation phase, where the Spt4/Spt5 complex binds to RNAP to increase transcriptional processivity (Chen et al. 2009; Hirtreiter et al. 2010a). Moreover, the transcript cleavage factor TFS binds to the active site to resolve transcriptional backtracking events and maintain fidelity (Lange and Hausner 2004). Orthologs of these archaeal transcription factors are conserved in the eukaryotic Pol II transcription system, where they perform analogous functions in promoter recognition, elongation control, and transcript proofreading (Table 1.2). Recent structural analyses, including the open-clamp structure of *Thermococcus kodakarensis* RNAP (PDB: 4QIW) (Jun et al. 2014), the RNAP structure of *Sulfolobus solfataricus* (PDB: 3HKZ) (Hirata et al. 2008b), and the complete Pol II structure from *Saccharomyces cerevisiae* (PDB: 1WCM) (Armache et al. 2005), have revealed the mechanistic conservation and evolutionary adaptation of the transcription machinery. These structural features show that the dynamic movements of the clamp domain and the peripheral stalk work together to regulate transcription initiation and elongation. In summary, archaeal RNAP represents an evolutionary bridge between simple bacterial transcription machinery to the highly regulated

Table 1.2 Transcription initiation and elongation factors in archaeal RNAP and eukaryotic Pol II

	Archaea	Eukaryotic Pol II
Initiation	TBP	TBP (TFIID)
	TFB	TFIIB
	TFE*	TFIIE
		TFIIF
		TFIIA
		TFIIH
Elongation	Spt4	Spt4
	Spt5	Spt5
	TFS	TFIIS

Note: *This list does not include all transcription factors known in eukaryotic systems*
**Most archaeal TFE proteins are homologous to the α subunit of eukaryotic TFIIE; however, in some species, TFE also includes a β-like subunit*

eukaryotic transcription system. Its structurally conserved architecture and fundamental regulatory features provide insight into the evolutionary origins of complex gene expression programs.

1.5 Structural Conservation and Functional Diversification of Rpo3

1.5.1 Overview of Rpo3 and Rpb3 Subunits

Archaeal RNAP and eukaryotic Pol II share a fundamentally conserved core architecture. Rpo3 (in archaea) and Rpb3 (in eukaryotes) are orthologous subunits critical for RNAP assembly and structural integrity. Rpo3 forms a dimer with Rpo11, analogous to the Rpb3-Rpb11 dimer in Pol II. This heterodimer stabilizes the RNAP structure and is critical for its proper function.

1.5.2 Structural Conservation and Divergence

Structures of archaeal RNAPs from *Thermococcus kodakarensis* (Tko, PDB: 4QIW) and *Sulfolobus solfataricus* (Sso, PDB: 3HKZ), and eukaryotic Pol II from *Saccharomyces cerevisiae* (Sce, PDB: 1WCM), reveal that Rpo3 and Rpb3 maintain a conserved overall fold (Figs. 1.3a–c). These structures show a core dimerization domain and a second conserved domain (domain 2) that together interface with the RNAP core. In domain 2, Sso Rpo3 features four conserved cysteine residues forming two disulfide bonds, contributing to structural stabilization, whereas Sce Rpb3 coordinates a zinc ion through conserved cysteine and histidine residues,

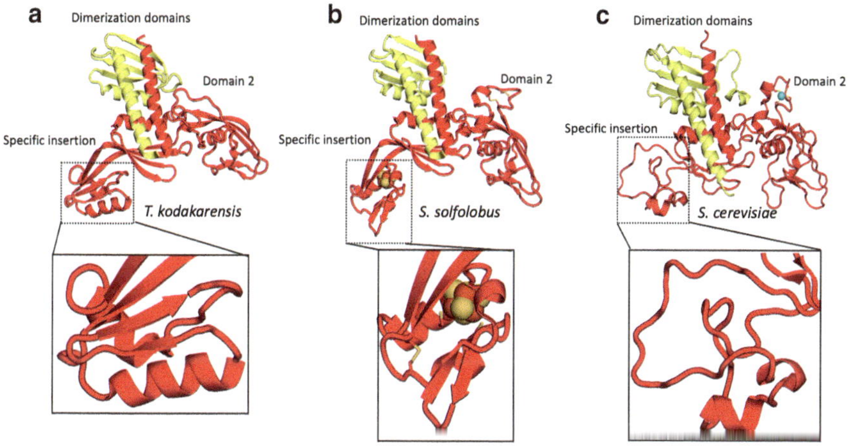

Fig. 1.3 Ribbon diagrams of the Rpo3/Rpo11 (**a**, *Thermococcus kodakarensis*, PDB: 4QJV; **b**, *Sulfolobus solfataricus*, PDB: 2PA8) and Rpb3/11 (**c**, *Saccharomyces cerevisiae*, PDB: 1WCM) subcomplexes. Rpo3/Rpb3 is colored red and Rpo11/Rpb11 yellow, with all domain boundaries labeled. A close-up view highlights the species-specific insertion, revealing distinct structural arrangements in each complex. Metal ions are shown as balls: Zn (cyan), Fe (brown), and S (gold)

reinforcing the domain architecture. Tko Rpo3 lacks these metal-binding features but retains the basic fold necessary for RNAP assembly. Supplementary Figs. 1.1, 1.2, and 1.3 highlight the striking conservation of the overall fold between archaeal and eukaryotic Rpo3/Rpb3.

1.5.3 Specific Insertion: Domain Architecture

Although overall conservation is maintained, each Rpo3/Rpb3 exhibits a specific insertion at a similar position, with distinct structural features.

- In Tko, the specific insertion forms a small loop-like structure composed of two α-helices and four β-strands (Fig. 1.3a, Supplementary Fig. 1.1). It is a compact domain without an iron-sulfur (Fe-S) cluster and is mainly surface-exposed.
- In Sso, the specific insertion forms a ferredoxin-like domain coordinating a [4Fe-4S] cluster (likely present *in vivo*, although crystallographic data suggest [3Fe-4S]) (Fig. 1.3b, Supplementary Figs. 1.1 and 1.4). Four conserved cysteine residues coordinate the cluster, supporting the notion that this insertion stabilizes the Rpo3/Rpo11 dimer and contributes to the structural stabilization of the entire RNAP architecture. Importantly, the formation of the [4Fe-4S] cluster is essential for the structural integrity, while two additional conserved cysteine residues are involved in forming a disulfide bond, which further stabilizes the specific insertion region

- In Sce, the specific insertion of Rpb3 forms a loosely packed loop-like structure composed of β-strands and α-helices (Fig. 1.2c, Supplementary Figs. 1.2 and 1.3). It does not contain an Fe-S cluster and is completely surface-exposed, indicating a potential role as an interaction platform for transcriptional regulators.

Supplementary Figs. 1.4 and 1.5 further highlight the variations in cysteine conservation and iron-sulfur cluster coordination across archaeal and eukaryotic species (Hirata and Murakami 2009). Supplementary Fig. 1.5 represents a phylogenetic tree based on the motifs shown in Supplementary Fig. 1.4, and the Rpo3 and Rpb3 proteins can be classified into four categories: (1) those without an iron-sulfur cluster motif, (2) those with two [4Fe-4S] clusters, (3) those with one [4Fe-4S] cluster, and (4) those lacking a specific insertion altogether. In *Sulfolobus* species, six conserved cysteine residues are present: four coordinate a [4Fe-4S] cluster and two form a disulfide bond, contributing to structural stabilization. In contrast, methanogenic archaea such as *Methanosarcina* species (highlighted in red boxes) preserve eight cysteine residues, enabling the coordination of two [4Fe-4S] clusters. This structural arrangement suggests increased oxygen sensitivity and the potential for functioning as an oxygen sensor involved in transcriptional regulation, as indicated by experimental and biochemical studies in *Methanosarcina acetivorans* (Lessner et al. 2012; Jennings et al. 2017). Importantly, proteins with a single [4Fe-4S] cluster motif, such as those in Sso, pathogenic microorganisms' AC40, and some Rpb3 orthologs, primarily contribute to structural stabilization and RNAP assembly. Interestingly, higher plants and certain pathogenic eukaryotic microorganisms maintain four conserved cysteine residues in their AC40 subunit of Pol I and III, implying the possibility of coordinating a single [4Fe-4S] cluster, paralleling the situation in archaeal systems.

1.5.4 Functional Diversification and Evolution

Supplemental Figs. 4 and 5 show that specific insertions vary among archaeal species: some coordinate Fe-S clusters (e.g., *Methanosarcina* species Rpo3), while others do not. Cross-linking experiments have also demonstrated that Rpb3 interacts with the Mediator complex through its Med17 subunit, implying that the specific insertion may contribute to this interaction (Soutourina et al. 2011). Furthermore, recent cryo-EM structures of the yeast PIC–Mediator complex revealed extensive contacts between Pol II and Mediator, particularly mediated by the C-terminal domain (CTD) of Pol II (Rengachari et al. 2021; Schilbach et al. 2021). Although direct structural evidence for the involvement of the Rpb3-specific insertion is lacking, its surface exposure suggests a potential role in facilitating interactions with Mediator. In *M. acetivorans*, the Rpo3-specific insertion coordinates two [4Fe-4S] clusters, which are crucial for the stability of the Rpo3/Rpo11 heterodimer and the overall assembly of RNAP. Disruption of the Fe-S clusters significantly impairs RNAP assembly, emphasizing the functional importance of these clusters in certain archaeal lineages. These findings suggest that the [4Fe-4S] clusters may act as oxygen sensors, linking environmental oxygen levels to transcriptional regulation in Methanosarcina. Thus, the specific insertion in archaeal Rpo3 initially contributed to structural stabilization, and in methanogenic archaea such as *M. acetivorans*, it may have evolved to function as an oxygen sensor involved in transcriptional regulation through Fe-S cluster coordination. In contrast, the eukaryotic Rpb3-specific insertion has diverged to serve as a versatile platform facilitating dynamic transcriptional regulation.

1.5.5 Concluding Remarks

The comparative analysis of Rpo3 in archaea and Rpb3 in eukaryotes reveals how a highly conserved subunit scaffold has accommodated lineage-specific innovations. Despite the preservation of a shared dimerization core with Rpo11/Rpb11, structural differences in the specific insertion region—especially the acquisition of iron-sulfur clusters in certain archaea—highlight adaptive mechanisms for environmental responsiveness and RNAP stability. In eukaryotes, the loss of Fe-S coordination in Rpb3 and its structural simplification coincided with the repurposing of the specific insertion domain into a surface-exposed platform that potentially engages regulatory factors, including components of the Mediator complex. This shift reflects an evolutionary trend in which structural elements originally contributing to enzyme stability are adapted to serve broader regulatory functions. Together, these observations highlight the evolutionary adaptability of Rpo3/Rpb3 and demonstrate how conserved structural cores evolve to fulfill diverse regulatory roles in increasingly complex transcriptional environments.

1.6 Structural and Functional Evolution of Rpo4/Rpo7 and Rpb4/Rpb7

1.6.1 Overview of Rpo4/Rpo7 in Archaea

In archaeal RNAP, the Rpo4/Rpo7 heterodimer forms a stalk-like extension that is critical for RNAP stability and function (Fig. 1.4a). Structural studies from *T. kodakarensis* and *S. solfataricus* have shown that this stalk facilitates RNA exit and contributes to the architectural integrity of the RNAP complex (Figs. 1.1a, b). The Rpo4 and Rpo7 subunits both adopt α/β folds typical of nucleic acid-binding proteins. As illustrated in Fig. 1.1a, they jointly form a prominent stalk extending from the main RNAP body. Based on sequence alignments shown in Supplementary Fig. 1.6, the *T. kodakarensis* Rpo4 possesses two defined structural domains: an N-terminal tip-associated domain and a C-terminal HRDC-like domain (Helicase

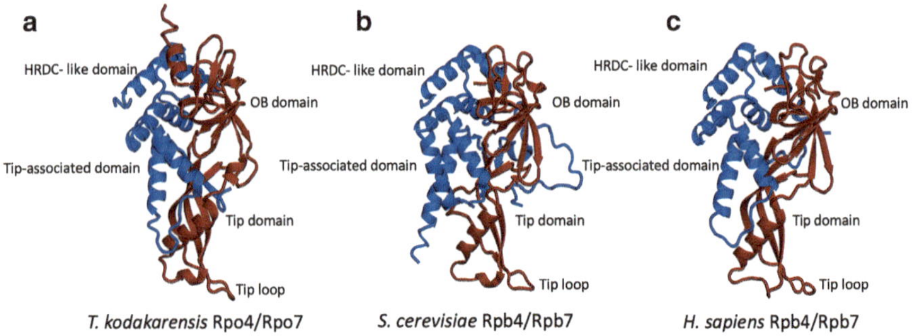

Fig. 1.4 Structural comparison of archaeal and eukaryotic RNAP stalk subcomplexes. Ribbon diagrams of the Rpo4/Rpo7 (**a**, PDB: 4QJF) from *Thermococcus kodakarensis,* the Rpb4/Rpb7 (**b**, PDB: 2C35) from *Saccharomyces cerevisiae*, and the Rpb4/Rpb7 (**c**, PDB: 1WCM) from *Homo sapiens* are shown. In each panel, Rpo4/Rpb4 is colored blue and Rpo7/Rpb7 brown. Key domains—the oligonucleotide/oligosaccharide-binding (OB) fold, Tip-associated domain, tip loop, tip domain, and HRDC-like domain—are labeled

and RNase D C-terminal-like domain) (Jun et al. 2014). The Tip-associated domain forms the distal region of the stalk, while the HRDC-like domain contributes to interactions with Rpo7 and to overall structural stability. Rpo7, meanwhile, consists of α-helical repeats and a central OB-fold domain—a five-stranded β-barrel structure commonly associated with RNA binding (Hirtreiter et al. 2010b). This OB-fold stabilizes the Rpo4/Rpo7 heterodimer through extensive contacts with Rpo4 and likely contributes to RNA handling near the exit channel. A functionally important element of Rpo7 is the tip loop, located at the distal end of the stalk and adjacent to the OB-fold. Structural studies have shown that this loop is disordered in isolation but folds upon interaction with the RNAP core, mediating contacts with subunits such as Rpo1 and Rpo6. Although the exact amino acid sequence of the tip loop is not strictly conserved among archaeal species, several positively charged or polar residues—critical for RNA binding and interaction—are functionally preserved (Supplementary Fig. 1.7). Although no direct interaction with transcription regulators has been confirmed in archaea, the evolutionary conservation of the stalk architecture implies a functional readiness for expansion, which is realized in eukaryotic systems. Interestingly, recent studies suggest that the Rpo4/Rpo7 heterodimer may also contribute to the recruitment or stabilization of general transcription factors such as TFE (Jun et al. 2020), especially under physiological stress conditions. These functional implications will be further explored in subsequent sections.

1.6.2 Rpb4/Rpb7 in Eukaryotes

In eukaryotic Pol II, the Rpb4 and Rpb7 subunits form a heterodimer analogous to the archaeal Rpo4/Rpo7 stalk. This subcomplex, known as the stalk domain, is located at the periphery of the Pol II core (Fig. 1.1c) and plays a crucial role not only in stabilizing the polymerase structure but also in regulating transcriptional responses under various environmental and developmental conditions (Garrido-Godino et al. 2022b). Structurally, Rpb4 and Rpb7 adopt similar folds to their archaeal counterparts. Rpb7 contains a conserved OB-fold domain, characteristic of RNA-binding proteins, while Rpb4 is composed predominantly of α-helical elements that interface tightly with Rpb7 (Supplementary Fig. 1.8). Cryo-EM and crystallographic studies have shown that this region is flexible and dynamic, potentially allowing interaction with various transcriptional regulators (Vos et al. 2018; Cramer 2021). Importantly, Rpb4/Rpb7 is not essential for basal transcription in vitro, but it is required for transcriptional responses to stress and during specific stages of cell differentiation. For example, in Saccharomyces cerevisiae, deletion of RPB4 results in temperature-sensitive phenotypes, defects in mRNA export, and impaired stress response gene expression (Harel-Sharvit et al. 2010; Garrido-Godino et al. 2022a, b). These phenotypes suggest that the Rpb4/Rpb7 complex contributes to the integration of transcription with post-transcriptional regulatory processes. The Rpb7 tip loop region, located adjacent to the OB-fold, has been shown to mediate interactions with the Pol II core and is involved in RNA binding. Although the amino acid sequence of this loop varies among eukaryotic species, structural studies indicate a conserved functional role. Supplementary Fig. 1.7 shows the structure and sequence alignment of Rpb7, highlighting key residues in the OB-fold and tip loop that contribute to its RNA-binding capability and core polymerase interaction. Together, Rpb4/Rpb7 has evolved beyond a structural role to become a multifunctional regulatory platform within the transcription machinery. This functional diversification is examined further in the next sections through detailed structural and phylogenetic comparisons.

1.6.3 Structural Comparison of Rpo4/Rpo7 and Rpb4/Rpb7

The archaeal Rpo4/Rpo7 and eukaryotic Rpb4/Rpb7 heterodimers share a conserved overall architecture, forming a stalk-like extension from the core RNAP complex. Despite differences in sequence length and domain boundaries, both subcomplexes adopt a similar spatial organization that positions them adjacent to the RNA exit channel, as shown in Fig. 1.2 and Supplementary Figs. 1.6, 1.7, and 1.8. Structurally, the most prominent shared feature is the presence of an OB-fold domain in Rpo7 and Rpb7. This β-barrel motif mediates RNA binding and contributes to interactions with other subunits of the core polymerase. In both archaeal and eukaryotic systems, the OB-fold is positioned at the base of the stalk and forms extensive contacts with Rpo4 or Rpb4. This interaction stabilizes the stalk heterodimer and anchors it to the polymerase core (Armache et al. 2005; Jun et al. 2014). Notably, Rpb7 possesses a longer and more elaborate tip loop region than its archaeal counterpart, with several positively charged residues involved in interactions with the RNA transcript and with core Pol II subunits such as Rpb1 and Rpb6 (Armache et al. 2003; Meka et al. 2005). This structural difference likely reflects the increased regulatory complexity of eukaryotic transcription, where additional protein–protein and protein–RNA interactions are required. Rpo4 and Rpb4 also exhibit important structural differences. While both contain predominantly α-helical domains, Rpb4 includes additional surface features and charged patches that may facilitate interactions with post-transcriptional machinery, including components involved in mRNA export and decay (Harel-Sharvit et al. 2010). As illustrated in Supplementary Figs. 1.7 and 1.8, the surface charge distribution and secondary structure elements in Rpb4/Rpb7 are more diversified compared to their archaeal counterparts, suggesting an evolutionary adaptation toward broader regulatory functions. Taken together, the conserved OB-fold architecture and stalk orientation in both archaeal and eukaryotic systems point to a common evolutionary origin. However, structural elaborations—particularly in the Rpb7 tip loop and Rpb4 surface—demonstrate the evolutionary transition from a basal transcriptional scaffold in archaea to a multifunctional regulatory module in eukaryotes.

1.6.4 Functional Diversification of the Stalk Subunits

While the Rpo4/Rpo7 heterodimer in archaea and its eukaryotic counterpart Rpb4/Rpb7 share structural conservation, their functional roles have diverged significantly through evolution. In archaea, the stalk subcomplex primarily serves a structural role, stabilizing the RNAP core and directly interacting with the general transcription factor TFE (Jun et al. 2020). Structural studies have shown that the Rpo4/Rpo7 heterodimer forms a binding surface for TFE, facilitating its recruitment to the RNAP and contributing to transcription initiation under physiological conditions. In contrast, Rpb4/Rpb7 in eukaryotic Pol II has acquired additional regulatory functions that extend far beyond transcription initiation. In archaea, although the Rpo4/Rpo7 heterodimer is not essential for in vitro transcription activity, its biological significance becomes evident under physiological conditions. In *T. kodakarensis*, deletion of the gene encoding Rpo4 (*ΔrpoF*) results in a severe growth defect at elevated temperatures, indicating a critical role for Rpo4 in thermophilic adaptation (Hirata et al. 2008a). Furthermore, RNAP purified from the *ΔrpoF* strain lacks co-purification of the general transcription factor TFE, reinforcing structural evidence that Rpo4/Rpo7 directly mediates TFE recruitment and stabilization during transcription initiation. A key area of diversification lies in the regulation of gene expression under stress conditions. Studies in *S. cerevisiae* have shown that Rpb4 is dispensable for basal transcription but essential for transcriptional responses to heat shock, nutrient deprivation, and other environmental stimuli. Rpb7 also contributes to the stability and export of mRNA transcripts, serving as a bridge between transcription and post-transcriptional regulation such as mRNA transport, localization, and degradation (Garrido-Godino et al. 2022a, b). This functional expansion is reflected in the structural elaboration of the Rpb7 tip loop and the surface-exposed regions of Rpb4, which are thought to mediate interactions with regulatory factors. For example, the Rpb7 OB-fold domain is involved not only in RNA binding but also in interactions with the Pol II core and possibly the Mediator complex, as suggested by cryo-EM studies of the transcription pre-initiation complex (Abascal-Palacios et al. 2021; Cramer 2021). Similarly, Rpb4 has been implicated in the recruitment of mRNA decay factors and in the coordination of transcription-coupled mRNA turnover. Moreover, Rpb4/Rpb7 has been shown to shuttle between the nucleus and the cytoplasm, especially during stress conditions. This behavior is consistent with its role in mRNA lifecycle regulation, and illustrate a distinct evolutionary divergence from the more static role of Rpo4/Rpo7 in archaea. These differences highlight how a conserved structural module has adapted to support the increased complexity of eukaryotic gene expression. Collectively, these findings suggest that the stalk subcomplex has evolved from a core structural component in archaea into a versatile hub that integrates transcription with RNA processing and stress responses in eukaryotes. This transition reflects a broader theme in molecular evolution: structural conservation paired with functional innovation.

1.6.5 Evolutionary Implications of the Stalk Subunits

The structural and functional comparison of Rpo4/Rpo7 in archaea and Rpb4/Rpb7 in eukaryotes highlights a striking example of molecular evolution where a conserved architecture has been repurposed to meet increasingly complex cellular needs. The stalk subcomplex in both systems originates from a shared ancestral configuration, with the OB-fold–containing subunit (Rpo7/Rpb7) interfacing tightly with its partner (Rpo4/Rpb4) to form a stable protrusion adjacent to the RNA exit channel. Despite this conserved framework, it has diverged markedly at both structural and functional levels. As shown in Sects. 1.6.1, 1.6.3 and 1.6.4, the archaeal Rpo4/Rpo7 heterodimer primarily contributes to RNAP stabilization and transcription factor recruitment, most notably providing a docking interface for the general transcription factor TFE. This function is essential in thermophilic archaea such as *T. kodakarensis,* where loss of Rpo4 impairs growth at high temperatures and disrupts TFE association with RNAP. These findings suggest that even in archaea, the stalk is more than a passive structural feature—it plays an active role in maintaining transcriptional competence under environmental stress. In contrast, as described in Sects. 1.6.2 and 1.6.4, the eukaryotic Rpb4/Rpb7 complex has evolved to support a broader range of regulatory processes. Beyond its conserved role in stabilizing the Pol II core and interacting with nascent RNA, Rpb4/Rpb7 now functions in mRNA export, degradation, and cytoplasmic shuttling during stress. These functions are enabled by structural elaborations—such as the extended tip loop of Rpb7 and the charged surface of Rpb4—which provide interaction surfaces for multiple co-factors involved in RNA metabolism. This functional expansion of the stalk subunits reflects a broader evolutionary trend: from a core transcriptional module in archaea to a dynamic regulatory interface in eukaryotes. The increased complexity of eukaryotic gene expression, including coordinated control of transcription, RNA processing, and degradation, required the recruitment and specialization of existing structural units. Rpb4/Rpb7 exemplifies this transition, integrating environmental responsiveness, post-transcriptional regulation, and cellular signaling into the transcriptional apparatus. Comparative structural analyses and sequence conservation across archaeal and eukaryotic stalk subunits suggest that the Pol II stalk evolved from its archaeal ancestor without major topological change but with newly acquired regulatory functions.

1.7 TFE and the Origin of General Transcription Factors

1.7.1 Overview of TFE in Archaea and TFIIE in Eukaryotes

In archaeal transcription, the general transcription factor TFE (transcription factor E) is critical for transcription initiation by facilitating promoter opening and stabilizing the transcription pre-initiation complex (PIC) (Ouhammouch et al. 2004; Grünberg et al. 2007; Grohmann et al. 2011). TFE is typically composed of a single α-subunit (TFEα), although a β-subunit (TFEβ) is also present in some archaeal lineages (Blombach et al. 2015). TFEα contains an N-terminal extended winged-helix domain and a C-terminal zinc ribbon domain, enabling interactions with both DNA and RNAP components. The structural and functional characteristics of TFEα are considered homologous to the eukaryotic general transcription factor TFIIEα, which, along with TFIIEβ, participates in the assembly of the eukaryotic PIC. TFIIE contributes to promoter melting, the recruitment of TFIIH, and the stabilization of the open complex (Schilbach et al. 2021). Both TFEα and TFIIEα are involved in the recruitment and positioning of RNAP (or Pol II in eukaryotes) at the transcription start site. Cryo-EM and crystallographic studies have revealed notable structural similarities between archaeal TFEα and eukaryotic TFIIEα (Fig. 1.5) (Jun et al. 2020; Schilbach et al. 2021). Both proteins interact with the polymerase stalk formed by Rpo4/Rpo7 in Archaea and Rpb4/Rpb7 in Eukarya, respectively, highlighting their evolutionary continuity. These interactions are essential for positioning TFE on the surface of RNAP and modulating its engagement with downstream DNA. The conservation of this binding surface suggests that TFE represents the evolutionary precursor of TFIIEα. Figure 1.6 shows a comparative structural alignment of TFEα from *T. kodakarensis* and TFIIEα from *S. cerevisiae*, based on the crystal and cryo-EM structures of the transcription pre-initiation complexes (PDB: 6KF9, 7O75). The overall architecture is preserved, including the location of the winged-helix and zinc ribbon domains, although domain orientations vary to accommodate species-specific interactions with RNAP and DNA. These findings position TFEα as a structural and functional ancestor of TFIIEα, providing insights into how transcriptional complexity has expanded from Archaea to Eukaryotes.

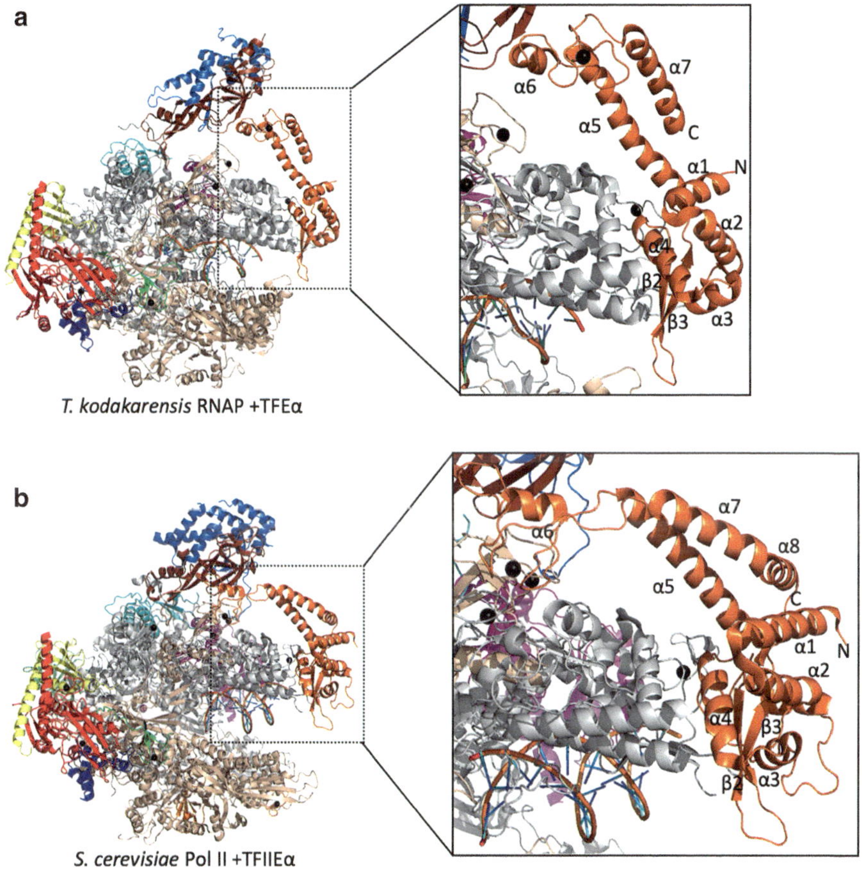

Fig. 1.5 Ribbon diagrams of the RNAP–TFEα/TFIIEα complexes from Archaea and Eukaryotes. *Thermococcus kodakarensis* RNA polymerase (RNAP) bound to the archaeal transcription factor TFEα (**a**, PDB: 6KF9). *Saccharomyces cerevisiae* RNA polymerase II complexed with the eukaryotic TFIIEα subunit (**b**, PDB: 7O75). TFEα/TFIIEα is depicted in light blue, and the RNAP core is depicted in light gray. The "Close-up view" highlights conserved features and structural differences in the interaction between RNAP and TFEα/TFIIEα across domains of life. Secondary structure elements of TFEα/TFIIEα (α-helices and β-strands) are labeled

1.7.2 Structural Conservation Between TFEα and TFIIEα

Recent structural studies have revealed that the archaeal transcription factor TFEα and its eukaryotic counterpart TFIIEα share a remarkable degree of structural conservation, despite differences in subunit composition and regulatory complexity between the two domains of life. TFEα consists of two main domains: an N-terminal extended winged-helix (eWH) domain and a C-terminal zinc ribbon (ZnR) domain. These domains are also observed in TFIIEα, highlighting a conserved core structure likely inherited from a common ancestor. As shown in Fig. 1.6, structural alignment

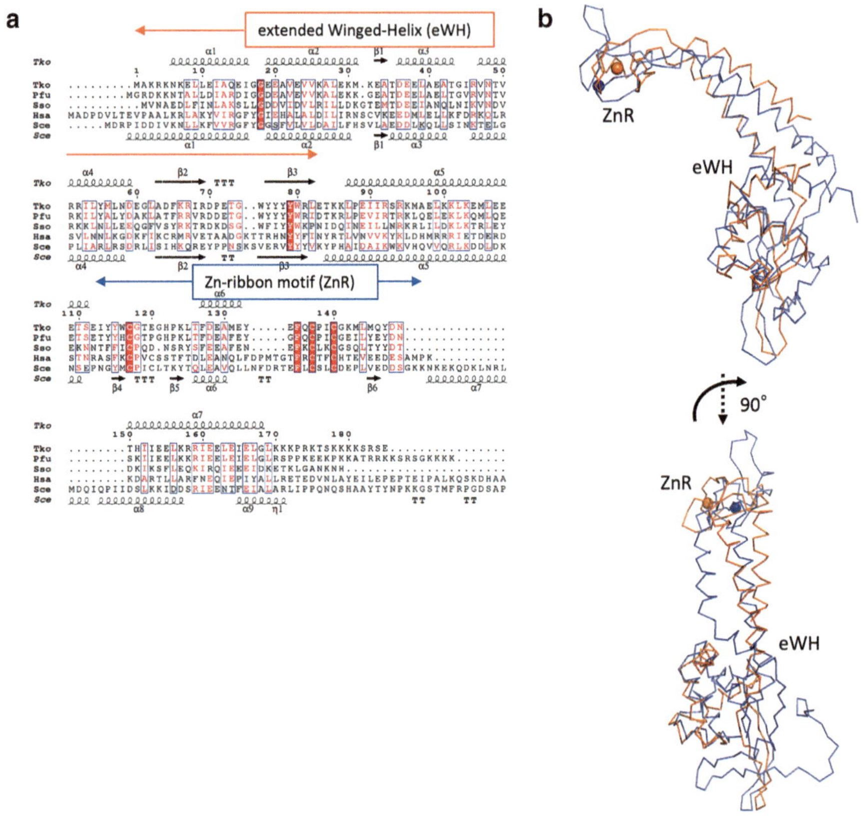

Fig. 1.6 Amino acid sequence alignment of TFEα/TFIIEα from archaeal and eukaryotic species. (**a**) Sequence alignment of TFEα from *Thermococcus kodakarensis* (*Tko*), *Pyrococcus furiosus* (*Pfu*), and *Sulfolobus solfataricus* (*Sso*), and TFIIEα from *Homo sapiens* (*Hsa*) and *Saccharomyces cerevisiae* (*Sce*). Identical and similar residues are boxed. Secondary structure elements derived from the crystal structures of *Tko* TFEα (PDB: 6KF9) and *Sce* TFIIEα (PDB: 7O75) are indicated above and below the alignment, respectively. (**b**) Ribbon diagram showing the superposition of 123 Cα atoms of *Tko* TFEα (orange) and *Sce* TFIIEα (blue), with a root-mean-square deviation (RMSD) of 5.45 Å. Zinc atoms are depicted as spheres. The lower panel shows the same structure rotated 90° clockwise relative to the top view

of TFEα from *T. kodakarensis* and TFIIEα from *S. cerevisiae* demonstrates that the eWH domains adopt a similar α-helical fold, with conservation of the key DNA-binding elements. Although the orientation of these domains differs slightly, reflecting species-specific interactions with RNAP and DNA, the topological architecture is largely maintained. This conservation suggests functional equivalency in their ability to stabilize the pre-initiation complex. The ZnR domain, located at the C-terminus of both proteins, exhibits an even higher level of structural preservation. In both TFEα and TFIIEα, the ZnR is involved in protein-protein interactions, including binding to the RNAP stalk (Rpo4/7 in Archaea; Rpb4/7 in Eukarya)

(Fig. 1.5). The positioning of the ZnR domain at the periphery of the polymerase complex allows for flexible interaction with downstream transcription factors and potentially with nucleic acids. Collectively, these findings support the notion that TFEα and TFIIEα are structurally conserved scaffolds that perform analogous roles in their respective domains. The conservation of core domains, especially the ZnR motif and the winged-helix topology, suggests a shared evolutionary origin and functional continuity across the archaeal–eukaryotic boundary.

1.7.3 Sequence Conservation and Divergence of TFEα

Sequence alignment of archaeal and eukaryotic transcription factors reveals that the core elements of TFEα and TFIIEα are structurally conserved, particularly within the eWH domain and the ZnR motif. As illustrated in Fig. 1.6, multiple archaeal species—including *T. kodakarensis* (Tko), *P. furiosus* (Pfu), and *S. solfataricus* (Sce)—share high sequence identity within the α-helical regions corresponding to α1 through α8. These α-helices align well with those found in TFIIEα from Sce and *Homo sapiens* (Hsa), suggesting that the eWH fold is an ancient and conserved structural element (Meinhart et al. 2003). The ZnR domain, located near the C-terminus, is characterized by a conserved cysteine pattern that coordinates a zinc ion, typically in a CXXC–Xn–CXXC motif. This motif is preserved in both archaeal TFEα and eukaryotic TFIIEα, indicating that the zinc-binding functionality has been evolutionarily maintained. Despite minor variations in the length and sequence of inter-helix loops, the core residues responsible for structural stability and metal coordination are well-conserved across domains. Notably, the C-terminal region exhibits greater variability, particularly among archaeal lineages. For example, *S. solfataricus* TFEα contains an extended acidic tail that is absent in *T. kodakarensis*, suggesting potential lineage-specific functional adaptations. Similarly, TFIIEα in eukaryotes often contains additional motifs or interaction regions that are absent in archaeal homologs, consistent with the expanded regulatory complexity of the eukaryotic transcription machinery. These sequence-level comparisons support the structural observations in Figs. 1.5 and 1.6, reinforcing the idea that the fundamental transcriptional machinery is built upon a conserved architectural framework, with species-specific insertions contributing to regulatory specialization.

1.7.4 Zinc Ribbon Domain (ZnR): A Structural and Functional Key

The ZnR domain, located at the C-terminal region of both archaeal TFEα and eukaryotic TFIIEα, plays a pivotal role in protein–protein interactions during transcription initiation. Structurally, the ZnR domain adopts a compact fold stabilized

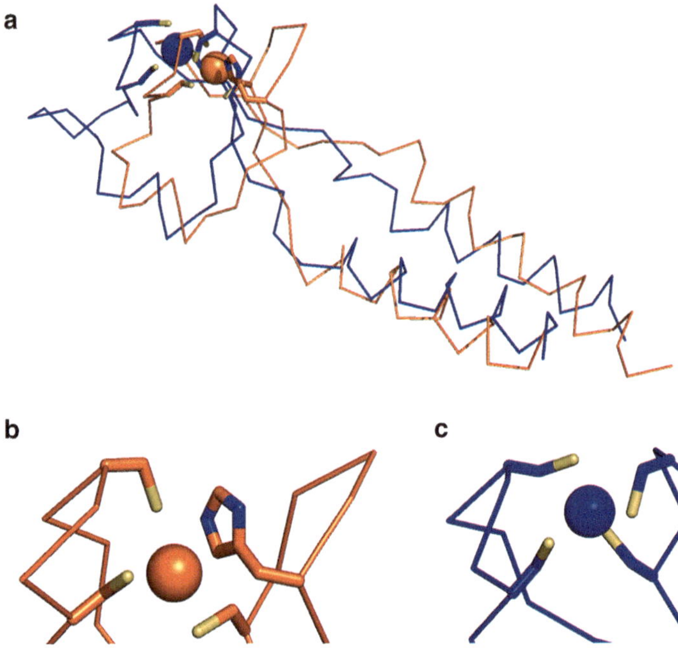

Fig. 1.7 Structural comparison of the zinc ribbon motifs of archaeal TFEα and eukaryotic TFIIEα. (**a**) Superposition of the C-terminal regions of *Thermococcus kodakarensis* (*Tko*) TFEα (residues 93–166) and *Saccharomyces cerevisiae* (*Sce*) TFIIEα (residues 100–180), highlighting the structural similarity of their zinc ribbon motifs. The alignment was performed using 50 Cα atoms, yielding a root-mean-square deviation (RMSD) of 3.76 Å. Ribbon diagrams showing the zinc ribbon motifs of *Tko* TFEα (**b**) and *Sce* TFIIEα (**c**), respectively. Zinc atoms and their coordinating amino acid residues are depicted in ball-and-stick representation

by coordination of a zinc ion via conserved cysteine residues, typically in a CXXC–Xn–CXXC configuration (Fig. 1.7). This domain is involved in anchoring TFEα or TFIIEα to the polymerase stalk and mediating interactions with other components of the pre-initiation complex (Fig. 1.5). Figure 1.7 shows a structural superposition of the ZnR domains from *T. kodakarensis* TFEα and *S. cerevisiae* TFIIEα, highlighting the strong conservation of the metal-binding core. The root-mean-square deviation (RMSD) between the two structures is minimal in the region encompassing the four coordinating cysteines, indicating that this motif has been preserved with high fidelity across the archaeal–eukaryotic divide. Despite this conservation, minor differences are observed in the orientation of peripheral loops and surface electrostatic properties, suggesting possible divergence in interaction specificity with transcriptional partners. Functionally, the ZnR domain contributes to stabilizing the association of TFEα/TFIIEα with RNAP and may facilitate conformational changes during promoter melting or transcription bubble stabilization. In Archaea, this domain interacts with the RNAP stalk formed by Rpo4/Rpo7, whereas in Eukarya, it binds to the analogous Rpb4/Rpb7 subcomplex. These contacts are essential for the correct positioning of general transcription factors within the PIC and may regulate the timing and efficiency of open complex formation. Recent structural studies have shown that in Archaea, TFEα binds simultaneously to the polymerase stalk (Rpo4/Rpo7) and the tip of the RNAP clamp, stabilizing the clamp in a closed conformation, which is essential for transcription initiation. In contrast, the absence of TFE leads to an open clamp state, as observed in isolated Rpo4/Rpo7-containing RNAP structures (Hirata et al. 2008a; Jun et al. 2014, 2020). This mechanism ensures that TFE not only stabilizes the PIC but also promotes DNA engagement by regulating clamp dynamics. A comparable interaction is observed in eukaryotic Pol II, where TFIIEα associates with the Rpb4/Rpb7 stalk and the clamp head of Pol II, thereby contributing to clamp closure. This evolutionarily conserved mode of action supports the hypothesis that the Zn ribbon domain of TFE/TFIIE serves as a molecular hinge, facilitating the transition from an open to a closed clamp configuration during transcription initiation.

1.7.5 Functional Diversification of TFE and the Origin of TFIIE

While the archaeal transcription factor TFE is generally composed of a single subunit, TFEα, a second subunit—TFEβ—has been identified in only a subset of archaeal lineages. In most archaeal species, TFEα is sufficient to support transcription initiation through its dual interaction with the RNAP clamp and stalk, as discussed in the previous sections. The occasional presence of TFEβ in some archaea, such as *Sulfolobus* species, appears to enhance the stability of the PIC under extreme environmental conditions, but it is not universally conserved. In certain archaeal lineages, such as the thermophilic and acidophilic crenarchaeon *S. solfataricus*, the transcription factor TFE functions as a heterodimer composed of TFEα and TFEβ. Structural and functional studies have shown that the presence of TFEβ enhances the ability of TFEα to promote open complex formation and transcription initiation (Blombach et al. 2015). TFEβ contains an iron–sulfur cluster that is essential for stable dimerization with TFEα and is thought to contribute to the formation and stabilization of the pre-initiation complex under environmental stress. Notably, *Sulfolobus* cells are unable to survive without TFEβ, highlighting its essential role in transcription and stress adaptation. These findings support the view that the acquisition of TFEβ in certain Archaea represents an adaptive innovation to stabilize transcription under extreme conditions such as high temperature and low pH. The fact that TFEβ is not universally conserved across Archaea further suggests that it evolved specifically in lineages requiring additional transcriptional robustness. This also provides an evolutionary parallel to the emergence of TFIIEβ in eukaryotes, which further diversified the transcription machinery by expanding its modular and regulatory potential. In contrast to the typically monomeric archaeal TFE, the eukaryotic transcription factor TFIIE has undergone clear structural and functional expansion, consisting of two subunits: TFIIEα and TFIIEβ. TFIIEα retains structural homology with TFEα, including the extended winged-helix and Zn ribbon domains, while TFIIEβ introduces additional domains involved in protein–protein interactions and possibly in transcriptional regulation. This subunit expansion likely reflects the increasing complexity of transcription regulation in eukaryotes, including the need for more dynamic control over promoter melting, TFIIH recruitment, and transition to elongation. The evolution of TFIIEβ may have allowed the decoupling of structural stabilization from regulatory interactions, thereby enhancing functional modularity within the PIC. Indeed, TFIIEβ has been shown to interact with TFIIH and influence DNA melting activity, suggesting that its appearance is linked to the elaboration of promoter unwinding mechanisms (Schilbach et al. 2021). These findings indicate that the two-subunit organization of TFIIE is a eukaryotic innovation that builds upon the ancestral TFEα framework. The retention of TFEα-like domains in TFIIEα, coupled with the addition of a specialized regulatory partner (TFIIEβ), exemplifies a common evolutionary strategy of functional diversification through modular assembly.

1.8 Transcription Elongation Factors Spt4/5 and TFS

1.8.1 Overview of Spt4/5 in Archaea and Eukaryotes

Transcription elongation is a crucial and tightly regulated phase of gene expression. In all domains of life, the progression of RNAP along the DNA template is modulated by various elongation factors. Among these, the universally conserved Spt5 protein is essential for facilitating transcriptional elongation. Both archaeal and eukaryotic organisms possess a structurally and functionally homologous elongation factor complex, Spt4/5, which is referred to as DSIF (5,6-Dichloro-1-β-D--ribofuranosylbenzimidazole (DRB) Sensitivity-Inducing Factor) in eukaryotes (Grohmann et al. 2011; Smollett et al. 2017; Blombach et al. 2021). In Archaea, Spt5 functions in association with Spt4 as a heterodimeric complex. Archaeal Spt5 typically consists of an N-terminal NGN (NusG N-terminal) domain and a single KOW (Kyprides–Ouzounis–Woese) domain. Spt4, a zinc finger protein, stabilizes the NGN domain and promotes its interaction with RNAP. Together, the Spt4/5 complex enhances transcriptional fidelity and is shown to stabilize the transcription bubble during elongation (Tarău et al. 2024). By contrast, eukaryotic Spt5 has undergone substantial domain expansion. Eukaryotic DSIF comprises Spt4 and a significantly larger Spt5 protein that contains multiple KOW domains (KOW1 to KOW5) and a C-terminal repeat (CTR) region (Bernecky et al. 2017). These additional domains have enabled eukaryotic DSIF to acquire diverse regulatory functions, including control of promoter-proximal pausing, recruitment of RNA capping enzymes, and interactions with elongation and RNA export factors. As illustrated in Fig. 1.8, the Spt4/5 complex features a conserved NGN domain in both Archaea and

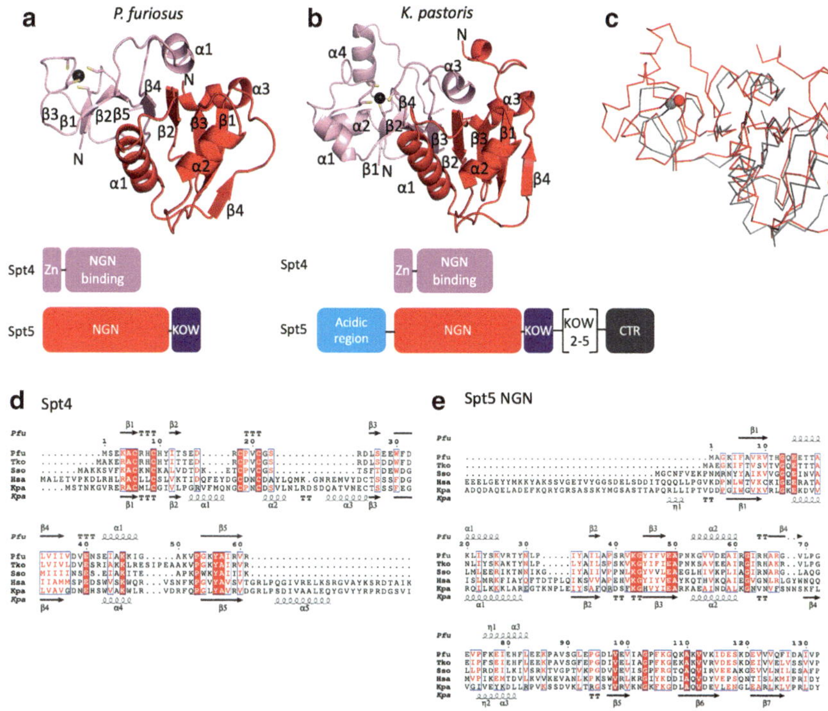

Fig. 1.8 Structural comparison of the Spt4/Spt5 transcription elongation factors from Archaea and Eukaryotes. Ribbon diagrams of the NGN domains of Spt5 and Spt4 from *Pyrococcus furiosus* (**a**) and *Komagataella pastoris* (**b**) are shown. The domain organization of the Spt4/Spt5 complex in both species highlights conserved regions, including the NGN domain, KOW domains, and the acidic C-terminal region (CTR). Secondary structure elements (α-helices and β-strands) are labeled to illustrate overall conservation and species-specific variations. Zinc atoms are depicted as spheres (black). Structural superposition of the NGN domains based on 48 Cα atoms from both structures (**a** and **b**), with a root-mean-square deviation (RMSD) of 1.69 Å, indicating a high degree of structural similarity (**c**). All domain boundaries and structural features are annotated. Amino acid sequence alignments of the Spt4 and NGN domain of Spt5 from *Pyrococcus furiosus* (*Pfu*), *Thermococcus kodakarensis* (*Tko*) *Sulfolobus solfataricus* (*Sso*), *Homo sapiens* (Hsa) and *Komagataella pastoris* (*Kpa*) (**d** and **e**). The two domain regions of *Tko* Rpo4 (extended Winged-Helix (eWH) domain and Zn-ribbon motif (ZnR) domain) are highlighted. Identical and similar residues are boxed. Secondary structures of the Spt4 and NGN domain of Spt5 from *Pfu* and *K*pa are indicated above and below the alignments

Eukaryotes, with notable structural complexity in eukaryotic Spt5. Archaeal Spt5 generally contains a single KOW domain, while its eukaryotic homologs have evolved multiple KOW domains along with a CTR. These structural innovations facilitate expanded regulatory functions, including interactions with the RNA exit tunnel and a variety of transcriptional regulators (Ehara et al. 2017).

1.8.2 Structural Architecture of the Archaeal Spt4/5 Complex

Recent cryo-EM analyses have elucidated the three-dimensional architecture of the archaeal Spt4/5 complex in association with RNAP, offering insight into its role in transcription elongation. In *P. furiosus*, the Spt4 and Spt5 proteins form a stable heterodimer that binds across the RNAP clamp and protrusion domains, anchoring the complex at the upstream edge of the transcription bubble (Fig. 1.9a) (Tarău et al. 2024). The core interaction is mediated by the NGN domain of Spt5, which is stabilized by its binding partner, Spt4. The NGN domain inserts between the clamp

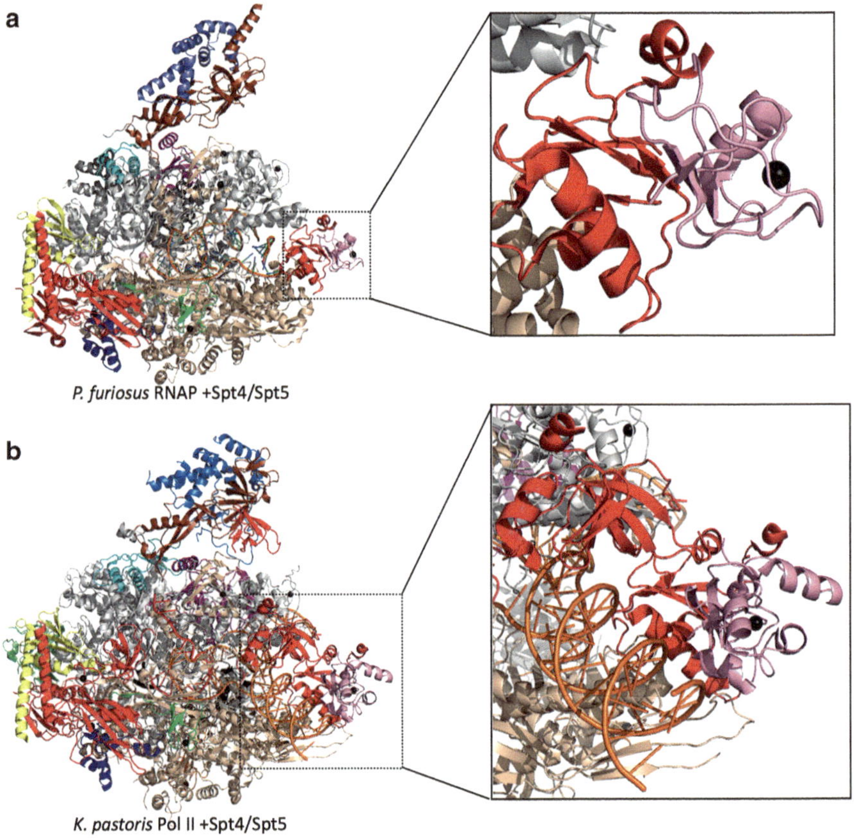

Fig. 1.9 Ribbon diagrams of cryo-EM structures of the RNAP–Spt4/Spt5 elongation complexes. *Pyrococcus furiosus* RNAP core bound to its archaeal Spt4–Spt5 heterodimer (**a**, PDB: 3P8B). *Komagataella pastoris* (formerly *Pichia pastoris*) Pol II complexed with the eukaryotic Spt4/Spt5 orthologue (DSIF) (**b**, PDB: 5XON). Spt4 is colored pink and Spt5 red, with the RNAP core shown in light gray. Zinc atoms are represented as black spheres. The “Closed-up view” highlights the interface between Spt4/Spt5 and RNAP, with all major domains of Spt5 (NGN) and the corresponding RNAP regions labeled to illustrate their points of contact

and protrusion of RNAP, forming direct contacts with conserved β-like elements. This structural configuration supports stabilization of the transcription bubble and facilitates efficient elongation. The stabilizing contribution of Spt4 is especially evident in its zinc-binding ribbon, which wraps around the NGN domain, forming a compact and functionally essential interface. In contrast to its eukaryotic counterpart, archaeal Spt5 contains only a single KOW domain. This domain is located distal to the NGN core and is oriented toward the RNA exit tunnel. This structural simplicity contrasts with the eukaryotic Spt5, which possesses multiple KOW domains and a CTR to support a broader range of regulatory interactions. As shown in Fig. 1.9, comparison of the archaeal and eukaryotic Spt4/5–RNAP complex structures reveals high conservation in the positioning of the NGN domain, while the spatial organization of the KOW domains shows notable divergence. In Archaea, the single KOW domain remains peripheral, whereas in Eukaryotes, the extended KOW domains fold into and around the RNA exit tunnel, forming an enlarged interface with RNAP. This structural distinction underscores the streamlined function of archaeal Spt4/5 in elongation and the regulatory complexity acquired by eukaryotic DSIF through domain expansion. Together, these findings highlight both the conservation and divergence of the Spt4/5 complex across domains of life. In Archaea, Spt4/5 acts as a structural stabilizer of the RNAP elongation complex, whereas in Eukaryotes, the elaborated architecture facilitates multifunctional roles in RNA processing and transcriptional regulation.

1.8.3 Structural and Functional Comparison with Eukaryotic DSIF (Spt4/5)

The Spt4/5 elongation factor complex is a universally conserved transcriptional regulator found in both Archaea and Eukaryotes. While the archaeal version exhibits a compact architecture with a minimal domain composition, the eukaryotic homolog—known as DSIF (DRB Sensitivity-Inducing Factor)—has undergone substantial domain expansion and functional diversification. Structurally, the core architecture of Spt5 is conserved, particularly the NGN domain, which interfaces with the RNA polymerase clamp and protrusion in both archaeal and eukaryotic systems. This conserved interaction supports transcriptional processivity by stabilizing the transcription bubble and facilitating the progression of RNA polymerase along the DNA template. However, the architecture of the KOW domains diverges markedly. Archaeal Spt5 contains only a single KOW domain (KOW1), flexibly tethered and located near the RNA exit channel. In contrast, eukaryotic Spt5 features five tandem KOW domains (KOW1–KOW5) and a CTR, allowing the protein to wrap around the RNA exit tunnel of Pol II. These additional domains interact with the polymerase, the emerging RNA transcript, and RNA-processing machinery. Cryo-EM structures of human and yeast DSIF reveal that KOW4 and KOW5 are especially critical for shaping the RNA exit channel, guiding the nascent RNA, and providing a platform for the recruitment of RNA capping enzymes and export factors (Ehara et al. 2017; Bernecky et al. 2017). Functionally, these structural differences translate into enhanced regulatory capabilities in eukaryotes. Archaeal Spt4/5 primarily enhances elongation processivity and prevents backtracking but lacks the ability to mediate transcriptional pausing or engage with RNA-processing complexes. In contrast, eukaryotic DSIF not only promotes elongation but also regulates promoter-proximal pausing—a key transcriptional checkpoint in higher eukaryotes. This process is further modulated by the phosphorylation of the Spt5 CTR, which recruits the positive transcription elongation factor b (P-TEFb) complex and facilitates the transition into productive elongation (Yamaguchi et al. 2013). To summarize, while archaeal Spt4/5 maintains a streamlined and efficient structure suitable for simple elongation control, eukaryotic DSIF has diversified to accommodate the increasingly complex regulatory needs of gene expression, coupling transcription with co-transcriptional RNA processing and chromatin-associated regulation.

1.8.4 Functional Implications of KOW Domain Expansion

The expansion of KOW domains in eukaryotic Spt5 represents a major evolutionary innovation that distinguishes DSIF from its archaeal counterpart. While archaeal Spt5 typically contains only a single KOW domain, eukaryotic Spt5 features up to five KOW domains (KOW1–KOW5) and a CTR region. These additions are not only structural; they are closely associated with the functional diversification of transcriptional regulation. Structural analyses of Pol II elongation complexes have demonstrated that the expanded KOW domains of eukaryotic Spt5 wrap around the RNA exit tunnel and establish multiple contacts with the nascent RNA and RNA-processing factors. For example, KOW4 and KOW5 form a surface that facilitates recruitment of the RNA capping enzyme complex during early elongation. This interaction functionally links transcript emergence to co-transcriptional RNA processing. Furthermore, the modular architecture of KOW domains enhances Spt5's ability to participate in broader transcriptional regulatory networks. KOW1–KOW3 mediate interactions with additional elongation factors, including the negative elongation factor (NELF) and the polymerase-associated factor 1 (PAF1) complex, which facilitates transcriptional elongation and coordinates histone modifications and mRNA processing, thereby contributing to promoter-proximal pausing and productive elongation (Chen et al. 2009; Yamaguchi et al. 2013). In contrast, the single KOW domain in archaeal Spt5 is structurally conserved but functionally limited. It is believed to interact with the RNA exit channel, potentially stabilizing the emerging RNA transcript and guiding it away from the transcription complex. However, due to the absence of a CTR and additional KOW modules, archaeal Spt5 lacks the capacity to recruit RNA-processing machinery or regulate transcriptional pausing. From an evolutionary standpoint, the expansion of the KOW domain architecture reflects the growing complexity of gene regulation in eukaryotic systems. These added domains not only enhance protein–protein and protein–RNA interactions but also serve as platforms for regulatory phosphorylation and allosteric modulation. Thus, the evolution of the KOW domain family represents a molecular adaptation that enables DSIF to act as a central scaffold for coordinating elongation, RNA processing, and gene expression.

1.8.5 Evolutionary Significance and Integration into Transcription Regulation

The evolution of the Spt4/5 complex, particularly the structural expansion of Spt5 in eukaryotes, demonstrates a broader evolutionary trend toward increasing complexity in transcriptional regulation across the domains of life. While the archaeal Spt4/5 complex exhibits a conserved and streamlined architecture, its eukaryotic counterpart—DSIF—has evolved into a multifunctional regulator that serves as a key hub in the transcription elongation machinery. From an evolutionary standpoint, the core Spt4/5 module likely originated in the last common ancestor of Archaea and Eukaryotes. In Archaea, this module is sufficient to stabilize the transcription elongation complex and enhance processivity. However, the relatively simple genomic architecture of archaeal organisms may have reduced the need for sophisticated control mechanisms during transcription (Bell and Jackson 2001; Werner and Grohmann 2011). In contrast, the increasing complexity of eukaryotic genomes—including the presence of introns, chromatin remodeling, and co-transcriptional RNA processing—has driven the structural and functional elaboration of Spt5. The acquisition of multiple KOW domains and a CTR region has enabled DSIF to interact with a wide variety of transcription-associated factors, such as NELF, the PAF1 complex, and RNA capping enzymes. These interactions allow DSIF to integrate transcription elongation with regulatory checkpoints and RNA processing. Notably, DSIF's role in promoter-proximal pausing represents a critical regulatory feature absent in Archaea. This pausing serves as a checkpoint that responds to environmental and developmental signals. Phosphorylation of the Spt5 CTR by the P-TEFb complex facilitates the transition from a paused to a productive elongation state (Yamaguchi et al. 2013). Thus, DSIF functions not merely as a structural element but as a regulatory platform coordinating elongation and gene expression control. To summarize, the divergence of the Spt4/5 complex—from a minimal archaeal elongation factor to a highly integrated regulatory component in eukaryotes—exemplifies how conserved molecular frameworks can evolve to meet the demands of complex transcriptional networks.

1.8.6 Role of TFS in Archaeal Transcription Elongation

Transcription Factor S (TFS) is an archaeal elongation factor that shares structural and functional homology with the eukaryotic transcription factor TFIIS (Hausner et al. 2000). TFS stimulates the intrinsic RNA cleavage activity of RNAP, enabling stalled elongation complexes to resume transcription (Lange and Hausner 2004). This proofreading mechanism supports high transcriptional fidelity.

TFS is composed of two conserved zinc ribbon motifs. The C-terminal ribbon inserts into the secondary channel of RNAP and positions acidic residues near the catalytic magnesium ions to promote RNA cleavage. In contrast, eukaryotic TFIIS includes an additional N-terminal domain and a longer linker that contacts not only the RNAP active site but also the Pol II-specific Rpb9 subunit—absent in archaeal RNAP (Kettenberger et al. 2003). This interaction stabilizes TFIIS binding and contributes to active site remodeling. Because archaeal RNAP lacks Rpb9, TFS functions independently of this auxiliary contact and instead relies solely on conserved core RNAP elements. Although archaeal TFS factors are closely related at the sequence level to Rpb9-like RNAP subunits, they differ functionally by reversibly associating with RNAP and stimulating transcript cleavage in a manner analogous to eukaryotic TFIIS (Hausner et al. 2000; Ruan et al. 2011). Cryo-EM analyses have shown that TFIIS induces structural changes in the clamp and trigger loop regions of RNAP, further facilitating active site rearrangement (Ehara et al. 2017). While TFS is dispensable under optimal conditions, its function becomes essential in response to environmental stress or transcription–replication conflicts that increase the frequency of RNAP stalling (Sanders et al. 2019). TFS thus represents a minimal yet effective strategy for transcript proofreading in Archaea. Its conserved function with TFIIS, despite architectural simplification, illustrates the evolutionary conservation and adaptation of transcriptional regulation across domains of life.

1.9 Summary and Perspectives

In this chapter, an integrated evolutionary view of the archaeal RNAP and its transformation into eukaryotic Pol II has been presented. Through structural and comparative analyses of core subunits (e.g., Rpo3, Rpo4/7) and regulatory factors (TFE, Spt4/5, TFS), it has been shown that a conserved enzymatic scaffold was elaborated across lineages to support transcriptional complexity in eukaryotes. Key evolutionary innovations—including modular domain acquisition, redox-responsive Fe–S clusters, and the coupling of elongation with RNA processing—have been highlighted as examples of lineage-specific adaptation. These modifications occurred while preserving the integrity of the catalytic core, thereby enabling the diversification of regulatory networks. Future insights are expected from the study of transitional archaeal systems and dynamic transcription complexes, which may further clarify the molecular logic underlying the evolution of gene regulation.

Acknowledgements The author wishes to thank Professor. Akio Kanai for his kind recommendation and continuous encouragement throughout the preparation of this chapter. Valuable discussions with colleagues in the fields of archaeal molecular biology and structural enzymology are also gratefully acknowledged. This work was supported in part by a JSPS KAKENHI grant JP24K09352 (to A.H). Assistance with language and editorial refinement was provided in part by OpenAI's ChatGPT, which was used to support grammar correction, phrasing, and the structuring of technical English. All scientific interpretations and conclusions remain the sole responsibility of the author. The author sincerely apologizes for any unintentional omissions of relevant references and acknowledges that it may not have been possible to cite every important contribution in this rapidly developing field.

Competing Interests The author declares no competing interests.

References

Abascal-Palacios G, Schilbach S, Cramer P (2021) Structure of the human Mediator–RNA polymerase II pre-initiation complex. Nature 594(7861):129–133. https://doi.org/10.1038/s41586-021-03223-9

Armache K-J, Kettenberger H, Cramer P (2003) Architecture of initiation-competent 12-subunit RNA polymerase II. Proc Natl Acad Sci 100:6964–6968. https://doi.org/10.1073/pnas.1030608100

Armache KJ, Mitterweger S, Meinhart A, Cramer P (2005) Structures of complete RNA polymerase II and its subcomplex, Rpb4/7. J Biol Chem 280:7131

Bell SD, Jackson SP (2001) Mechanism and regulation of transcription in archaea. Curr Opin Microbiol 4:208–213. https://doi.org/10.1016/S1369-5274(00)00190-9

Bell SD, Kosa PL, Sigler PB, Jackson SP (1999) Orientation of the transcription preinitiation complex in archaea. Proc Natl Acad Sci USA 96:13662

Bernecky C, Plitzko JM, Cramer P (2016) Structure of a transcribing RNA polymerase II–DSIF complex reveals a multidentate DNA-RNA clamp. Nat Struct Mol Biol 23:899–904. https://doi.org/10.1038/nsmb.3273

Bernecky C, Plitzko JM, Cramer P (2017) Structure of a transcribing RNA polymerase II–DSIF complex reveals a multidentate DNA–RNA clamp. Nat Struct Mol Biol 24:809–815. https://doi.org/10.1038/nsmb.3465

Bernecky C, Plitzko JM, Cramer P (2020) Cryo-electron microscopy of RNA polymerase II. Methods. https://doi.org/10.1016/j.ymeth.2020.01.001

Blombach F, Werner F (2022) The evolutionary conservation and divergence of the archaeal transcription apparatus. Curr Opin Microbiol 65:96–104

Blombach F, Salvadori E, Fouqueau T et al (2015) Archaeal TFEα/β is a hybrid of TFIIE and the RNA polymerase III subcomplex hRPC62/39. elife 4:e08378. https://doi.org/10.7554/eLife.08378

Blombach F, Fouqueau T, Matelska D et al (2021) Promoter-proximal elongation regulates transcription in archaea. Nat Commun 12:5524. https://doi.org/10.1038/s41467-021-25669-2

Chen Y et al (2009) DSIF, the Paf1 complex, and Tat-SF1 have nonredundant, cooperative roles in RNA polymerase II elongation. Genes Dev 23:2765–2777. https://doi.org/10.1101/gad.1834709

Cramer P (2021) Structure and function of RNA polymerase II transcription machinery: recent advances and future prospects. Nat Rev Mol Cell Biol. https://doi.org/10.1038/s41580-021-00366-8

Ehara H, Yokoyama T, Shigematsu H et al (2017) Structure of the complete elongation complex of RNA polymerase II with basal factors. Science 357:921–924. https://doi.org/10.1126/science.aan8552

Eick D, Geyer M (2013) The RNA polymerase II carboxy-terminal domain (CTD) code. Nat Rev Mol Cell Biol 14:264–274. https://doi.org/10.1038/nrm3547
Garrido-Godino AI, Cuevas-Bermúdez A, Gutiérrez-Santiago F et al (2022a) The association of Rpb4 with RNA Polymerase II depends on CTD Ser5P Phosphatase Rtr1 and influences mRNA decay in Saccharomyces cerevisiae. Int J Mol Sci 23:2002. https://doi.org/10.3390/ijms23042002
Garrido-Godino AI, Martín-Expósito M, Gutiérrez-Santiago F et al (2022b) Rpb4/7, a key element of RNA pol II to coordinate mRNA synthesis in the nucleus with cytoplasmic functions in Saccharomyces cerevisiae. Biochim Biophys Acta 1865:194846. https://doi.org/10.1016/j.bbagrm.2022.194846
Geiger SR (2010) RNA polymerase II pre-initiation complex architecture and dynamics. Nat Struct Mol Biol 17:1232–1239. https://doi.org/10.1038/nsmb.1903
Grohmann D, Nagy J, Chakraborty A et al (2011) The initiation factor TFE and the elongation factor Spt4/5 compete for the RNAP clamp during transcription initiation and elongation. Mol Cell 43:263–274. https://doi.org/10.1016/j.molcel.2011.05.030
Grünberg S, Bartlett MS, Naji S, Thomm M (2007) Transcription factor E is a part of transcription elongation complexes. J Biol Chem 282:35482
Harel-Sharvit L, Eldad N, Haimovich G et al (2010) RNA polymerase II subunits link transcription and mRNA decay to translation. Cell 143:552–563. https://doi.org/10.1016/j.cell.2010.10.033
Hausner W, Wettach J, Hethke C, Thomm M (1996) Two transcription factors related with the eukaryotic transcription factor TFIIB interact with the archaeal RNA polymerase. J Biol Chem 271:30144
Hausner W, Lange U, Musfeldt M (2000) Transcription factor S, a cleavage induction factor of the archaeal RNA polymerase. J Biol Chem 275:12393–12399. https://doi.org/10.1074/jbc.275.17.12393
Hirata A, Murakami KS (2009) Structural studies of archaeal RNA polymerase and its similarity to eukaryotic Pol II. Curr Opin Struct Biol 19:724–731. https://doi.org/10.1016/j.sbi.2009.09.001
Hirata A, Kanai T, Santangelo TJ et al (2008a) Archaeal RNA polymerase subunits E and F are not required for transcription *in vitro*, but a *Thermococcus kodakarensis* mutant lacking subunit F is temperature-sensitive. Mol Microbiol 70:623–633. https://doi.org/10.1111/j.1365-2958.2008.06430.x
Hirata A, Klein BJ, Murakami KS (2008b) The X-ray crystal structure of RNA polymerase from archaea. Nature 451:851–854. https://doi.org/10.1038/nature06532
Hirtreiter A, Damsma GE, Cheung AC et al (2010a) Spt4/5 stimulates transcription elongation through the RNA polymerase clamp coiled-coil motif. Nucleic Acids Res 38:4040–4051. https://doi.org/10.1093/nar/gkq129
Hirtreiter A, Grohmann D, Werner F (2010b) Molecular mechanisms of RNA polymerase—the F/E (RPB4/7) complex is required for high processivity in vitro. Nucleic Acids Res 38:585–596. https://doi.org/10.1093/nar/gkp928
Jennings ME, Lessner FH, Karr EA, Lessner DJ (2017) The [4Fe-4S] clusters of Rpo3 are key determinants in the post Rpo3/Rpo11 heterodimer formation of RNA polymerase in *Methanosarcina acetivorans*. Microbiology 6(1):e00399. https://doi.org/10.1002/mbo3.399
Jun S-H, Hirata A, Kanai T et al (2014) The X-ray crystal structure of the euryarchaeal RNA polymerase in an open-clamp configuration. Nat Commun 5:5132. https://doi.org/10.1038/ncomms6132
Jun S-H, Hyun J, Cha JS et al (2020) Direct binding of TFEα opens DNA binding cleft of RNA polymerase. Nat Commun 11:6123. https://doi.org/10.1038/s41467-020-19998-x
Kettenberger H, Armache K-J, Cramer P (2003) Architecture of the RNA polymerase II-TFIIS complex and implications for mRNA cleavage. Cell 114:347–357. https://doi.org/10.1016/S0092-8674(03)00598-1
Koonin EV (2021) Origin of eukaryotes from within archaea: genome reduction and function transfer. Nat Rev Genet 22:455–468. https://doi.org/10.1038/s41576-021-00340-3

Korkhin Y, Unligil UM, Littlefield O et al (2009) Evolution of complex RNA polymerases: the complete archaeal RNA polymerase structure. PLoS Biol 7:e1000102. https://doi.org/10.1371/journal.pbio.1000102

Lange U, Hausner W (2004) Transcription regulation in Archaea: concerted action of transcription factors and histones. Mol Microbiol 52:1133–1142. https://doi.org/10.1111/j.1365-2958.2004.04058.x

Lessner FH, Jennings ME, Hirata A et al (2012) Subunit D of RNA polymerase from Methanosarcina acetivorans contains two oxygen-labile [4Fe-4S] clusters: implications for oxidant-dependent regulation of transcription. J Biol Chem 287(22):18510–18523. https://doi.org/10.1074/jbc.M111.331199

Malik S, Roeder RG (2010) The metazoan Mediator co-activator complex as an integrative hub for transcriptional regulation. Nat Rev Genet 11:761–772. https://doi.org/10.1038/nrg2901

Meinhart A, Blobel J, Cramer P (2003) An extended winged helix domain in general transcription factor E/IIEα. J Biol Chem 278:48267–48274. https://doi.org/10.1074/jbc.M307874200

Meka H, Werner F, Cordell SC et al (2005) Crystal structure and RNA binding of the Rpb4/Rpb7 subunits of human RNA polymerase II. Nucleic Acids Res 33:6435–6444. https://doi.org/10.1093/nar/gki945

Mushegian A, Koonin EV (2020) Evolution of the RNA polymerase catalytic core. Biochimica et Biophysica Acta 1863:194529. https://doi.org/10.1016/j.bbagrm.2020.194529

Ouhammouch M, Werner F, Weinzierl RO, Geiduschek EP (2004) A set of bacterial-type promoters in an archaeal organism. J Biol Chem 279:51719

Rengachari S, Schilbach S, Aibara S et al (2021) Structure of the human Mediator–RNA polymerase II pre-initiation complex. Nature 594:129–133. https://doi.org/10.1038/s41586-021-03555-7

Ruan W, Lehmann E, Thomm M et al (2011) Evolution of two modes of intrinsic RNA polymerase transcript cleavage. J Biol Chem 286:18701–18707. https://doi.org/10.1074/jbc.M111.222273

Sanders TJ, Lammers M, Marshall CJ et al (2019) TFS and Spt4/5 accelerate transcription through archaeal histone-based chromatin. Mol Microbiol 111:784–797. https://doi.org/10.1111/mmi.14191

Schilbach S, Aibara S, Dienemann C et al (2021) Structure of RNA polymerase II pre-initiation complex at 2.9 Å defines initial DNA opening. Cell 184:4064–4072.e28. https://doi.org/10.1016/j.cell.2021.05.012

Smollett K, Blombach F, Reichelt R et al (2017) A global analysis of transcription reveals two modes of Spt4/5 recruitment to archaeal RNA polymerase. Nat Microbiol 2:17021. https://doi.org/10.1038/nmicrobiol.2017.21

Soutourina J, Wydau S, Ambroise Y et al (2011) Direct interaction of RNA polymerase II and mediator required for transcription in vivo. Science 331:1451–1454. https://doi.org/10.1126/science.1200188

Tarău D, Grünberger F, Pilsl M et al (2024) Structural basis of archaeal RNA polymerase transcription elongation and Spt4/5 recruitment. Nucleic Acids Res 52:6017–6035. https://doi.org/10.1093/nar/gkae282

Vos SM, Farnung L, Boehning M et al (2018) Structure of activated transcription complex Pol II–DSIF–PAF–SPT6. Nature 560:607–612. https://doi.org/10.1038/s41586-018-0440-4

Werner F, Grohmann D (2011) Evolution of multisubunit RNA polymerases in the three domains of life. Nat Rev Microbiol 9:85–98. https://doi.org/10.1038/nrmicro2507

Yamaguchi Y, Shibata H, Handa H (2013) Transcription elongation factors DSIF and NELF: promoter-proximal pausing and beyond. Biochimica et Biophysica Acta 1829:98–104. https://doi.org/10.1016/j.bbagrm.2012.11.007

Zillig W (1981) The archaebacterium Thermoproteus tenax: growth, morphology, and a DNA-dependent RNA polymerase insensitive to rifampicin. Arch Microbiol 129:275–280. https://doi.org/10.1007/BF00417159

Zillig W (1989) Transcription in archaea. Syst Appl Microbiol 11:265–270. https://doi.org/10.1016/S0723-2020(89)80051-7

Chapter 2
Ancient Protein Folds in RNA Polymerase and Ribosomes

Sota Yagi and Shunsuke Tagami

Abstract Modern cellular RNA polymerases and ribosomes are gigantic, multiple-subunit complexes, and they mutually synthesize each other at the core of the gene expression system. How such complicated molecules and systems emerged remains a fundamental mystery in molecular biology. Conservation of protein structures provides essential information to elucidate such ancient evolutionary events, as structures are more conserved than sequences. In this review, we showcase the conserved protein structures within RNA polymerase and ribosomes and present our recent experimental findings on the evolutionary relationships between ancient β-barrels, thereby shedding light on the common origin of RNA polymerase and multiple ribosomal proteins.

Keywords RNA polymerase · Ribosomes · Ancient proteins · Protein evolution · Protein folds · The origin of life

S. Yagi
RIKEN Center for Integrative Medical Sciences, Yokohama, Kanagawa, Japan

Faculty of Human Sciences, Waseda University, Tokorozawa, Saitama, Japan
e-mail: sota.yagi@aoni.waseda.jp

S. Tagami (✉)
RIKEN Center for Integrative Medical Sciences, Yokohama, Kanagawa, Japan

International Institute for Sustainability with Knotted Chiral Meta Matter (WPI-SKCM2), Hiroshima University, Higashi-Hiroshima, Hiroshima, Japan
e-mail: shunsuke.tagami@riken.jp

A. Kanai, A. W. Nicholson (eds.), *Molecular Evolution of RNA Regulatory Enzymes and Their Systems*, Nucleic Acids and Molecular Biology 38,
https://doi.org/10.1007/978-3-032-03300-0_2

2.1 Basic Protein Folds Conserved in the Gene Expression System

At the core of our life system, RNA is synthesized by proteins, and proteins are synthesized by RNA. Specifically, RNA polymerase and ribosomes synthesize each other. Both are massive macromolecular complexes composed of multiple polymer components. Even a simple bacterial RNA polymerase contains five protein subunits and more than 3000 amino acid residues (Zhang et al. 1999). While ribosomes are composed primarily of RNA, they also contain more than fifty ribosomal proteins (Ban et al. 2000; Korobeinikova et al. 2012; Shin et al. 2004). How such a vast number of proteins initially emerged and eventually assembled into the huge complexes that are responsible for the modern gene expression system remains an unsolved mystery.

Structural similarity between proteins is considered the most reliable indicator of ancient evolutionary relationships between different proteins, because structures are more strongly conserved than sequences or functions (Illergård et al. 2009). Although there is no universally accepted evolutionary classification system for proteins, they are generally grouped based on their shared sequences, functions, and structures. Proteins with clear similarities in sequence, function and structure are classified as a *family*. Families sharing functional and structural similarities without strong sequence conservation form a *superfamily*. Superfamilies with the same fundamental structural framework but uncertain sequence and functional similarities are categorized into a *fold*. Proteins within the same family are clearly derived from a common ancestor. Although there is less certainty, proteins within the same superfamily or fold are also often thought to have diversified from a common ancestor. In some rare cases, evolutionary relationships between different folds with distinct topologies are indicated, for example, by partial sequence similarities (*meta-fold*) (Alva et al. 2008; Kolodny 2021; Kolodny et al. 2021). Regardless of evolutionary relationships, folds with similar secondary structure components are also classified into an *architecture*, such as α-bundles and β-barrels.

SCOP, CATH, and ECOD are the major protein structure classification databases, and each employs slightly different classification methods (Andreeva et al. 2014; Cheng et al. 2014; Fox et al. 2014; Sillitoe et al. 2019). To date, more than 1000 independent folds or meta-folds have been classified (e.g., X-groups in ECOD). However, the origins of these diverse folds remain unclear. Even within RNA polymerase and ribosomes, a wide variety of distinct protein folds can be found. The protein folds in the archaeal ribosomal proteins of *Thermococcus kodakarensis* (PDB ID 6TH6) are summarized in Table 2.1. In this chapter, we highlight the representative folds conserved in RNA polymerase and ribosomal proteins, mainly focusing on ancient β-barrels (Table 2.2), and explain the results of our recent research on their evolutionary relationships.

Table 2.1 Protein folds conserved in ribosomal proteins. Each domain of the proteins in the *Thermococcus kodakarensis* ribosome structure (PDB ID 6TH6) was compared against the ECOD database by using Foldseek (van Kempen et al. 2024) and categorized into the corresponding X-groups. The X-group names (or T-group names if no X-group name is assigned) are listed as the fold names. Universally conserved ribosomal proteins are indicated in bold, whereas archaeal/eukaryal-specific proteins are shown in regular font

Architectures	Fold	ECOD_X	Protein	Domain	PDB_chain
α arrays	**HTH**	**101**	**uS15**	1	Q [auth Aq]
			eS17	1	T [auth At]
			eS19	1	V [auth Av]
	HhH/H2TH	**102**	**uS13**	1	P [auth Ap]
	Ribosomal protein uS7	**167**	**uS7**	1	I [auth Ai]
	S15/NS1 RNA-binding domain	**616**	**uS15**	2	Q [auth Aq]
	Ribosomal protein eL19, N-terminal domain	**637**	eL19	1	SA [auth BS]
α bundles	**Long α-hairpin**	**192**	**uL29**	1	ZA [auth BZ]
	Ribosomal protein eL19, C-terminal domain	**3050**	eL19	2	SA [auth BS]
α + β two layers	**Ribosomal protein uS5 domain 2-like**	**212**	**uS5**	2	G [auth Ag]
			uS9	1	L [auth Al]
	Enolase-N/ribosomal protein	**218**	**uL22**	1	VA [auth BV]
	β-Grasp	**221**	**uS4**	1	E [auth Ae]
			eS4	1	F [auth Af]
	T-fold	**230**	**uS3**	2	C [auth Ac]
			eS1 (S3Ae)	2	D [auth Ad]
	Ribosomal protein eL31-like	**259**	eL31	1	CB [auth Bc]
	α-β plaits	**304**	**uS10**	1	M [auth Am]
			eS24	1	W [auth Aw]
			uL23	1	WA [auth BW]
			eL15	1	PA [auth BP]
	Ribosomal protein uL6	**318**	**uL6**	1	GA [auth BG]
			uL6	2	GA [auth BG]
	R3H domain-like	**320**	**uS8**	1	J [auth Aj]

(continued)

Table 2.1 (continued)

Architectures	Fold	ECOD_X	Protein	Domain	PDB_chain
	α-Lytic protease prodomain-like	**327**	**uS3**	1	C [auth Ac]
	dsRBD-like	**330**	**uS5**	1	G [auth Ag]
	DNA polymerase III C-terminal domain	**585**	eL39	1	HB [auth Bi]
	Ribosomal protein uS19	**838**	**uS19**	1	U [auth Au]
	Ribosomal protein uL5	**882**	**uL5**	1	FA [auth BF]
	Ribosomal protein uL30	**887**	**uL30**	1	AB [auth Ba]
	GP2-like	**3115**	eL20 (L18Ae)	1	TA [auth BT]
	Ribosomal protein uS8, C-terminal domain	**4988**	**uS8**	2	J [auth Aj]
α + β three layers	**Bacillus chorismate mutase-like**	**301**	eL30	1	BB [auth Bb]
			L7Ae (eL30-like)	1	HA [auth BH], IA [auth BI]
α + β complex topology	**α/β-Hammerhead/ Barrel-sandwich hybrid**	**325**	**uL16**	1	JA [auth BJ]
Mixed α + β and α/β	**Ribonuclease H-like**	**2484**	**uS11**	1	N [auth An]
			uL18	1	QA [auth BQ]
α/β three-layered sandwiches	**Flavodoxin-like**	**2007**	**uS2**	1	B [auth Ab]
	Ribosomal protein uL13/uL15/eL18/ eL32	**2490**	**uL13**	1	KA [auth BK]
			uL15	1	OA [auth BO]
			eL18	1	RA [auth BR]
			eL32	1	DB [auth Bd]
	Ribosomal protein uL4	**7559**	**uL4**	1	EA [auth BE]

(continued)

Table 2.1 (continued)

Architectures	Fold	ECOD_X	Protein	Domain	PDB_chain
β barrels	**Cradle loop barrel**	**1**	**uL3**	1	DA [auth BD]
	OB-fold	**2**	**uS12**	1	O [auth Ao]
			uS17	1	S [auth As]
			eS4	2	F [auth Af]
			eS28	1	Y [auth Ay]
			uL2	1	CA [auth BC]
	SH3	**4**	eS4	3	F [auth Af]
			uL2	2	CA [auth BC]
			uL24	1	XA [auth BX]
			eL14	1	MA [auth BM], NA [auth BN]
			eL21	1	UA [auth BU]
	Ribosomal protein bL25-like	**239**	eS1 (S3Ae)	1	D [auth Ad]
			eS6	1	H [auth Ah]
	Ribosomal protein uL14-like	**3174**	**uL14**	1	LA [auth BL]
			eS8	1	K [auth Ak]
Few secondary structure elements (β barrel-like)	**Rubredoxin-like**	**375**	eS27	1	X [auth Ax]
			eL37	1	GB [auth Bh]
			eL40	1	IB [auth Bj]
			eL42 (eL44)	1	KB [auth Bl]
			eL43	1	FB [auth Bg]
			Predicted	1	Z [auth Az]
	Glucocorticoid receptor-like	**377**	**uS14**	1	R [auth Ar]
			eL24	1	YA [auth BY]
			eL34	1	EB [auth Be]

Table 2.2 Structural and functional classification of ancient β-barrel folds

Fold	b-strand pattern	Amino acid length	Biological function	Representative proteins
Double-ψ β-barrel (DPBB)	$\downarrow_6\uparrow_1\uparrow_5\downarrow_2\downarrow_4\uparrow_3$	~80	Transcription/ replication Metabolism/ protein degradation	Cellular RNA polymerase Archaeal replicative DNA polymerase PolD VCP chaperone
RIFT	$\downarrow_6\uparrow_1\downarrow_2\uparrow_5\downarrow_4\uparrow_3$	~80	Translation/ metabolism Protein degradation	Ribosomal protein uL3 Elongation factor Tu Riboflavin synthase
Oligonucleotide/ oligosaccharide binding fold (OB)	$\uparrow_5\downarrow_4\uparrow_1\downarrow_2\uparrow_3$	~70	Transcription/ translation Replication/ metabolism DNA repair	Ribosomal protein uL2 Lysyl- and aspartyl-tRNA synthase Single-stranded DNA binding (SSB) protein
SRC homology 3 (SH3)	$\uparrow_5\downarrow_1\uparrow_2\downarrow_3\uparrow_4$	~70	Translation Signal transduction	Ribosomal protein uL2 RNaseP, Rpp29 Phosphoinositide 3-kinase
Rubredoxin	$\downarrow_1\uparrow_2\downarrow_3\uparrow_4$	~50	Replication/ translation Metabolism/ electron transfer	Rubredoxin Archaeal replicative DNA polymerase PolD Ribosomal protein eL44

2.1.1 DPBB Fold Conserved in RNA Polymerase

The catalytic core of cellular RNA polymerases is composed of two double-psi β-barrel domains (DPBBs); one (DPBB-A) coordinates the catalytic magnesium ions with the well-conserved motif "DFDGD," and the other (DPBB-B) interacts with DNA by two invariant lysine residues (Fouqueau et al. 2017; Iyer et al. 2003; Lane and Darst 2010; Werner and Grohmann 2011) (Fig. 2.1a). These two DPBB domains are usually encoded in different subunits (e.g., β′ and β subunits in bacterial RNA polymerase). In some exceptional cases, these subunits are fused into one polypeptide containing two DPBB domains (e.g., in *Helicobacter pylori*) (Zakharova et al. 1999). Most single subunit RNA polymerases (e.g., T7 RNA polymerase) are evolutionarily unrelated to cellular RNA polymerases, and have different catalytic domains. However, single subunit RNA polymerases containing two DPBB domains have also been identified from some bacteriophages (Chaban et al. 2024; Drobysheva et al. 2021; Forrest et al. 2017). The arrangement of two DPBB domains is also conserved in the archaeal replicative DNA polymerase PolD (Raia et al. 2019a, b; Sauguet et al. 2016) (Fig. 2.1b). These observations suggest that the evolution of (deoxy-) ribonucleic acid polymerases started by the dimerization and fusion of DPBB domains, followed by the acquisition of other domains and, in the case of

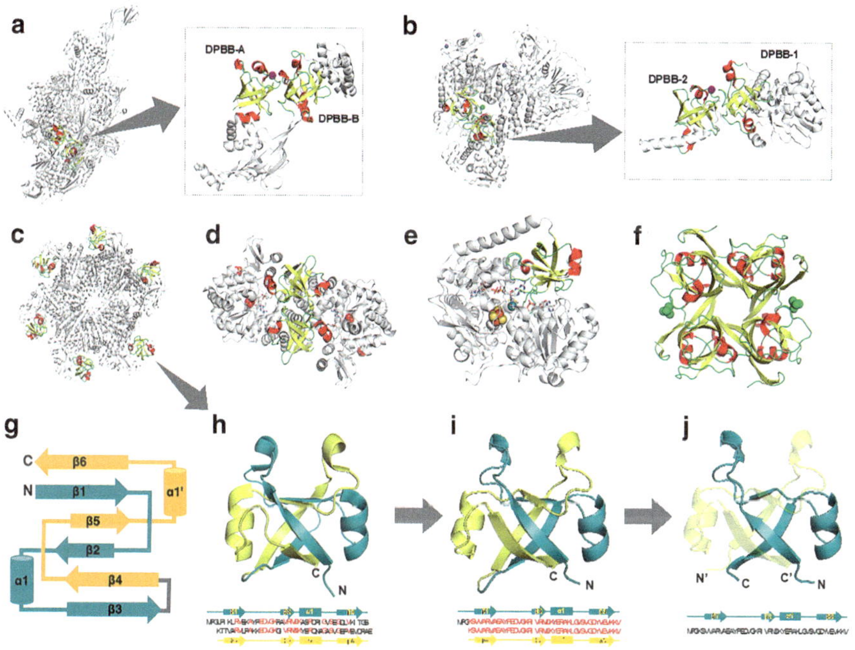

Fig. 2.1 Structures of the DPBB fold and its experimental simplification. (**a**–**f**) The DPBB domains in (**a**) cellular RNA polymerases (PDB ID: 4G7H), (**b**) archaeal replicative DNA polymerase PolD (PDB ID: 8PPT), (**c**) hexameric VCP chaperone (PDB ID: 5G4F), (**d**) dimeric AMP phosphorylase (PDB ID: 4GA6), (**e**) dimethylsulfoxide reductase (PDB ID: 1FDO), and (**f**) homotetrameric aspartate decarboxylase (PDB ID: 1PPY) are colored by secondary structure (α-helix: red; β-sheet: yellow; loop: green). (**g**) Schematic topology of the DPBB fold. The N- and C-terminal ββαβ motifs are colored cyan and yellow, respectively. (H–J) The crystal structures of (**h**) the isolated DPBB domain from *Methanopyrus kandleri* VCP chaperone (PDB ID: 7DG7), (**i**) a designed DPBB protein with perfect sequence repeats (PDB ID: 7DU6), and (**j**) half-sized homo-dimeric DPBB peptides (PDB ID: 7DXT) are shown with the same color scheme as in panel G. The amino acid sequences and secondary structures are presented below each panel. The conserved residues in the N- and C-terminal halves are highlighted in red

cellular RNA polymerases, separation into multiple subunits to perform sophisticated gene expression control.

The DPBB fold is also found in distinct protein superfamilies, such as VCP chaperone (Fig. 2.1c), AMP phosphorylase (Fig. 2.1d), dimethylsulfoxide reductase (Fig. 2.1e), and aspartate decarboxylase (Fig. 2.1f), indicating its versatility to realize diverse enzymatic functions (Castillo et al. 1999).

The DPBB fold is a six-stranded β-barrel capped by two alpha-helices, in which a ββαβ motif is repeated twice (Fig. 2.1g, h). The N- and C-terminal motifs are interlaced to form the barrel part with partially parallel and antiparallel orientations $(\downarrow_6\uparrow_1\uparrow_5\downarrow_2\downarrow_4\uparrow_3)$. The two loops between β1(β4) and β2(β5) cross over the loop between β5(β2) and β6(β3). As this irregular loop configuration resembles the Greek letter ψ, the two loops are called ψ loops (Suguna et al. 1987). The pseudo-two-fold symmetry of the DPBB fold implies that DPBB originally emerged as a homodimer of a short peptide with a single $\beta\beta\alpha\beta$ motif and then evolved into modern pseudo-symmetrical domains by gene duplication and fusion events (Coles et al. 1999).

We experimentally reconstructed the ancient homo-dimeric DPBB fold to validate the above evolutionary scenario (Yagi et al. 2021). The DPBB domains of the chaperone-like protein VCP, from the hyper-thermophilic archaea *Methanopyrus kandleri* and *Aeropyrum pernix*, exhibited high internal sequence symmetry: the sequence identities between their N- and C-terminal fragments are 42 and 45%, respectively (Fig. 2.1h). This suggests that proteins from hyper-thermophilic species have retained ancestral features, consistent with the hypothesis that the last universal common ancestor (LUCA) was a hyper-thermophile (Akanuma et al. 2013; Nasir et al. 2014; Weiss et al. 2016a, b). By engineering these hyper-thermophilic proteins, we reconstructed DPBB domains with perfect sequence repeats (Fig. 2.1i) and half-sized homo-dimeric DPBB peptides with 43 amino acid residues (Fig. 2.1j). Furthermore, we could simplify the sequence of the homo-dimeric DPBB peptides to include only limited sets of amino acid types. The most simplified peptide comprises only seven amino acid types (Ala, Gly, Val, Asp, Glu, Lys, and Arg) and forms random-coil structures in standard phosphate buffer. However, it adopted the DPBB fold when crystallized under conditions with a high malonate or malic acid concentration. This study demonstrates that the DPBB structure could emerge even within primitive translation systems restricted to a limited number of amino acids in early evolutionary periods.

2.1.2 RIFT Fold Conserved in Ribosomal Proteins and Translational Factors

The RIFT fold is one of the widespread small β-barrels and, like DPBB, comprises two α-helices and six β-strands with two $\beta\beta\alpha\beta$ motifs (Fig. 2.2a) (Alva et al. 2008; Coles et al. 2006). However, its β-strand arrangement is all antiparallel ($\downarrow_6\uparrow_1\downarrow_2\uparrow_5\downarrow_4\uparrow_3$), differing from that of DPBB ($\downarrow_6\uparrow_1\uparrow_5\downarrow_2\downarrow_4\uparrow_3$). The loops linking β1 (β4) and β2 (β5) form simple β-hairpin structures in the RIFT fold, instead of the ψ-loops in DPBB. The similarity between DPBB and RIFT is also apparent in their sequences, with two well conserved GD-box motifs at the N-termini of β3 and β6 (Alva et al. 2009). Because of these similarities, DPBB and RIFT are considered sister proteins with a common ancestor (Alva et al. 2008; Coles et al. 2006). Thus, both have been classified in the same meta-fold, denoted as cradle loop barrel (ECOD X-group 1).

The RIFT fold is conserved across various proteins and named after its occurrence in riboflavin synthase (Fig. 2.2b), F1 ATP synthase (Fig. 2.2c), and translation proteins such as the universal ribosomal protein uL3 (Fig. 2.2d), archaea- and eukarya-specific ribosomal protein aL35 (Fig. 2.2e), EF-Tu (Fig. 2.2f), EF-G (Fig. 2.2g), and eIF5B (Fig. 2.2h) (Ammelburg et al. 2007; Coles et al. 2006). uL3

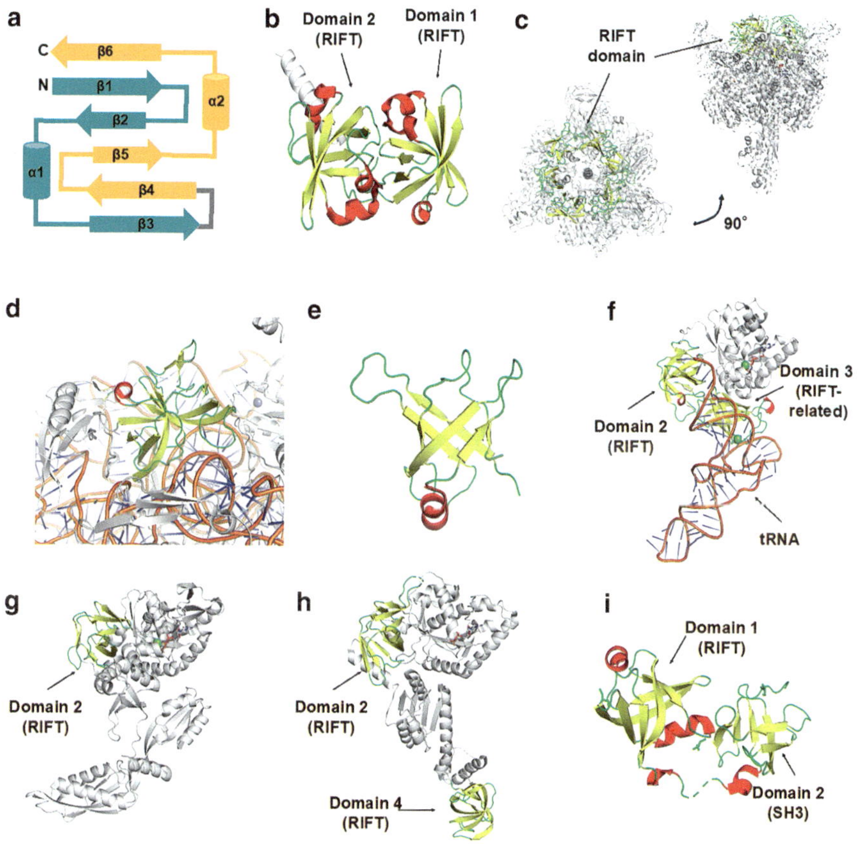

Fig. 2.2 Structure of the RIFT fold. (**a**) Schematic topology of the RIFT fold. The N- and C-terminal ββαβ motifs are colored cyan and yellow, respectively. The structures of (**b**) riboflavin synthase (PDB ID: 1I8D), (**c**) F1 ATP synthase (PDB ID: 3W3A), (**d**) ribosomal protein uL3 (PDB ID: 6TH6), (**e**) eL35 (PDB ID: 1SQR), (**f**) EF-Tu (PDB ID: 1TTT), (**g**) EF-G (PDB ID: 2DY1), (**h**) eIF5B (PDB ID: 1G7S), and (**i**) ribosome maturation factor M (RimM; PDB ID: 2F1L). EF-Tu in panel F and RimM in panel I are fused with other β-barrel domains, RIFT-related and SH3, respectively

plays roles in tRNA binding to the ribosome and the peptidyl-transferase activity (Meskauskas and Dinman 2007). EF-Tu incorporates tRNA into the ribosome, while EF-G and IF2/eIF5B facilitate the translation cycle (Abel and Jurnak 1996; Hartman and Smith 2010; Kuhle and Ficner 2014; Roll-Mecak et al. 2000; Savelsbergh et al. 2009; Sergiev et al. 2005). Furthermore, ribosome maturation factor M (RimM; Fig. 2.2i) also has a RIFT domain (Guo et al. 2013; Maksimova et al. 2022). Thus, the RIFT fold is one of the essential components in multiple stages of translation.

Some other β-barrel folds in the cradle loop barrel group function as RNA binding domains. EF-Tu has not only the RIFT barrel fold but also a different

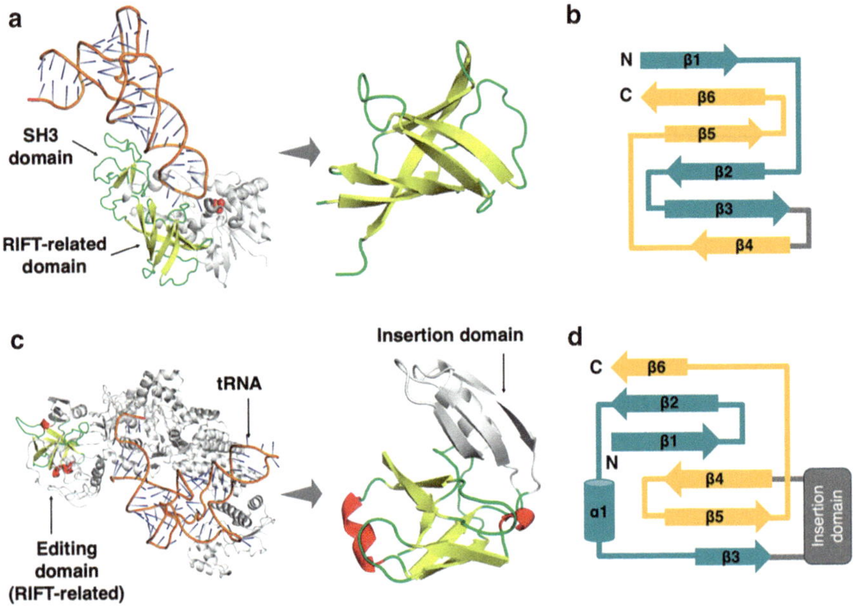

Fig. 2.3 Structures of RIFT-related folds. (**a**) The structure of the tRNA sulfuration enzyme MnmA and its aminomethyltransferase β-barrel domain (PDB ID: 2DET). (**b**) The topology of the aminomethyltransferase β-barrel domain is shown in the same manner as Fig. 2.2a. (**c**) The structure of leucyl-tRNA synthase and its acid protease domain (PDB ID: 1WZ2). (**d**) The topology of the acid protease domain is shown in the same manner as Fig. 2.2a

RIFT-related barrel fold, assigned as "Aminomethyltransferase β-barrel domain" T-group in ECOD (ECOD ID 1.1.8.9) (Fig. 2.3a) (Kjeldgaard et al. 1993). Its β-sheet topology has the $\uparrow_1\downarrow_6\uparrow_5\downarrow_2\uparrow_3\downarrow_4$ arrangement (Fig. 2.3b). The same β-barrel structure is found in the EF-Tu homolog proteins eIF2γ (Roll-Mecak et al. 2004) and SelB (Leibundgut et al. 2005). Additionally, the tRNA binding domain of MnmA, which modifies the uridine at the first anticodon base of the tRNAs for glutamate, lysine, and glutamine by introducing sulfur, is composed of this type of β-barrel fold (Numata et al. 2006). Therefore, along with the DPBB and RIFT folds, this β-barrel structure serves as an essential component behind the translation system.

Leucyl-, isoleucyl-, and valyl-tRNA synthetases share a proofreading domain that removes incorrectly attached amino acids from the tRNA. This editing domain adopts a six-stranded β-barrel structure with a mixed parallel and antiparallel β-strand orientation ($\uparrow_6\downarrow_2\uparrow_1\downarrow_4\uparrow_5\downarrow_3$; ECOD ID 1.1.1.31) and includes specific insertion domains (Fig. 2.3c, d) (Fukunaga and Yokoyama 2006; Liu et al. 2006). Notably, this structure is classified within the "Acid Protease" T-group in ECOD, suggesting a possible evolutionary link between its hydrolytic activity in tRNA editing and the proteolytic function of acid proteases.

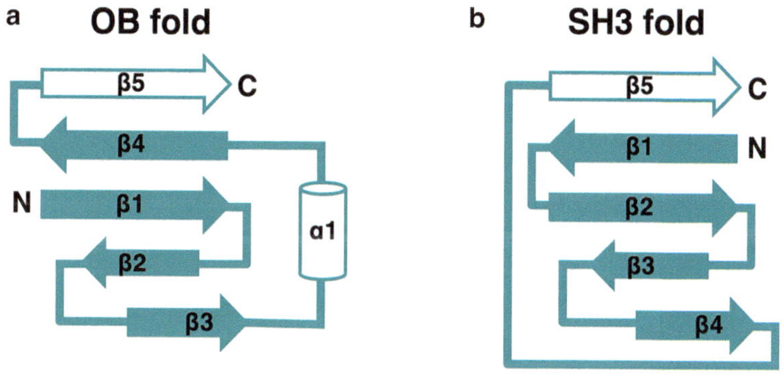

Fig. 2.4 Schematic topologies of the (**a**) OB and (**b**) SH3 folds. The four strands of the core part in both folds are filled in cyan

2.1.3 OB and SH3 Folds in Translational Machinery and its Relatives

The oligonucleotide/oligosaccharide binding fold (OB-fold; ECOD X-group 2) and SRC homology three fold (SH3-fold; ECOD X-group 4) are also considered as ancient small β-barrel structures and are found in many transcriptional and translational proteins (Agrawal and Kishan 2001; Youkharibache et al. 2019). These versatile structural motifs play crucial roles in RNA binding and enzymatic functions, making them essential for pivotal biological processes. While both folds adopt a five-stranded β-barrel structure, their topologies differ: the β-sheet patterns of OB and SH3 are $\uparrow_5\downarrow_4\uparrow_1\downarrow_2\uparrow_3$ (Fig. 2.4a) and $\uparrow_5\downarrow_1\uparrow_2\downarrow_3\uparrow_4$ (Fig. 2.4b), respectively. The conformation of the fifth β-strand is highly variable between and within OB and SH3 proteins. Some homologs, such as ribosomal protein L2, even lack this strand (Alvarez-Carreño et al. 2021). Therefore, it is thought that the original OB and SH3 folds formed four-stranded core structures, and then their fifth strands emerged independently to adopt the five-stranded structures observed today.

2.1.3.1 OB and SH3 Folds in Ribosomal Proteins

Many ribosomal proteins possess OB and/or SH3 domains (Fig. 2.5a). The OB structures are present in uL2, eS4, uS12, uS17, and eS28 (Fig. 2.5b–d, k, l), and the SH3 structures are components of uL2, eS4, bL19, uL14, eL14, and eL21 (Fig. 2.5e–l) (Alvarez-Carreño et al. 2021). While the peptidyl-transferase activity of the ribosome is catalyzed by rRNA (Ban et al. 2000; Harms et al. 2001; Nissen et al. 2000), ribosomal proteins enhance the folding and precise conformations of rRNAs (Nikolay et al. 2015). For example, the uS12 protein interacts with tRNA and rRNA at the decoding center. The removal of uS12 from the reconstructed ribosome system substantially decreased the protein synthesis activity (Shimojo et al. 2020).

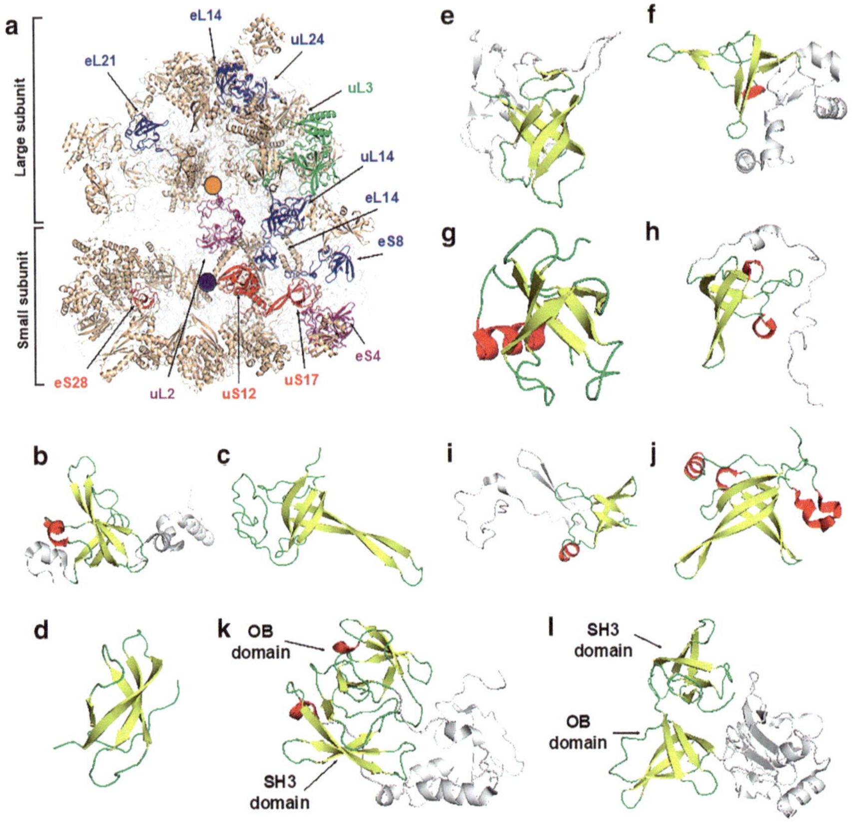

Fig. 2.5 β-barrel proteins in the ribosome. (**a**) The ribosome structure of the archaeon, *Thermococcus kodakarensis* (PDB ID: 6TH6). Its peptidyl-transfer center (PTC) and decoding center (DC) are indicated by orange and purple circles. The proteins with the OB, SH3, OB + SH3, and RIFT folds are colored red, blue, pink, and green, respectively. (**b–d**) The structures of OB-fold proteins, uS12 (**b**), uS17 (**c**), and eS28 (**d**). (E–J) The structures of SH3-fold proteins, uL14(**e**), uL24(**f**), eL14(**g**), eL21(**h**), eS28(**i**), and bL19(**j**). The structures of ribosomal proteins containing both OB and SH3 folds: uL2(**k**) and eS4(**l**)

Additionally, the translation accuracy was affected by mutations in uS12, indicating its significant importance in the decoding system (Demirci et al. 2013; Sharma et al. 2007). Other ribosomal proteins with the OB and SH3 folds reportedly correspond to the assembly and stabilization of rRNA, mRNA, and tRNA molecules (Delvillani et al. 2011; Espinar-Marchena et al. 2018; Ramaswamy and Woodson 2009; Zhang 2024).

Furthermore, two ribosome biogenesis factors, RimP (Garaeva et al. 2024) and RsgA/YjeQ (Rocchio et al. 2019; Shin et al. 2004), have SH3 and OB domains, respectively. These proteins enhance the assembly of the small ribosomal subunit through interactions with their β-barrel domains. Indeed, *in vitro* ribosome reconstitution experiments confirmed that these proteins improve the assembly of the small ribosomal subunit (Tamaru et al. 2018).

2.1.3.2 Translational Factors

Bacterial translation initiation factor IF1 and its archaeal and eukaryal homolog, eIF1a, are OB-fold proteins and bind to the A-site of the small subunit through interactions with mRNA and another OB-fold protein, uS12 (Fig. 2.6a, b) (Schmitt et al. 2020). One of the archaeal initiation factors with the OB fold, eIF2a, stabilizes the complex of initiator methionyl-tRNA and eIF2γ, which contain RIFT and RIFT-related β-barrel folds (Fig. 2.6b) (Naveau et al. 2013). Small protein B (SmpB), an essential component in the tmRNA-SmpB system, also contains a single OB domain and rescues the stalled ribosome from truncated mRNA (Fig. 2.6c) (Guyomar et al. 2021; Neubauer et al. 2012). It binds at a similar site to that bound by bIF1 and eIF1a.

Escherichia coli bS1 comprises six-repeat OB folds (Fig. 2.6d). Unlike other typical ribosomal proteins, bS1 is not an integral component of the ribosome. It recruits mRNA onto the small subunit of the ribosome by binding to the ribosome surface with its N-terminal OB domains and interacting with mRNA via its C-terminal OB domains (Lauber et al. 2012; Sengupta et al. 2001). Recently, structural models of a giant complex with RNA polymerase, ribosome, and bS1 have been reported, providing a mechanistic framework for their cooperative action in translation initiation (Fig. 2.6d) (Webster et al. 2024).

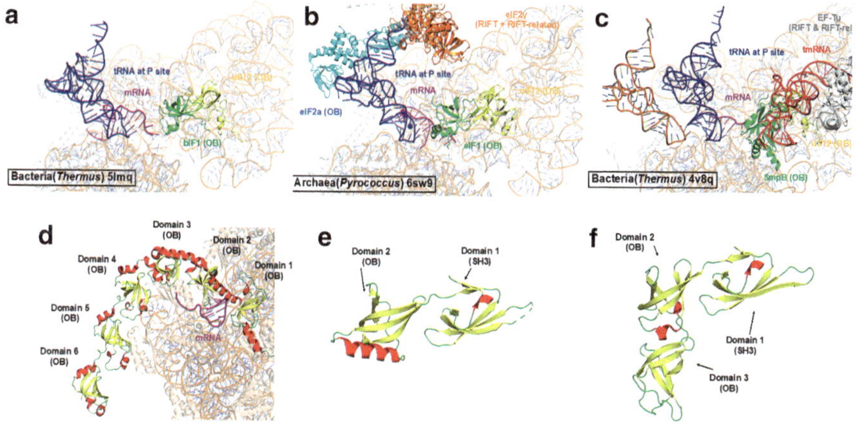

Fig. 2.6 OB and SH3 folds in translational factors. (**a**, **b**) Structures of bIF1 and eIF1a (green) bound to the ribosomal small subunit along with mRNA (pink) and tRNA (blue) (PDB IDs: 5LMQ and 6SW9). S12, located near the decoding center, is highlighted in yellow. In panel B, aIF2a and eIF2γ are shown in cyan and orange, respectively. (**c**) Complex structure of SmpB, tmRNA, and EF-Tu bound to the ribosomal small subunit (PDB ID: 4V8Q). (**d**) Complex structure of ribosomal protein bS1 and mRNA on the ribosomal small subunit (PDB ID: 9GUT). (**e**, **f**) Structures of eIF5a and EF-P (PDB IDs: 8QZV and 1UEB)

Archaeal and eukaryal IF5a (eIF5a) were initially considered to function in the initiation stage of translation. However, later studies discovered that it acts more in the elongation and termination stages (Guo and Zhou 2024; Schuller et al. 2017; Yao et al. 2003). Its bacterial homolog is elongation factor P (EF-P). eIF5a and EF-P share conserved SH3 and OB domains, and EF-P possesses another OB domain at its C-terminus (Fig. 2.6e, f). These proteins bind to the E site of a ribosome stalled at proline-rich sequences to relieve the ribosome from arrest (Doerfel et al. 2013; Guo and Zhou 2024; Schuller et al. 2017).

2.1.3.3 tRNA Modification

In class-IIb aminoacyl tRNA synthases, including lysyl- and aspartyl-tRNA synthases, the OB domain contributes to tRNA anticodon recognition (Fig. 2.7a) (Moulinier et al. 2001; Onesti et al. 1995). The bacterial Trbp111 protein comprises

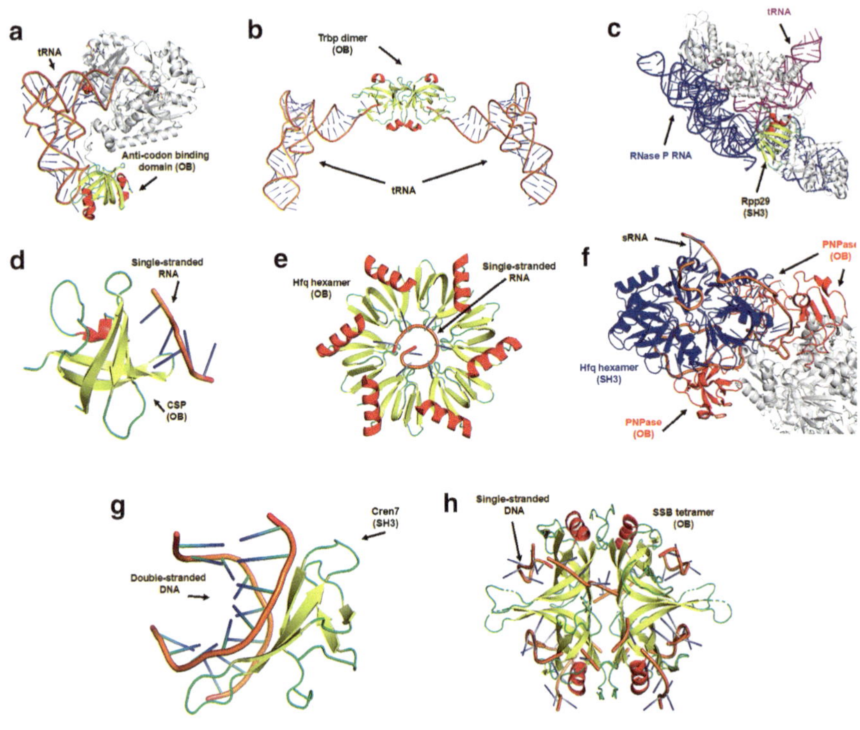

Fig. 2.7 OB and SH3 folds in proteins functioning in tRNA modification, RNA folding, and DNA binding. The structures of (**a**) lysyl-tRNA synthase (PDB ID: 1C0A), (**b**) dimeric Trbp (PDB ID: 1VU0), (**c**) Rpp29 in the complex of RNase P (PDB ID: 6K0B), (**d**) CSP with single-stranded RNA (ssRNA) (PDB ID: 2ES2), (**e**) homo-hexameric Hfq with ssRNA (PDB ID: 1KQ2), (**f**) the complex of Hfq-PNPase-ssRNA (PDB ID: 9GUT), (**g**) cren7 with double-stranded DNA (PDBID: 4R56), and (**h**) homo-tetrameric SSB with ssDNA (PDB ID: 3A5U)

a dimeric OB structure and facilitates tRNA aminoacylation and folding (Fig. 2.7b) (Umuhire Juru et al. 2024). Rpp29, an essential subunit of the RNase P enzyme in the archaea and eukarya, is composed of an SH3 domain, interacts with the RNA subunit (Fig. 2.7c) (Boomershine et al. 2003; Honda et al. 2010; Wan et al. 2019), and plays a crucial role in tRNA maturation by cleaving the precursor tRNA to create the mature 3′-end.

2.1.3.4 RNA Chaperone

Beyond their roles in translation, the OB domains in some proteins such as cold shock proteins (Fig. 2.7d) (Giaquinto et al. 2007; Giuliodori et al. 2023; Max et al. 2006), bacterial ribosomal protein S1 (Duval et al. 2013; Qu et al. 2012; Qureshi et al. 2018), and *E. coli* IF1 (Croitoru et al. 2006; Phadtare et al. 2007; Weber et al. 2001) also function as RNA chaperones. These proteins assist in unwinding and modifying RNA structures. For example, the RNA melting activity of bS1 unfolds mRNA with secondary structures and facilitates its accommodation into the decoding center of the ribosome. Furthermore, like the OB fold, the SH3 domain in the abundant bacterial protein Hfq is a crucial post-transcriptional regulator (Fig. 2.7e). Hfq binds various types of RNA, such as small RNA (sRNA), mRNA, tRNA, and rRNA, and then promotes their proper base-pairing as an RNA matchmaker (Dos Santos et al. 2019; Hoekzema et al. 2019; Massé and Gottesman 2002; Møller et al. 2002; Sledjeski et al. 2001; Zhang et al. 2002). Recently, it was reported that Hfq forms a complex with polynucleotide phosphorylase (PNPase) via sRNA (Dendooven et al. 2021). Although PNPase primarily functions to degrade sRNA, in the complex with Hfq and sRNA, it protects sRNA from degradation by other ribonucleases and acts as an "RNA carrier" in the cell. In this complex, PNPase associates with sRNA through its C-terminal OB domain. As a result, the sRNA is wrapped by the SH3 domain of Hfq and the OB domain of PNPase (Fig. 2.7f).

2.1.3.5 DNA Binding Proteins

Some DNA-binding proteins also contain OB or SH3 folds. For example, SH3 domains are present in chromosomal proteins. In certain archaeal species, such as the *Sulfolobus* group, SH3-containing proteins bind genomic DNA, providing structural protection similar to the histones observed in the eukaryal nucleosome (Fig. 2.7g) (Driessen et al. 2013; Geethika et al. 2024; L. Guo et al. 2008; Kalichuk et al. 2016; Laursen et al. 2021; Zhang et al. 2010). The single-stranded DNA binding (SSB) protein with the OB domain is believed to be an essential component for DNA replication, recombination, and repair processes (Bochkarev et al. 1997; Fedorov et al. 2006; Kerr et al. 2003; Raghunathan et al. 1997; Suksombat et al. 2015). Homo-tetrameric SSB wraps and stabilizes single-stranded DNA through interactions between DNA bases and invariant aromatic residues (Fig. 2.7h).

2.1.4 Rubredoxin Fold

The rubredoxin (zinc ribbon) fold (ECOD X-group 375) is another interesting β-barrel fold shared by numerous ribosomal proteins. It consists of four β-strands ($\downarrow_1\uparrow_2\downarrow_3\uparrow_4$) with two hairpins, each containing two cysteine residues, which cluster together to coordinate a single zinc ion. It is conserved in the universal ribosomal protein uL33 (Kovacs et al. 2018), the bacterial-specific ribosomal protein bL32, and the archaeal and eukaryotic-specific proteins eL37, eL40, eL43, eS27, and eS31. Interestingly, the rubredoxin fold is also observed in the DNA binding site of archaeal DNA polymerase PolD (Mayanagi et al. 2020). In the PolD structure complexed with template DNA, it contacts the backbone phosphate groups of double-stranded DNA via several positively-charged residues.

2.1.5 αβ-Protein Folds in Ribosomal Proteins Shared with Metabolic Enzymes

Several αβ-protein folds found in ribosomal proteins, such as the ferredoxin-like fold, RNase H-like fold, and flavodoxin-like fold, are also among the most frequently observed folds in metabolic enzymes (Caetano-Anollés et al. 2007; Ma et al. 2008). The ferredoxin-like fold (also called RNA recognition motif or RRM) is classified within the meta-fold called α-β plaits in ECOD (X-group 304), along with some other αβ-protein folds. Ribosomal protein uS10 exhibits the typical RRM topology, while uL23, eS24, and eL15 adopt slightly different topologies in the same X-group. The catalytic cores of many DNA polymerases and viral-type RNA polymerases also contain the RRM fold (Koonin et al. 2020), suggesting that some ribosomal proteins might share a common ancestor with these polymerases.

Ribosomal proteins uS11 and uL18 contain the RNase H-like fold (X-group 2484), which is also found, for example, in glycerol kinase, acetate kinase, DNA transposase, many nucleases, and eRF1. Ribosomal protein uS2 is composed of the flavodoxin-like fold (X-group 2007), a highly conserved structure present in a wide variety of metabolic enzymes, such as phosphoglycerate kinase, succinyl-CoA synthetase, and flavodoxin. The flavodoxin-like fold likely has an evolutionary relationship with the TIM barrel fold, another well-conserved fold in metabolic enzymes (Farías-Rico et al. 2014).

2.1.6 α- and αβ-Protein Folds in Ribosomal Proteins Shared with Transcription Regulators

α-Protein folds are relatively rare in ribosomal proteins (Table 2.1). Still, the helix-turn-helix (HTH) fold (X-group 101) is conserved among a few ribosomal proteins (uS15, eS17, and eS19). HTH is a well-known DNA-binding motif present in numerous transcription regulators, including the λ repressor.

Another well-known protein fold shared between ribosomal proteins and transcription regulators is the type II KH fold (X-group 327). The KH domains were first identified by the conserved GxxG motif (Siomi et al. 1993), and were subsequently shown to adopt three distinct αβ topologies (type I–III) as a meta-fold (Grishin 2001; Pereira and Lupas 2018). The type II KH fold is found in uS3 and the universally conserved transcription factor NusA (Pereira and Lupas 2018), indicating that it originated before LUCA. Some archaea possess a homolog of NusA (NusA2), which retains the same basic topology but lacks the GxxG motif. Structural and phylogenetic analyses indicated that NusA2 is the ancestor of ribosomal protein eS7 (Phung et al. 2025).

2.1.7 Unique Folds in Ribosomal Proteins

Although many protein folds in ribosomal proteins are also conserved in non-ribosomal proteins, as seen above, some unique folds are found exclusively in ribosomal proteins. For example, ribosomal protein uL6 comprises two domains with the same αβ fold and a unique topology (X-group 318). Ribosomal proteins uL13, uL15, eL18, and eL32 share the same fold (X-group 2490), strongly indicating that they diversified from a common ancestor by gene duplication and subsequent mutations. It is not clear why these folds are used only in ribosomes. They might have adapted to the special RNA-rich environment in the ribosomes.

2.2 Evolutionary Relationships Between Distinct β-Barrel Folds

So far, we have discussed the fundamental protein folds conserved in the gene expression system. The same folds are utilized as domains in various proteins, highlighting their versatility and evolutionary significance. Understanding how these prototypical proteins emerged is a major challenge in protein science and studies of the origins of life. Research into their evolution is ongoing, and here we present some of the key recent findings.

2.2.1 *Transition Between DPBB and RIFT via an Evolutionary Missing Link*

As the DPBB and RIFT folds have similar tertiary structures with a highly conserved sequence motif, the GD box, both folds are regarded as sister protein folds. Two plausible diversification pathways linking the DPBB and RIFT folds have been proposed. One hypothesis is that an ancestral homo-dimeric RIFT protein became a monomeric form by gene duplication and fusion, and subsequently underwent strand swapping twice, resulting in the modern DPBB architecture (Alva et al. 2008; Coles et al. 2006). The other hypothesis is that homo-dimers of the common ancestor initially diverged to the homo-dimeric DPBB and RIFT proteins, followed by the duplication and fusion of individual genes independently, leading to the emergence of the modern monomeric DPBB and RIFT proteins (Coles et al. 1999; Lupas and Koretke 2008).

These hypotheses have been difficult to test, based only on sequence and structure comparisons. However, more compelling evidence has been provided through experimental reconstructions of ancestral proteins. As described above, the simplified DPBB peptide with only seven amino acid types adopted the homo-dimeric DPBB fold in the presence of malonate or malic acid (Fig. 2.8a) (Yagi et al. 2021). This peptide formed a new homo-dimeric β-barrel structure, named the double-Z β-barrel (DZBB), at a high sulfate ion concentration, instead of the original DPBB fold (Fig. 2.8b) (Yagi and Tagami 2024). Such a metamorphic property may have served as an intermediate state in drastic structural evolution and connected the sequence spaces for distinct folds (Das et al. 2021; Madhurima et al. 2021; Murzin 2008; Yadid et al. 2010). In the DZBB structure, the loop connecting β1 and β2 forms a long, coil-like conformation. Notably, in the DPBB structure, the corresponding loop adopts the ψ-loop, whereas in the case of the RIFT structure, the loop forms the simple β-hairpin. Additionally, the strand directions of DZBB are completely antiparallel (↓↑↓↑↓↑) like those of the RIFT fold, implying that it might be the evolutionary intermediate form between DPBB and RIFT.

This potential evolutionary role of DZBB has been examined through further experimental verifications, as detailed below (Yagi and Tagami 2024). Replacing

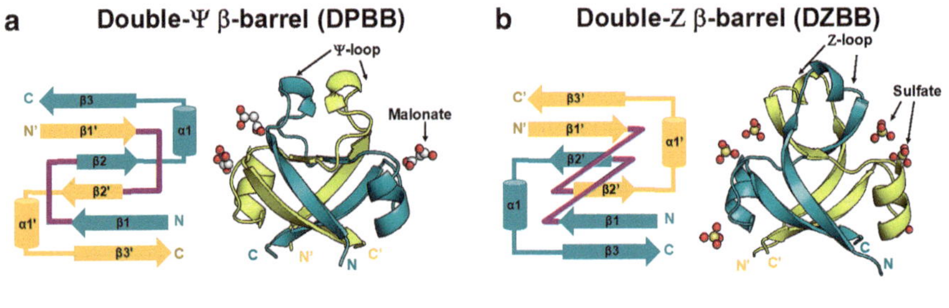

Fig. 2.8 The topologies and crystal structures of the simplified peptide with (**a**) DPBB and (**b**) DZBB structures

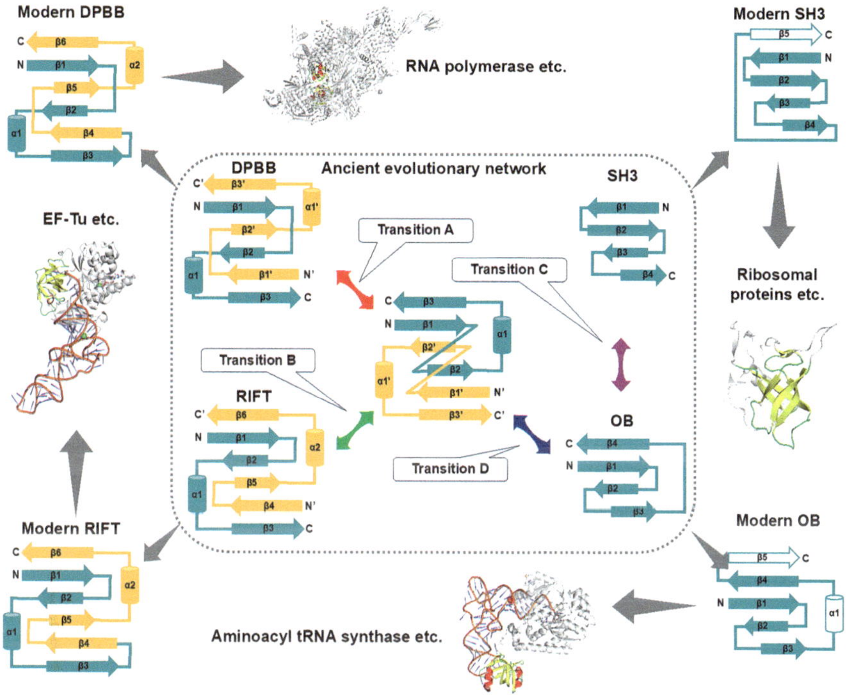

Fig. 2.9 The proposed evolutionary network for four distinct β-barrels. The topologies of the modern DPBB, RIFT, OB, and SH3 folds and their structures in representative proteins are shown outside the dashed box. The experimentally supported evolutionary relationships between distinct β-barrels are shown inside the dashed box. The fold conversion between DPBB and DZBB is indicated as a red arrow (transition A). The fold conversion between DZBB and RIFT is shown as a green arrow (transition B). The fold conversion between OB and SH3 is shown as a pink arrow (transition C). The fold conversion between DZBB and OB is shown as a blue arrow (transition D)

only five residues in the metamorphic peptide with those conserved in Phs018, a simple RIFT protein with presumed ancestral features, stabilized the DZBB structure. This demonstrated that the complete fold transition from DPBB to DZBB can be achieved by just a few amino acid substitutions (Transition A in Fig. 2.9).

Subsequently, the sequence connecting β1 and β2 in this DZBB variant was engineered to form the β-hairpin found in the RIFT fold, leading to the formation of the homo-dimeric RIFT structure (Transition B in Fig. 2.9). These experimental results strongly support the evolutionary hypothesis that the early homo-dimeric DPBB and RIFT proteins diverged from a common ancestral one, via the DZBB fold intermediate. The DZBB fold is not found in extant proteins, and thus is likely to be an evolutionary missing link between the DPBB and RIFT folds.

2.2.2 Transition Between OB and SH3

While the OB and SH3 folds are classified into different categories in structure-based databases such as SCOP, CATH, and ECOD, it has been hypothesized that both originated from a common ancestral protein and subsequently diverged (Agrawal and Kishan 2001). By carefully comparing their structures and sequences, Alvarez-Carreño and colleagues suggested that the ancient core of the OB fold, composed of four β-strands, was converted into the core of the SH3 fold by a simple circular permutation (and vice versa) (Alvarez-Carreño et al. 2021). They also computationally reconstructed the permutated sequences of the ancestral OB and SH3 cores, which were predicted to adopt SH3- and OB-like structures, respectively, by AlphaFold2. Recently, this fold conversion has been experimentally verified (Yagi and Tagami 2024). The artificially designed protein with a four-stranded OB structure has been engineered by transferring its fourth strand to the N-terminal end with two additional residues, to form an ideal β-turn. The resultant protein adopted the four-stranded SH3 fold (Transition C in Fig. 2.9). Therefore, the two major building blocks comprising many proteins involved in the gene expression system likely share a common origin.

2.2.3 Transition Between DZBB and OB

The GD box mentioned above as a conserved sequence motif between DPBB and RIFT has also been detected in other protein folds, including OB and SH3. In addition to this shared GD-box motif across RIFT and OB, Alvarez-Carreño and colleagues detected the significant cross-fold sequence similarity between the RIFT and OB folds (Alvarez-Carreño et al. 2022). These observations imply the evolutionary link between the six stranded β-barrels, DPBB and RIFT, and the five stranded β-barrels, OB and SH3. However, the exact mechanism by which such distinct structures diverged was not clearly explained by those analyses.

Interestingly, structural and sequence similarities were detected between the pseudo-monomeric region of DZBB and the OB domain of the ribosomal protein L2 (Yagi and Tagami 2024). Furthermore, structure-based protein engineering revealed that the interconversion between DZBB and the four-stranded core of OB required only a single insertion of 7 amino acids, corresponding to the β2 strand in OB, along with five point substitutions in the sequence. (Transition D in Fig. 2.9). The fact that only a few mutations can enable a fold conversion implies that such an event may have taken place at a certain point in early evolutionary history. As the DZBB structure can mediate the fold transition between DPBB and RIFT, it may have served as an evolutionary hub connecting multiple ancient β-barrel folds.

2.2.4 Limited Evolvability of the DZBB Fold

Given the significant role of DZBB as an evolutionary hub linking DPBB, RIFT, and OB, its absence in modern proteins is puzzling. One possible explanation is the low thermostability and poor functional evolvability of the DZBB structure. Contemporary DPBB and RIFT proteins are present as monomers, resulting from the duplication and fusion of the genes encoding their precursor homo-dimeric short peptides. Such tandem duplication and fusion into a monomer generally improve thermostability and liberate structural constraints imposed by symmetry, facilitating the evolution of asymmetric scaffolds that enable various specific interactions with other molecules (Blaber et al. 2012; Kawakami et al. 2017). Indeed, the thermostabilities of TIM-barrels and β-propellers were reportedly improved through tandem duplications of their original half fragments (Huang et al. 2016; Romero-Romero et al. 2021; Voet et al. 2014). Given the substantial evidence suggesting that LUCA was a highly thermophilic organism (Akanuma et al. 2013; Brooks and Gaucher 2007; Moody et al. 2024; Weiss et al. 2016b), we can infer that those ancient proteins underwent at least one selection process for heat resistance.

On the one hand, the N- and C-termini of the dimeric DPBB and RIFT peptides are close to each other. This proximity likely facilitated gene fusion, allowing the evolution into monomers without requiring additional sequence optimization. On the other hand, the N- and C-termini of the two chains of DZBB are far apart, making the transition from homo-dimer to monomer by gene fusion unrealistic. Therefore, DZBB may have become extinct during natural selection, due to its limited evolvability into a monomer protein with high thermostability and functionality.

2.2.5 Implications for the Early Evolution of Life

Collectively, the analyses described above suggest that ancient four-stranded core barrels (OB and SH3) and six-stranded homo-dimeric barrels (DPBB and RIFT) form a large evolutionary group, likely originating from a common ancestral protein (Fig. 2.9) (Yagi and Tagami 2024). Furthermore, the fold conversion experiments demonstrated that a few mutations and the insertion or deletion of short sequences can induce drastic structural changes across distinct barrel folds (Yagi and Tagami 2024). This indicates that fold conversion, especially among RNA-regulatory β-barrels, could have occurred rapidly without the need to explore vast sequence space, as most of the amino acid sequence remains unchanged. In the early evolutionary stages of proteins, mutations generated by the immature central dogma machinery may also have accelerated the diversification of protein folds.

Such rapid evolution of RNA-regulatory proteins might have facilitated the transition from the RNA world to the modern protein world. The RNA world hypothesis, which proposes that RNA molecules served as both genetic storage and catalysts before the emergence of the modern life system with protein coding (Gilbert 1986; Neveu et al. 2013), is a widely accepted scenario for explaining the origin of life. If we assume that the life system evolved from the RNA world to the early protein-RNA world, the most ancient proteins (peptides) likely functioned as cofactors or auxiliary factors for RNA (Codispoti et al. 2024; Li et al. 2022; Tagami 2023; Tagami et al. 2017; Tagami and Li 2023). The ability of small β-barrels to modify and stabilize RNA structures suggests that they represent one of the most primitive protein functions. Therefore, understanding the evolution of the ancient proteins discussed in this review will provide insights into the transition from the RNA world to the RNA-protein world.

2.3 Conclusion

This review highlighted the ancient protein folds that are highly conserved in proteins involved in the gene expression system. In particular, small β-barrel folds are prevalent in diverse proteins and play critical roles in stabilizing RNA conformations and guiding structural assembly. This indicates that there has been a long co-evolutionary process between β-barrel proteins and RNA molecules. We have also outlined the evolutionary network of diverse β-barrel structures, DPBB, RIFT, OB, and SH3. Further elucidation of the early evolution of protein structures and functions will be a key to unraveling the origins of the gene expression system. Given the possibility that missing-link proteins like the DZBB fold are essential to bridge evolutionary gaps between different protein folds, advanced computational approaches as well as comprehensive experimental validations remain indispensable.

Acknowledgement This work is supported by Japan Science and Technology Agency, Fusion Oriented Research for Disruptive Science and Technology (FOREST) program [grant no. 24012213 (S.T.)].

References

Abel K, Jurnak F (1996) A complex profile of protein elongation: translating chemical energy into molecular movement. Structure (London, England: 1993) 4(3):229–238

Agrawal V, Kishan RK (2001) Functional evolution of two subtly different (similar) folds. BMC Struct Biol 1:5

Akanuma S, Nakajima Y, Yokobori S-I, Kimura M, Nemoto N, Mase T, Miyazono K-I, Tanokura M, Yamagishi A (2013) Experimental evidence for the thermophilicity of ancestral life. Proc Natl Acad Sci USA 110(27):11067–11072

Alva V, Koretke KK, Coles M, Lupas AN (2008) Cradle-loop barrels and the concept of metafolds in protein classification by natural descent. Curr Opin Struct Biol 18(3):358–365

Alva V, Dunin-Horkawicz S, Habeck M, Coles M, Lupas AN (2009) The GD box: a widespread noncontiguous supersecondary structural element. Protein Sci 18(9):1961–1966

Alvarez-Carreño C, Penev PI, Petrov AS, Williams LD (2021) Fold evolution before LUCA: common ancestry of SH3 domains and OB domains. Mol Biol Evol 38(11):5134–5143
Alvarez-Carreño C, Gupta RJ, Petrov AS, Williams LD (2022) Creative destruction: new protein folds from old. Proc Natl Acad Sci USA 119(52):e2207897119
Ammelburg M, Hartmann MD, Djuranovic S, Alva V, Koretke KK, Martin J, Sauer G, Truffault V, Zeth K, Lupas AN, Coles M (2007) A CTP-dependent archaeal riboflavin kinase forms a bridge in the evolution of cradle-loop barrels. Structure 15(12):1577–1590
Andreeva A, Howorth D, Chothia C, Kulesha E, Murzin AG (2014) SCOP2 prototype: a new approach to protein structure mining. Nucleic Acids Res 42(Database issue):D310–D314
Ban N, Nissen P, Hansen J, Moore P, Steitz T (2000) The complete atomic structure of the large ribosomal subunit at 2.4 Å resolution. Science 289(5481):905–920
Blaber M, Lee J, Longo L (2012) Emergence of symmetric protein architecture from a simple peptide motif: evolutionary models. Cell Mol Life Sci 69(23):3999–4006
Bochkarev A, Pfuetzner RA, Edwards AM, Frappier L (1997) Structure of the single-stranded-DNA-binding domain of replication protein a bound to DNA. Nature 385(6612):176–181
Boomershine WP, McElroy CA, Tsai H-Y, Wilson RC, Gopalan V, Foster MP (2003) Structure of Mth11/Mth Rpp29, an essential protein subunit of archaeal and eukaryotic RNase P. Proc Natl Acad Sci USA 100(26):15398–15403
Brooks DJ, Gaucher EA (2007) A thermophilic last universal ancestor inferred from its estimated amino acid composition. In: Ancestral Sequence Reconstruction. Oxford University Press, Oxford, pp 200–207
Caetano-Anollés G, Kim HS, Mittenthal JE (2007) The origin of modern metabolic networks inferred from phylogenomic analysis of protein architecture. Proc Natl Acad Sci USA 104(22):9358–9363
Castillo RM, Mizuguchi K, Dhanaraj V, Albert A, Blundell TL, Murzin AG (1999) A six-stranded double-psi beta barrel is shared by several protein superfamilies. Structure 7(2):227–236
Chaban A, Minakhin L, Goldobina E, Bae B, Hao Y, Borukhov S, Putzeys L, Boon M, Kabinger F, Lavigne R, Makarova KS, Koonin EV, Nair SK, Tagami S, Severinov K, Sokolova ML (2024) Tail-tape-fused virion and non-virion RNA polymerases of a thermophilic virus with an extremely long tail. Nat Commun 15(1):317
Cheng H, Schaeffer RD, Liao Y, Kinch LN, Pei J, Shi S, Kim B-H, Grishin NV (2014) ECOD: an evolutionary classification of protein domains. PLoS Comput Biol 10(12):e1003926
Codispoti S, Yamaguchi T, Makarov M, Giacobelli VG, Mašek M, Kolář MH, Sanchez Rocha AC, Fujishima K, Zanchetta G, Hlouchová K (2024) The interplay between peptides and RNA is critical for protoribosome compartmentalization and stability. Nucleic Acids Res 52(20):12689–12700
Coles M, Diercks T, Liermann J, Gröger A, Rockel B, Baumeister W, Koretke KK, Lupas A, Peters J, Kessler H (1999) The solution structure of VAT-N reveals a “missing link” in the evolution of complex enzymes from a simple betaalphabetabeta element. Curr Biol 9(20):1158–1168
Coles M, Hulko M, Djuranovic S, Truffault V, Koretke K, Martin J, Lupas AN (2006) Common evolutionary origin of swapped-hairpin and double-psi beta barrels. Structure 14(10):1489–1498
Croitoru V, Semrad K, Prenninger S, Rajkowitsch L, Vejen M, Laursen BS, Sperling-Petersen HU, Isaksson LA (2006) RNA chaperone activity of translation initiation factor IF1. Biochimie 88(12):1875–1882
Delvillani F, Papiani G, Dehò G, Briani F (2011) S1 ribosomal protein and the interplay between translation and mRNA decay. Nucleic Acids Res 39(17):7702–7715
Demirci H, Murphy F 4th, Murphy E, Gregory ST, Dahlberg AE, Jogl G (2013) A structural basis for streptomycin-induced misreading of the genetic code. Nat Commun 4(1):1355
Das M, Chen N, LiWang A, & Wang LP. (2021) Identification and characterization of metamorphic proteins: Current and future perspectives. Biopolymers, 112(10), e23473.
Dendooven T, Sinha D, Roeselová A, Cameron TA, De Lay NR, Luisi BF, Bandyra KJ (2021) A cooperative PNPase-Hfq-RNA carrier complex facilitates bacterial riboregulation. Mol Cell 81(14):2901–2913.e5

Doerfel LK, Wohlgemuth I, Kothe C, Peske F, Urlaub H, Rodnina M (2013) EF-P is essential for rapid synthesis of proteins containing consecutive proline residues. Science 339:85–88
Dos Santos RF, Arraiano CM, Andrade JM (2019) New molecular interactions broaden the functions of the RNA chaperone Hfq. Curr Genet 65(6):1313–1319
Driessen RPC, Meng H, Suresh G, Shahapure R, Lanzani G, Priyakumar UD, White MF, Schiessel H, van Noort J, Dame RT (2013) Crenarchaeal chromatin proteins Cren7 and Sul7 compact DNA by inducing rigid bends. Nucleic Acids Res 41(1):196–205
Drobysheva AV, Panafidina SA, Kolesnik MV, Klimuk EI, Minakhin L, Yakunina MV, Borukhov S, Nilsson E, Holmfeldt K, Yutin N, Makarova KS, Koonin EV, Severinov KV, Leiman PG, Sokolova ML (2021) Structure and function of virion RNA polymerase of a crAss-like phage. Nature 589(7841):306–309
Duval M, Korepanov A, Fuchsbauer O, Fechter P, Haller A, Fabbretti A, Choulier L, Micura R, Klaholz BP, Romby P, Springer M, Marzi S (2013) Escherichia coli ribosomal protein S1 unfolds structured mRNAs onto the ribosome for active translation initiation. PLoS Biol 11(12):e1001731
Espinar-Marchena F, Rodríguez-Galán O, Fernández-Fernández J, Linnemann J, de la Cruz J (2018) Ribosomal protein L14 contributes to the early assembly of 60S ribosomal subunits in Saccharomyces cerevisiae. Nucleic Acids Res 46(9):4715–4732
Farías-Rico JA, Schmidt S, Höcker B (2014) Evolutionary relationship of two ancient protein superfolds. Nat Chem Biol 10(9):710–715
Fedorov R, Witte G, Urbanke C, Manstein DJ, Curth U (2006) 3D structure of Thermus aquaticus single-stranded DNA-binding protein gives insight into the functioning of SSB proteins. Nucleic Acids Res 34(22):6708–6717
Forrest D, James K, Yuzenkova Y, Zenkin N (2017) Single-peptide DNA-dependent RNA polymerase homologous to multi-subunit RNA polymerase. Nat Commun 8:15774
Fouqueau T, Blombach F, Werner F (2017) Evolutionary origins of two-barrel RNA polymerases and site-specific transcription initiation. Ann Rev Microbiol 71:331–348
Fox NK, Brenner SE, Chandonia J-M (2014) SCOPe: Structural Classification of Proteins—extended, integrating SCOP and ASTRAL data and classification of new structures. Nucleic Acids Res 42(Database issue):D304–D309
Fukunaga R, Yokoyama S (2006) Structural basis for substrate recognition by the editing domain of isoleucyl-tRNA synthetase. J Mol Biol 359(4):901–912
Garaeva N, Fatkhullin B, Murzakhanov F, Gafurov M, Golubev A, Bikmullin A, Glazyrin M, Kieffer B, Jenner L, Klochkov V, Aganov A, Rogachev A, Ivankov O, Validov S, Yusupov M, Usachev K (2024) Structural aspects of RimP binding on small ribosomal subunit from Staphylococcus aureus. Structure (London, England: 1993) 32(1):74–82.e5
Geethika K, Verma A, Mondal P, Mandal SS (2024) Molecular analysis of changes in DNA binding affinity and bending extent induced by the mutations on the aromatic amino residues in Cren7. Chemphyschem 25(24):e202400545
Giaquinto L, Curmi PMG, Siddiqui KS, Poljak A, DeLong E, DasSarma S, Cavicchioli R (2007) Structure and function of cold shock proteins in archaea. J Bacteriol 189(15):5738–5748
Gilbert W (1986) Origin of life: the RNA world. Nature 319(6055):618–618
Giuliodori AM, Belardinelli R, Duval M, Garofalo R, Schenckbecher E, Hauryliuk V, Ennifar E, Marzi S (2023) Escherichia coli CspA stimulates translation in the cold of its own mRNA by promoting ribosome progression. Front Microbiol 14:1118329
Grishin NV (2001) KH domain: one motif, two folds. Nucleic Acids Res 29(3):638–643
Guo K, Zhou J (2024) Insights into eukaryotic translation initiation factor 5A: its role and mechanisms in protein synthesis. Biochimt Biophys Acta Mol Cell Res 1871(8):119849
Guo L, Feng Y, Zhang Z, Yao H, Luo Y, Wang J, Huang L (2008) Biochemical and structural characterization of Cren7, a novel chromatin protein conserved among crenarchaea. Nucleic Acids Res 36(4):1129–1137

Guo Q, Goto S, Chen Y, Feng B, Xu Y, Muto A, Himeno H, Deng H, Lei J, Gao N (2013) Dissecting the in vivo assembly of the 30S ribosomal subunit reveals the role of RimM and general features of the assembly process. Nucleic Acids Res 41(4):2609–2620

Guyomar C, D'Urso G, Chat S, Giudice E, Gillet R (2021) Structures of tmRNA and SmpB as they transit through the ribosome. Nat Commun 12(1):4909

Harms J, Schluenzen F, Zarivach R, Bashan A, Gat S, Agmon I, Bartels H, Franceschi F, Yonath A (2001) High resolution structure of the large ribosomal subunit from a mesophilic eubacterium. Cell 107:679–688

Hartman H, Smith TF (2010) GTPases and the origin of the ribosome. Biol Direct 5(1):36

Hoekzema M, Romilly C, Holmqvist E, Wagner EGH (2019) Hfq-dependent mRNA unfolding promotes sRNA-based inhibition of translation. EMBO J 38(7):e101199

Honda T, Hara T, Nan J, Zhang X, Kimura M (2010) Archaeal homologs of human RNase P protein pairs Pop5 with Rpp30 and Rpp21 with Rpp29 work on distinct functional domains of the RNA subunit. Biosci Biotechnol Biochem 74(2):266–273

Huang P-S, Feldmeier K, Parmeggiani F, Velasco DAF, Höcker B, Baker D (2016) De novo design of a four-fold symmetric TIM-barrel protein with atomic-level accuracy. Nat Chem Biol 12(1):29–34

Illergård K, Ardell DH, Elofsson A (2009) Structure is three to ten times more conserved than sequence--a study of structural response in protein cores. Proteins 77(3):499–508

Iyer LM, Koonin EV, Aravind L (2003) Evolutionary connection between the catalytic subunits of DNA-dependent RNA polymerases and eukaryotic RNA-dependent RNA polymerases and the origin of RNA polymerases. BMC Struct Biol 3:1

Kalichuk V, Béhar G, Renodon-Cornière A, Danovski G, Obal G, Barbet J, Mouratou B, Pecorari F (2016) The archaeal "7 kDa DNA-binding" proteins: extended characterization of an old gifted family. Sci Rep 6:37274

Kawakami E, Adachi N, Senda T, Horikoshi M (2017) Leading role of TBP in the establishment of complexity in eukaryotic transcription initiation systems. Cell Rep 21(13):3941–3956

Kerr ID, Wadsworth RIM, Cubeddu L, Blankenfeldt W, Naismith JH, White MF (2003) Insights into ssDNA recognition by the OB fold from a structural and thermodynamic study of Sulfolobus SSB protein. EMBO J 22(11):2561–2570

Kjeldgaard M, Nissen P, Thirup S, Nyborg J (1993) The crystal structure of elongation factor EF-Tu from Thermus aquaticus in the GTP conformation. Structure (London, England: 1993) 1(1):35–50

Kolodny R (2021) Searching protein space for ancient sub-domain segments. Curr Opin Struct Biol 68:105–112

Kolodny R, Nepomnyachiy S, Tawfik DS, Ben-Tal N (2021) Bridging themes: short protein segments found in different architectures. Mol Biol Evol 38(6):2191–2208

Koonin EV, Krupovic M, Ishino S, Ishino Y (2020) The replication machinery of LUCA: common origin of DNA replication and transcription. BMC Biol 18(1):61

Korobeinikova AV, Garber MB, Gongadze GM (2012) Ribosomal proteins: structure, function, and evolution. Biochemistry (Mosc) 77(6):562–574

Kovacs NA, Penev PI, Venapally A, Petrov AS, Williams LD (2018) Circular permutation obscures universality of a ribosomal protein. J Mol Evol 86(8):581–592

Kuhle B, Ficner R (2014) eIF5B employs a novel domain release mechanism to catalyze ribosomal subunit joining. EMBO J 33(10):1177–1191

Lane WJ, Darst SA (2010) Molecular evolution of multisubunit RNA polymerases: structural analysis. J Mol Biol 395(4):686–704

Lauber MA, Rappsilber J, Reilly JP (2012) Dynamics of ribosomal protein S1 on a bacterial ribosome with cross-linking and mass spectrometry. Mol Cel Proteomics 11(12):1965–1976

Laursen SP, Bowerman S, Luger K (2021) Archaea: the final frontier of chromatin. J Mol Biol 433(6):166791

Leibundgut M, Frick C, Thanbichler M, Böck A, Ban N (2005) Selenocysteine tRNA-specific elongation factor SelB is a structural chimaera of elongation and initiation factors. EMBO J 24(1):11–22

Li P, Holliger P, Tagami S (2022) Hydrophobic-cationic peptides modulate RNA polymerase ribozyme activity by accretion. Nat Commun 13(1):3050

Liu Y, Liao J, Zhu B, Wang E-D, Ding J (2006) Crystal structures of the editing domain of Escherichia coli leucyl-tRNA synthetase and its complexes with Met and Ile reveal a lock-and-key mechanism for amino acid discrimination. Biochem J 394(Pt 2):399–407

Lupas AN, Koretke KK (2008) Evolution of protein folds. In: Computational structural biology: methods and applications. World Scientific, Singapore, pp 131–152

Ma B-G, Chen L, Ji H-F, Chen Z-H, Yang F-R, Wang L, Qu G, Jiang Y-Y, Ji C, Zhang H-Y (2008) Characters of very ancient proteins. Biochem Biophys Res Commun 366(3):607–611

Madhurima K, Nandi B, Sekhar A (2021) Metamorphic proteins: the Janus proteins of structural biology. Open Biol 11(4):210012

Maksimova E, Kravchenko O, Korepanov A, Stolboushkina E (2022) Protein assistants of small ribosomal subunit biogenesis in bacteria. Microorganisms 10(4):747. https://doi.org/10.3390/microorganisms10040747

Massé E, Gottesman S (2002) A small RNA regulates the expression of genes involved in iron metabolism in Escherichia coli. Proc Natl Acad Sci USA 99(7):4620–4625

Max KEA, Zeeb M, Bienert R, Balbach J, Heinemann U (2006) T-rich DNA single strands bind to a preformed site on the bacterial cold shock protein Bs-CspB. J Mol Biol 360(3):702–714

Mayanagi K, Oki K, Miyazaki N, Ishino S, Yamagami T, Morikawa K, Iwasaki K, Kohda D, Shirai T, Ishino Y (2020) Two conformations of DNA polymerase D-PCNA-DNA, an archaeal replisome complex, revealed by cryo-electron microscopy. BMC Biol 18(1):152

Meskauskas A, Dinman JD (2007) Ribosomal protein L3: gatekeeper to the A site. Mol Cell 25(6):877–888

Møller T, Franch T, Højrup P, Keene DR, Bächinger HP, Brennan RG, Valentin-Hansen P (2002) Hfq: a bacterial Sm-like protein that mediates RNA-RNA interaction. Mol Cell 9(1):23–30

Moody ERR, Álvarez-Carretero S, Mahendrarajah TA, Clark JW, Betts HC, Dombrowski N, Szánthó LL, Boyle RA, Daines S, Chen X, Lane N, Yang Z, Shields GA, Szöllősi GJ, Spang A, Pisani D, Williams TA, Lenton TM, Donoghue PCJ (2024) The nature of the last universal common ancestor and its impact on the early Earth system. Nat Ecol Evol 8(9):1654–1666

Moulinier L, Eiler S, Eriani G, Gangloff J, Thierry JC, Gabriel K, McClain WH, Moras D (2001) The structure of an AspRS-tRNA(Asp) complex reveals a tRNA-dependent control mechanism. EMBO J 20(18):5290–5301

Murzin AG (2008) Biochemistry. Metamorphic proteins. Science 320(5884):1725–1726

Nasir A, Kim KM, Caetano-Anollés G (2014) A phylogenomic census of molecular functions identifies modern thermophilic archaea as the most ancient form of cellular life. Archaea (Vancouver, BC) 2014:706468

Naveau M, Lazennec-Schurdevin C, Panvert M, Dubiez E, Mechulam Y, Schmitt E (2013) Roles of yeast eIF2α and eIF2β subunits in the binding of the initiator methionyl-tRNA. Nucleic Acids Res 41(2):1047–1057

Neubauer C, Gillet R, Kelley A, Ramakrishnan V (2012) Decoding in the absence of a codon by tmRNA and SmpB in the ribosome. Science 335:1366–1369

Neveu M, Kim H-J, Benner SA (2013) The "strong" RNA world hypothesis: fifty years old. Astrobiology 13(4):391–403

Nikolay R, van den Bruck D, Achenbach J, Nierhaus KH (2015) Ribosomal proteins: role in ribosomal functions. In: eLS. Wiley, pp 1–12. https://doi.org/10.1002/9780470015902.a0000687.pub4

Nissen P, Hansen J, Ban N, Moore P, Steitz T (2000) The structural basis of ribosome activity in peptide bond synthesis. Science 289(5481):920–930

Numata T, Ikeuchi Y, Fukai S, Suzuki T, Nureki O (2006) Snapshots of tRNA sulphuration via an adenylated intermediate. Nature 442(7101):419–424

Onesti S, Miller AD, Brick P (1995) The crystal structure of the lysyl-tRNA synthetase (LysU) from Escherichia coli. Structure (London, England: 1993) 3(2):163–176

Pereira J, Lupas AN (2018) The ancestral KH peptide at the root of a domain family with three different folds. Bioinf (Oxf) 34(23):3961–3965

Phadtare S, Kazakov T, Bubunenko M, Court DL, Pestova T, Severinov K (2007) Transcription antitermination by translation initiation factor IF1. J Bacteriol 189(11):4087–4093

Phung DK, Pilotto S, Matelska D, Blombach F, Pinotsis N, Hovan L, Gervasio FL, Werner F (2025) Archaeal NusA2 is the ancestor of ribosomal protein eS7 in eukaryotes. Structure (London, England: 1993) 33(1):149–159.e6

Qu X, Lancaster L, Noller HF, Bustamante C, Tinoco I Jr (2012) Ribosomal protein S1 unwinds double-stranded RNA in multiple steps. Proc Natl Acad Sci USA 109(36):14458–14463

Qureshi NS, Bains JK, Sreeramulu S, Schwalbe H, Fürtig B (2018) Conformational switch in the ribosomal protein S1 guides unfolding of structured RNAs for translation initiation. Nucleic Acids Res 46(20):10917–10929

Raghunathan S, Ricard CS, Lohman TM, Waksman G (1997) Crystal structure of the homo-tetrameric DNA binding domain of Escherichia coli single-stranded DNA binding protein determined by multiwavelength x-ray diffraction on the selenomethionyl protein at 2.9- Å resolution. Proc Natl Acad Sci USA 94(13):6652–6657

Raia P, Carroni M, Henry E, Pehau-Arnaudet G, Brûlé S, Béguin P, Henneke G, Lindahl E, Delarue M, Sauguet L (2019a) Structure of the DP1-DP2 PolD complex bound with DNA and its implications for the evolutionary history of DNA and RNA polymerases. PLoS Biol 17(1):e3000122

Raia P, Delarue M, Sauguet L (2019b) An updated structural classification of replicative DNA polymerases. Biochem Soc Trans 47(1):239–249

Ramaswamy P, Woodson SA (2009) Global stabilization of rRNA structure by ribosomal proteins S4, S17, and S20. J Mol Biol 392(3):666–677

Rocchio S, Santorelli D, Rinaldo S, Franceschini M, Malatesta F, Imperi F, Federici L, Travaglini-Allocatelli C, Di Matteo A (2019) Structural and functional investigation of the Small Ribosomal Subunit Biogenesis GTPase A (RsgA) from Pseudomonas aeruginosa. FEBS J 286(21):4245–4260

Roll-Mecak A, Cao C, Dever T, Burley S (2000) X-ray structures of the universal translation initiation factor IF2/eIF5B conformational changes on GDP and GTP binding. Cell 103:781–792

Roll-Mecak A, Alone P, Cao C, Dever TE, Burley SK (2004) X-ray structure of translation initiation factor eIF2gamma: implications for tRNA and eIF2alpha binding. J Biol Chem 279(11):10634–10642

Romero-Romero S, Costas M, Silva Manzano D-A, Kordes S, Rojas-Ortega E, Tapia C, Guerra Y, Shanmugaratnam S, Rodríguez-Romero A, Baker D, Höcker B, Fernández-Velasco DA (2021) The stability landscape of de novo TIM barrels explored by a modular design approach. J Mol Biol 433(18):167153

Sauguet L, Raia P, Henneke G, Delarue M (2016) Shared active site architecture between archaeal PolD and multi-subunit RNA polymerases revealed by X-ray crystallography. Nat Commun 7:12227

Savelsbergh A, Rodnina MV, Wintermeyer W (2009) Distinct functions of elongation factor G in ribosome recycling and translocation. RNA (New York, N.Y.) 15(5):772–780

Schmitt E, Coureux P-D, Kazan R, Bourgeois G, Lazennec-Schurdevin C, Mechulam Y (2020) Recent advances in archaeal translation initiation. Front Microbiol 11:584152

Schuller AP, Wu CC-C, Dever TE, Buskirk AR, Green R (2017) EIF5A functions globally in translation elongation and termination. Mol Cell 66(2):194–205.e5

Sengupta J, Agrawal RK, Frank J (2001) Visualization of protein S1 within the 30S ribosomal subunit and its interaction with messenger RNA. Proc Natl Acad Sci USA 98(21):11991–11996

Sergiev PV, Bogdanov AA, Dontsova OA (2005) How can elongation factors EF-G and EF-Tu discriminate the functional state of the ribosome using the same binding site? FEBS Lett 579(25):5439–5442

Sharma D, Cukras AR, Rogers EJ, Southworth DR, Green R (2007) Mutational analysis of S12 protein and implications for the accuracy of decoding by the ribosome. J Mol Biol 374(4):1065–1076
Shimojo M, Amikura K, Masuda K, Kanamori T, Ueda T, Shimizu Y (2020) In vitro reconstitution of functional small ribosomal subunit assembly for comprehensive analysis of ribosomal elements in E. coli. Commun Biol 3(1):142
Shin DH, Lou Y, Jancarik J, Yokota H, Kim R, Kim S-H (2004) Crystal structure of YjeQ from Thermotoga maritima contains a circularly permuted GTPase domain. Proc Natl Acad Sci USA 101(36):13198–13203
Sillitoe I, Dawson N, Lewis TE, Das S, Lees JG, Ashford P, Tolulope A, Scholes HM, Senatorov I, Bujan A, Ceballos Rodriguez-Conde F, Dowling B, Thornton J, Orengo CA (2019) CATH: expanding the horizons of structure-based functional annotations for genome sequences. Nucleic Acids Res 47(D1):D280–D284
Siomi H, Matunis M, Michael W, Dreyfuss G (1993) The pre-mRNA binding K protein contains a novel evolutionarily conserved motif. Nucleic Acids Res 21(5):1193–1198
Sledjeski DD, Whitman C, Zhang A (2001) Hfq is necessary for regulation by the untranslated RNA DsrA. J Bacteriol 183(6):1997–2005
Suguna K, Bott RR, Padlan EA, Subramanian E, Sheriff S, Cohen GH, David D (1987) Structure and refinement at 1.8 Å resolution of the aspartic proteinase from Rhizopus chinensis. J Mol Biol 196(4):877–900
Suksombat S, Khafizov R, Kozlov AG, Lohman TM, Chemla YR (2015) Structural dynamics of E. coli single-stranded DNA binding protein reveal DNA wrapping and unwrapping pathways. elife 4:e08193. https://doi.org/10.7554/eLife.08193
Tagami S (2023) Why we are made of proteins and nucleic acids: structural biology views on extraterrestrial life. Biophys Physicobiol 20(2):e200026
Tagami S, Li P (2023) The origin of life: RNA and protein co-evolution on the ancient Earth. Develop Growth Differ 65(3):167–174
Tagami S, Attwater J, Holliger P (2017) Simple peptides derived from the ribosomal core potentiate RNA polymerase ribozyme function. Nat Chem 9(4):325–332
Tamaru D, Amikura K, Shimizu Y, Nierhaus KH, Ueda T (2018) Reconstitution of 30S ribosomal subunits in vitro using ribosome biogenesis factors. RNA (New York, NY) 24(11):1512–1519
Umuhire Juru A, Ghirlando R, Zhang J (2024) Structural basis of tRNA recognition by the widespread OB fold. Nat Commun 15(1):6385
van Kempen M, Kim SS, Tumescheit C, Mirdita M, Lee J, Gilchrist CLM, Söding J, Steinegger M (2024) Fast and accurate protein structure search with Foldseek. Nat Biotechnol 42(2):243–246
Voet ARD, Noguchi H, Addy C, Simoncini D, Terada D, Unzai S, Park S-Y, Zhang KYJ, Tame JRH (2014) Computational design of a self-assembling symmetrical β-propeller protein. Proc Natl Acad Sci USA 111(42):15102–15107
Wan F, Wang Q, Tan J, Tan M, Chen J, Shi S, Lan P, Wu J, Lei M (2019) Cryo-electron microscopy structure of an archaeal ribonuclease P holoenzyme. Nat Commun 10(1):2617
Weber MH, Beckering CL, Marahiel MA (2001) Complementation of cold shock proteins by translation initiation factor IF1 in vivo. J Bacteriol 183(24):7381–7386
Webster MW, Chauvier A, Rahil H, Graziadei A, Charles K, Miropolskaya N, Takacs M, Saint-André C, Rappsilber J, Walter NG, Weixlbaumer A (2024) Molecular basis of mRNA delivery to the bacterial ribosome. Science (New York, NY) 386(6725):eado8476
Weiss MC, Neukirchen S, Roettger M, Mrnjavac N, Nelson-Sathi S, Martin WF, Sousa FL (2016a) Reply to "Is LUCA a thermophilic progenote?". Nat Microbiol 1:16230
Weiss MC, Sousa FL, Mrnjavac N, Neukirchen S, Roettger M, Nelson-Sathi S, Martin WF (2016b) The physiology and habitat of the last universal common ancestor. Nat Microbiol 1(9):16116
Werner F, Grohmann D (2011) Evolution of multisubunit RNA polymerases in the three domains of life. Nat Rev Microbiol 9(2):85–98
Yadid I, Kirshenbaum N, Sharon M, Dym O, Tawfik DS (2010) Metamorphic proteins mediate evolutionary transitions of structure. Proc Natl Acad Sci USA 107(16):7287–7292

Yagi S, Tagami S (2024) An ancestral fold reveals the evolutionary link between RNA polymerase and ribosomal proteins. Nat Commun 15(1):5938

Yagi S, Padhi AK, Vucinic J, Barbe S, Schiex T, Nakagawa R, Simoncini D, Zhang KYJ, Tagami S (2021) Seven amino acid types suffice to create the Core fold of RNA polymerase. J Am Chem Soc 143(39):15998–16006

Yao M, Ohsawa A, Kikukawa S, Tanaka I, Kimura M (2003) Crystal structure of hyperthermophilic archaeal initiation factor 5A: a homologue of eukaryotic initiation factor 5A (eIF-5A). J Biochem 133(1):75–81

Youkharibache P, Veretnik S, Li Q, Stanek KA, Mura C, Bourne PE (2019) The small β-barrel domain: a survey-based structural analysis. Structure 27(1):6–26

Zakharova N, Paster BJ, Wesley I, Dewhirst FE, Berg DE, Severinov KV (1999) Fused and overlapping rpoB and rpoC genes in helicobacters, campylobacters, and related bacteria. J Bacteriol 181(12):3857–3859

Zhang J (2024) Recognition of the tRNA structure: everything everywhere but not all at once. Cell Chem Biol 31(1):36–52

Zhang G, Campbell E, Minakhin L, Richter C, Severinov K, Darst S (1999) Crystal structure of Thermus aquaticus Core RNA polymerase at 3.3 Å resolution. Cell 98:811–824

Zhang A, Wassarman KM, Ortega J, Steven AC, Storz G (2002) The Sm-like Hfq protein increases OxyS RNA interaction with target mRNAs. Mol Cell 9(1):11–22

Zhang Z, Gong Y, Guo L, Jiang T, Huang L (2010) Structural insights into the interaction of the crenarchaeal chromatin protein Cren7 with DNA: crystal structures of Cren7-DNA complexes. Mol Microbiol 76(3):749–759

Part II
Evolution of RNA Processing, Modification, and Decay Mechanisms

Chapter 3
Regulation of Exon Recognition by Serine-Arginine Rich Splicing Factors and Heterogeneous Ribonucleoproteins in Higher Eukaryotes

Naoyuki Kataoka

Abstract MRNA splicing (splicing) is a process to that removes introns from pre-mRNAs and ligates remaining regions termed exons to produce mRNAs. This reaction must be accurate, since insertion/deletion of nucleotides causes frame shift, which results in production of non-functional proteins. Signals for splicing mostly reside in introns. However, higher eukaryotes that harbor long introns in their genomes require additional splicing signals to modulate splicing appropriately. The additional signals in exons are termed ESE and ESS, exonic splicing enhancer and exonic splicing silencer, respectively. ESEs are bound to Serine-Arginine rich splicing factors (SRSFs) and promote exon recognition by recruiting essential splicing factors adjacent to splice sites. In contrast, ESSs can block exon recognition through binding to heterogeneous ribonucleoproteins (hnRNPs). This system, termed exon recognition, is a main system for splicing involving long introns in higher eukaryotes. The diverse sequences of splice sites and the large numbers and lengths of introns may have served to facilitate evolution of many splicing regulators such as SRSF proteins and hnRNPs in higher eukaryotes.

Keywords Splicing · Exonic splicing enhancer · Exonic splicing silencer · Exon recognition · Serine-Arginine rich splicing factor · Heterogeneous ribonucleoprotein · Higher eukaryotes

---

N. Kataoka (✉)
Laboratory of Cellular Biochemistry, Department of Animal Resource Sciences, Graduate School of Agricultural and Life Sciences, The University of Tokyo, Tokyo, Japan
e-mail: akataoka@g.ecc.u-tokyo.ac.jp

A. Kanai, A. W. Nicholson (eds.), *Molecular Evolution of RNA Regulatory Enzymes and Their Systems*, Nucleic Acids and Molecular Biology 38,
https://doi.org/10.1007/978-3-032-03300-0_3

3.1 Pre-mRNA Splicing and Exon Recognition Model

In eukaryotic genomes, many of the protein coding genes contain intervening sequences, termed introns. Therefore, precursors of messenger RNA (pre-mRNAs) must undergo a step that removes introns and ligates exons. This step is called mRNA splicing (refers as splicing, hereafter), which is essential for protein expression (Hastings and Krainer 2001; Wahl et al. 2009; Kataoka 2017). Splicing requires accuracy in which nucleotides should be recognized for the reaction. Signals essential for splicing reactions mainly reside in introns (Hastings and Krainer 2001; Wahl et al. 2009; Kataoka 2017). At the 5′ ends of introns, GU dinucleotides are found, whereas AG dinucleotides exist at the 3′ end (Fig. 3.1). Those nucleotides are almost perfectly conserved. In addition to those sequences, a branch point sequence resides near 3′ end of intron (Fig. 3.1). Adenosine is mostly utilized as a branch nucleotide in higher eukaryotes. These signals are well conserved, from budding yeast to humans (Hastings and Krainer 2001; Wahl et al. 2009; Kataoka 2017). The consensus sequence of 5′ splice sites in humans is GURAGU (R stands for purine residue, such as A or G), whereas GUAAGU is the consensus in budding yeast (Fig. 3.1). These sequences are complementary to the 5′ end of U1 snRNA, and U1 snRNP recognizes 5′ splice site through RNA-RNA interactions (Fig. 3.2). As for 3′ splice sites, YAG (Y represents pyrimidine residues, such as C or U) is used in both species

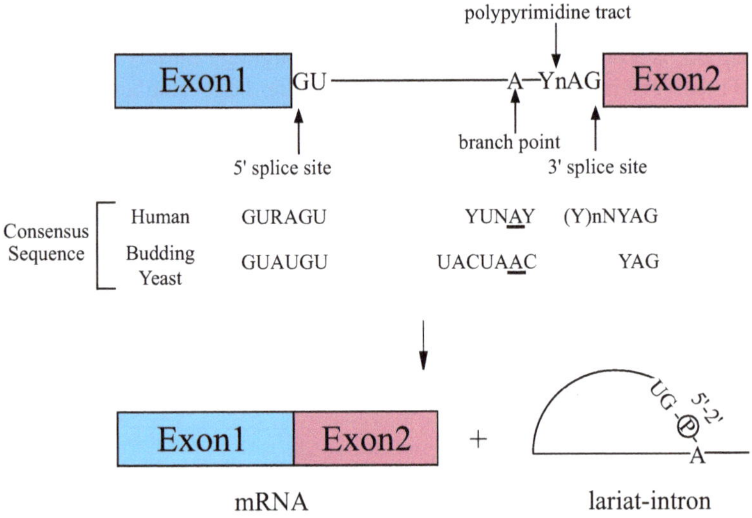

Fig. 3.1 A scheme for splicing reaction with splicing signals. Schematic representations of sequences required for splicing reaction. Boxes indicate exons and lines between them represent introns. Conserved sequence elements of human and budding yeast pre-mRNAs are shown. R and Y stand for purine and pyrimidine residues, respectively. Yn stands for a polypyrimidine tract. N indicates any nucleotide. Adenosine residue used as the branchpoint nucleotide is underlined. In the splicing reaction, a 2′–5′ phosphodiester bond is formed, creating a lariat intron structure, and the exons are ligated

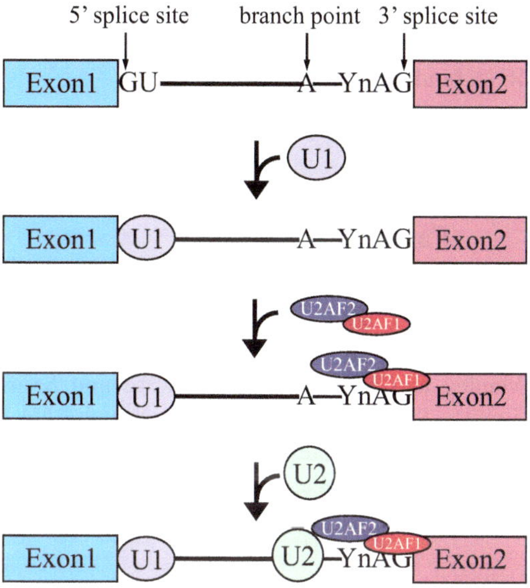

Fig. 3.2 Schematic representation of splicing signal recognition. Schematic representation of stepwise splicing signal recognition by splicing factors. Spliceosomal Uridine-rich small nuclear ribonucleoproteins (U snRNPs) are indicated as U1 and U2. U2AF denotes U2 snRNP auxiliary factor

(Fig. 3.1), although the pyrimidine residue is occasionally not utilized in humans. The branch point consensus sequence in almost all introns for budding yeast is UACUAAC (the underlined Adenosine is the branch point nucleotide) (Fig. 3.1). This sequence perfectly matches to the U2 snRNA, by which the U2 snRNP recognizes the branch point through RNA hybridization (Fig. 3.2). These mechanisms for splice site recognition are evolutionarily conserved from budding yeast to human (Hastings and Krainer 2001; Wahl et al. 2009; Kataoka 2017). However, in humans, the sequences around the branch point are highly degenerate, and the consensus sequence is YUNAY (Y: pyrimidine, N: any nucleotide) (Fig. 3.1) (Gao et al. 2008; Wahl et al. 2009). Thus, in humans a polypyrimidine tract, shown as (Y)n in Fig. 3.1 (Y stands for pyrimidine, 'n' means number of Y stretch), precedes the AG dinucleotide, which helps recognition of the 3′ splice site through binding of U2AF2 (Fig. 3.2). In addition, 5′ splice site sequences often do not match the consensus sequence. Therefore, splicing signals in introns are highly degenerate in humans (Fig. 3.1) (Wahl et al. 2009; Kataoka 2017). Moreover, the average lengths of introns are much longer in humans than those of budding yeast or other lower eukaryotes (Zhu et al. 2009). Taken together, short splice site consensus sequences at both ends of introns are essential but not sufficient for recognition of introns to be spliced out. Higher eukaryotes also employ additional regulatory elements in exons for exon recognition. Those elements are named exonic splicing enhancers (ESE) (Cartegni et al. 2002; Fu and Ares 2014). ESE promotes exon recognition, through binding to splicing regulators termed Serine-Arginine rich splicing factors (SRSFs) (Manley and Krainer 2010; Howard and Sanford 2015). On the other hand, there is an element that weakens exon recognition. This element is termed Exonic splicing silencer (ESS). ESSs are bound to another class of splicing regulators, heterogeneous nuclear ribonucleoproteins (hnRNPs). The hnRNP protein family consists of 20 proteins, from hnRNP A1 to hnRNP U, which have many different RNA binding domains.

3.2 Serine-Arginine Rich Splicing Factor (SRSF) Family Proteins

SRSFs are a family of proteins whose characteristic is the occurrence of Arginine-Serine (RS) repeats at the C-terminal region (Fig. 3.3) (Shepard and Hertel 2009; Manley and Krainer 2010; Howard and Sanford 2015). In addition to this RS domain, SRSF proteins harbor one or two RNA binding domains at N-terminal half. The first SRSF protein characterized, SRSF1, was termed Splicing Factor 2 (SF2). SF2 was biochemically purified and identified as an essential splicing factor, which was able to complement S100 cytoplasmic extracts for *in vitro* splicing reactions

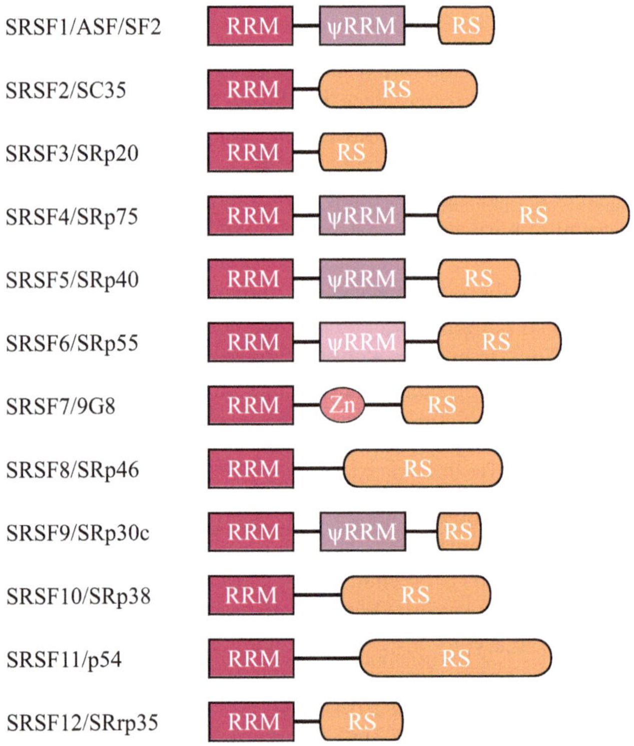

Fig. 3.3 Structure of human SRSF family proteins. Current specific names of SRSFs are shown at the left of the panel along with previous names. RRM denotes RNA binding domain, RS denotes arginine-serine dipeptide repeat rich domain. Zn and ψRRM indicate Zn-finger domain and RRM homology domain, respectively, which are also known to bind RNA

(Krainer et al. 1990; Krainer et al. 1991). At the same time, SRSF1 was also reported as Alternative Splicing Factor (ASF) that can modulate SV40 early pre-mRNA alternative splicing through 5′ splice site selection (Ge and Manley 1990; Ge et al. 1991). With these two simultaneous findings, it was established that SRSF1 has dual roles in splicing, as both an essential splicing factor and as an alternative splicing regulator. Another SR protein, originally termed SC35, was identified as an essential splicing factor (Fu and Maniatis 1990; Fu and Maniatis 1992), and which was also demonstrated to have an ability to regulate 5′ splice site selection (Fu et al. 1992; Tacke and Manley 1995). In the meantime, a monoclonal antibody against a nuclear protein, mAb104, was isolated and demonstrated to recognize a group of *Xenopus laevis* proteins that share a common epitope which is a phosphorylated RS domain (Roth et al. 1990; Zahler et al. 1992). Following the identification of these proteins, twelve different human SRSFs have been isolated to date. Canonical SRSFs are defined with two functions. One is an ability to complement cytoplasmic S100 extracts in *in vitro* splicing reactions, and the other is to modulate 5′ splice site selection *in vitro* and *in vivo* (Shepard and Hertel 2009; Manley and Krainer 2010; Howard and Sanford 2015).

SRSFs are conserved among metazoans, including frog, fly, nematode and plants. As described previously, mAb104 antibody recognizes *Xenopus laevis* SR proteins including SRp20, SRp40, SRp55 and SRp75 (SRSF3, SRSF5, SRSF6 and SRSF4 in humans) and it binds to active sites of RNA polymerase II transcription in lampbrush chromosome in *Xenopus* oocytes (Roth et al. 1990; Zahler et al. 1992). In *Drosophila melanogaster*, B52 protein, which has a similarity to human SRSF6, was identified as a protein binding to boundaries of transcriptionally active chromosomes (Ring and Lis 1994). Three more SRSFs were identified in *Drosophila melanogaster*, termed SWAP (suppressor of white apricot) (Mancebo et al. 1990), Tra (transformer) (Belote et al. 1989) and Tra2 (transformer 2) (Amrein et al. 1988; Goralski et al. 1989). Database searches also identified *Caenorhabditis elegans* SRSF proteins as rsp-1 to rsp-7, and it was shown that RNAi knockdown of rsp-3, which is assumed to be a homolog of human SRSF1, causes embryonic lethality, suggesting an essential role in development (Longman et al. 2000). SRSF proteins also have been identified in plants including carrots, tobacco and *Arabidopsis thaliana*, by cross-reactivity with mAb104 and anti-SRSF1 antibodies (Lopato et al. 1996). Genome database searches revealed that *Arabidopsis thaliana* and *Oryza sativa* encode 18 and 22 SRSF proteins, respectively, with some of them specific to plants (Shepard and Hertel 2009; Manley and Krainer 2010; Howard and Sanford 2015). In unicellular organisms such as fission yeast at least two SRSF protein homologs, Srp1 and Srp2, were reported (Gross et al. 1998; Lutzelberger et al. 1999). There are no apparent homologs of canonical SRSF proteins in budding yeast. However, one RNA binding protein, Npl3, has six 'SRGG' motifs at the C-terminus, and can promote splicing (Kress et al. 2008). Interestingly, the 'SRGG' motif is a combination of 'SR' dipeptide and 'RGG' tripeptide. 'RGG' is a characteristic sequence in heterogeneous ribonucleoprotein (hnRNP) A/B type proteins (Geuens et al. 2016). It has been well accepted that hnRNP A/B type proteins antagonize SRSFs in splice site selection (Hastings and Krainer 2001; Wahl et al. 2009;

Kataoka 2017). It would be quite interesting if Npl3 protein is a prototype of canonical SRSFs and hnRNP A/B type proteins.

3.3 The Role of SRSF Proteins in Exon Recognition

After completion of human genome project, it became evident that the average length of human introns is much longer than that of exons. The average size of exons is 163 nucleotides, whereas the average length of introns is 5849 nucleotides (Zhu et al. 2009). In the fly and nematode, introns are relatively short, and they can be recognized as discrete 'units' to be spliced out. This is called the 'Intron recognition model' (Fig. 3.4a) (Berget 1995). Compared to fly and nematode, humans have much longer introns, although exon sizes are not so different. In addition, internal exons humans are typically no longer than 300 nucleotides. Moreover, the exons, whose adjacent downstream 5′ splice sites are mutated, tend to be skipped in human diseases (Scotti and Swanson 2016). Based on these findings, an exon recognition model was proposed for higher eukaryotes that harbor long introns in their genomes. This model is called as 'Exon Recognition Model' (Fig. 3.4b) (Berget 1995). With this model, exons are recognized as a unit to be protected from removal during splicing, through an interaction involving U2 snRNP at the branch point, a U2AF heterodimer near 3′ splice site in the upstream intron, and a U1 snRNP at the

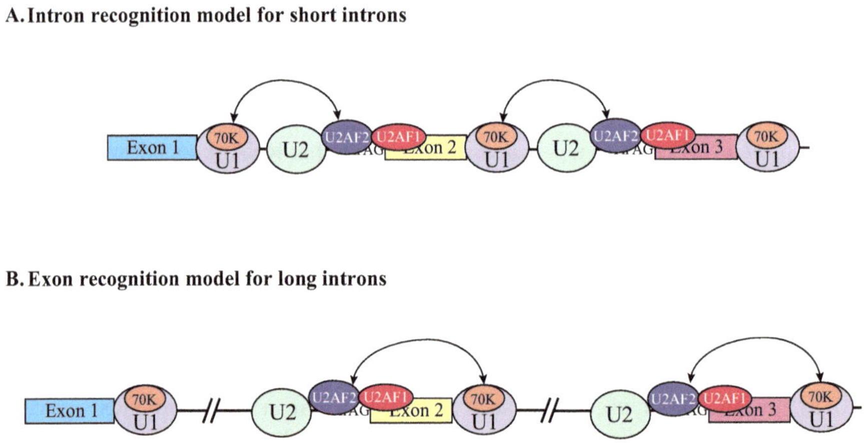

Fig. 3.4 Schematic representations of the intron recognition model and the exon recognition model for splicing. (**a**) Intron recognition model. In lower eukaryotes, intron lengths are short, and intron recognition is a major mode for splicing. Introns are recognized by a crosstalk between U170K protein in the U1 snRNP and the U2AF heterodimer that reside at the 5′ splice site and the branch point, respectively. (**b**) Exon recognition model. Exon recognition is dominant for splicing in humans, whose introns are long. In this mode, exons are recognized via interaction over an exon between the U2AF heterodimer and the U170K protein that are located at the region spanning the polypyrimidine tract and 3′ splice site, and the 5′ splice site, respectively

downstream 5′ splice site (Fig. 3.4b). Some of splice sites are ‘weak’, which means that their sequences show low similarity to the splice site consensus sequences. Low homology to the consensus sequences results in inefficient recognition by U2 snRNP, U2AF proteins for 3′ splice site region, and by U1 snRNP for 5′ splice sites. Under these circumstances, these exons are less efficiently recognized than exons with ‘strong’ adjacent splice sites. ESEs in those exons support exon recognition by employing SRSF proteins (Fig. 3.6a). The RS domain of SRSF proteins functions in protein-protein interactions with other RS domain-containing proteins. Specifically, SRSFs bind to ESEs through their RNA binding domain and recruit RS domain-containing splicing factors, such as the U2AF heterodimer and U170K protein (namely, U1 snRNP) via their RS domains to the polypyrimidine tract and 5′ splice site, respectively (Fig. 3.6a). The U2AF heterodimer assists in the association of the U2 snRNP with the branch point sequence. With these actions, recognition of the ESE-containing exons is promoted.

3.4 Heterogeneous Ribonucleoprotein Family Proteins (hnRNPs)

In eukaryotic cells, primary transcripts associate with many RNA binding proteins to form ribonucleoproteins (RNPs). The heterogeneous nuclear ribonucleoproteins (hnRNPs) assist in controlling the maturation of heterogeneous nuclear RNAs (hnRNAs/pre-mRNAs) into messenger RNAs (mRNAs), as well assist mRNA in cellular transport, and control mRNA translation. (Dreyfuss et al. 1993; Dreyfuss et al. 2002; Geuens et al. 2016). The hnRNP complex, isolated via pulldown and UV-cross-linking experiments (Dreyfuss et al. 1993; Dreyfuss et al. 2002; Geuens et al. 2016) consists of 20 major types of hnRNPs and minor hnRNPs (Dreyfuss et al. 1993; Dreyfuss et al. 2002; Geuens et al. 2016). These proteins are localized predominantly in the nucleus, since they harbor several different types of nuclear localization signals. Some of the hnRNP proteins can shuttle between the nucleus and the cytoplasm, which are likely to be important for their functions. By analyzing their amino acid sequences, four unique RNA binding domains (RBDs) were identified: the RNA recognition motif (RRM), the quasi-RRM, a glycine-rich domain constituting an RGG box, and a KH domain (Fig. 3.5). The most common RBD is the RNA recognition motif (RRM) with two conserved RNP1 and RNP2 motifs (Dreyfuss et al. 1993; Dreyfuss et al. 2002; Geuens et al. 2016).

The quasi(q)-RRMs are not conserved but have a similar structure to the RRM (Dominguez et al. 2011; Samatanga et al. 2013). The qRRMs interact with the G-tract RNA, which forms G-quadruplex(G4) (Hansel-Hertsch et al. 2017), in a specific way (Dominguez et al. 2011). The RGG box is characterized by repeats of the Arg-Gly-Gly amino acid sequence (Kiledjian and Dreyfuss 1992). RGG box functions include homologous and heterologous interactions with other hnRNPs (Cartegni et al. 1996), and it can interact with RNAs. The KH (K-Homology)

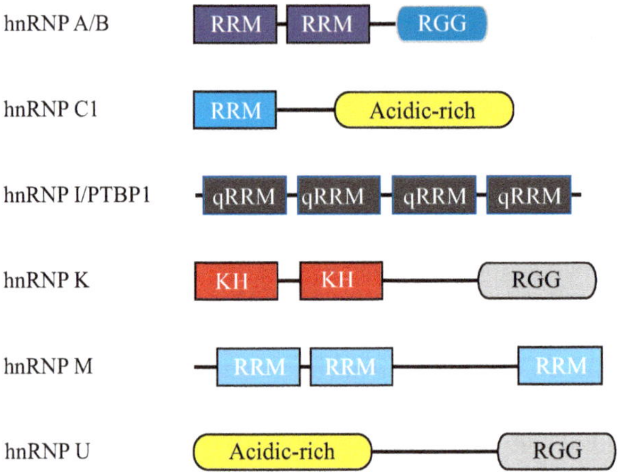

Fig. 3.5 Structure of hnRNPs. Schematic representations of hnRNPs. RRM: RNA recognition motif, RGG: arginine–glycine–glycine repeat-rich region; Acidic-rich: acidic amino acid residue-rich region, qRRM: non-canonical quasi-RRM; KH: hnRNP K homology domain

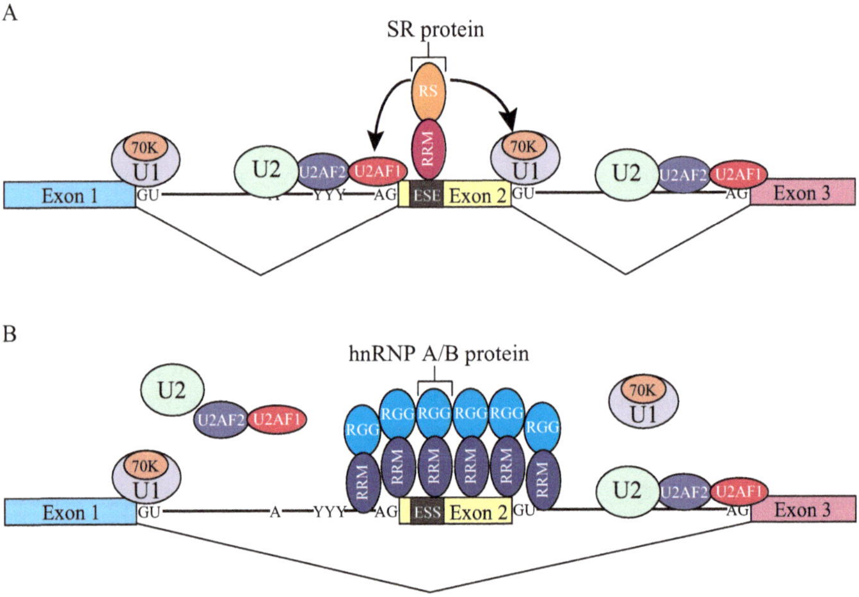

Fig. 3.6 Models for promotion or inhibition of exon recognition by SR proteins or hnRNP A/B type proteins, respectively. (**a**) SR proteins recognize and associate with exonic splicing enhancer (ESE) through their RNA binding domain (RRM). They interact with U2AF1 and U170K proteins via the Arginine-Serine-rich (RS) domains. This interaction promotes exon recognition. U1, U2: U1, U1 small nuclear ribonucleoprotein, U2 small nuclear ribonucleoprotein; U2AF1, U2AF2: U2 snRNP auxiliary factor 1, 2. (**b**) hnRNP A/B type proteins bind to the exonic splicing silencer (ESS) via their RRM and can associate and spread on the pre-mRNA in both the 5′ and 3′ directions. This results in sequestration of the exon and other regions on the pre-mRNA, even if an ESE exists, and prevents the association of splicing factors. Consequently, the exon is less recognized and tends to be excluded from the splicing pathway. RGG: Arginine-Glycine-Glycine repeat-rich domain

domain was originally found in hnRNP K protein as triple repeats (Siomi et al. 1993a) and later found in many RNA binding proteins, including FMR1 (Siomi et al. 1993b). The modularity created by the combination of these different domains including RBDs and auxiliary domains likely contributes to the diverse functions of hnRNPs (Fig. 3.5). The hnRNPs are assumed to be the first proteins to associatewith primary transcripts, to assist in packing and stabilization. This assembly step facilitates the downstream events such as constitutive and alternative splicing, resulting in the fully mature mRNAs which are then exported to the cytoplasm. Even after transport, shuttling hnRNP proteins remain bound to newly exported mRNAs, and regulate their translation, degradation and subcellular localization (Dreyfuss et al. 2002). The hnRNP complex was first identified in human cells, but it is also present in *Drosophila melanogaster* and *Xenopus laevis* oocytes and somatic cells (Matunis et al. 1992a; Matunis et al. 1992b; Marcu et al. 2001). Although the composition of the complex is slightly different among organisms, some of the components are well conserved (Dreyfuss et al. 1993; Dreyfuss et al. 2002; Geuens et al. 2016). The apparent homologs are not found in budding yeast. As described previously, Npl3 is a candidate for SRSF and hnRNP, and may be a prototype of hnRNP. In the budding yeast, only 2.5 percent of the protein-encoding genes contain introns. In addition, essentially only one intron exists in each gene. In contrast, almost all protein-coding genes in humans harbor introns, and the average number of introns per gene is more than seven. This huge difference could drive a vast evolution of many RNA binding proteins with different RNA binding domains found in hnRNPs.

3.5 The Role of hnRNP A/B Type Proteins in Exon Recognition

Some hnRNPs have functions in splicing. Among them, hnRNP A1 protein, one of the hnRNP family members, was demonstrated to represses exon recognition by binding initially to a high-affinity binding site (ESS) in the exon, which then promotes further hnRNP A1 association with the upstream and downstream regions of the exon, expanding to the intron (Zhu et al. 2001) (Fig. 3.6b). Thus, this results in inhibition of SR protein binding to ESEs. It also causes the concealment of splice sites by hnRNP A1 at both ends of exons (Fig. 3.6b).

The elements for promotion/inhibition of exon recognition are also found in introns, which, in turn, are termed Intronic Splicing Enhancer (ISE) and Intronic Splicing Silencer (ISS), respectively (Fu and Ares 2014). These regulatory elements and associated factors are involved not only in constitutive splicing but also in tissue- and stage-specific alternative splicing (Fu and Ares 2014). Both SR proteins and hnRNP proteins are involved in alternative splicing. SR proteins act to promote exon inclusion by binding to ESE and ISE, whereas hnRNP proteins, especially hnRNP A/B proteins, act to enhance exon skipping via binding to ESS and ISS (Fu and Ares 2014). The ratio of SR protein and hnRNP A/B protein varies among tissues (Hanamura et al. 1998), which likely contributes to tissue-specific alternative splicing. It is likely that combinations of differing levels of SR proteins, hnRNP proteins, and tissue-specific splicing regulators determine which alternative splicing pattern is used in a given tissue or cell type.

3.6 Conclusions and Perspectives

There are twelve SRSFs in humans, and other species also employ many different SRSF proteins. Since there are so many different ESEs with diverse sequences, it may come as no surprise that humans require a corresponding diversity of SRSF proteins. The amino acid sequences of the RRM of each SR protein are similar but not identical. This indicates that each SRSF has a unique RNA sequence preference. The target genes partially, but not completely overlap. This is most likely the reason why a single SRSF knockout causes embryonic lethality in the mouse (Wang et al. 2001; Xu et al. 2005). In addition, there are several SRSF protein kinases that phosphorylate serine residues in RS domains, and several of these can modulate alternative splicing through phosphorylation of SRSF proteins. Thus, different combinations of SRSF kinases and SRSF proteins may be able to regulate alternative splicing of their specific target genes (Yomoda et al. 2008; Shkreta et al. 2024). It is likely that the presence of numerous long introns and the occurrence of splice site sequence diversity in higher eukaryotes in turn require multiple SRSF proteins to regulate splicing properly. These circumstances may facilitate evolution of many different SRSF proteins in higher eukaryotes.

In addition to SRSFs, higher eukaryotes evolved another class of RNA binding proteins, hnRNPs. The hnRNPs associate with the long intronic regions and exons. The different ratio between SRSFs and hnRNP proteins in protein expression level may regulate alternative splicing differently among tissues (Hanamura et al. 1998). Multicellular organisms likely evolved these protein systems with expansion of gene numbers and transcriptome diversity.

Acknowledgements I would like to thank members of Laboratory of Cellular Biochemistry for critical reading and discussion. I am thankful to Drs. Akio Kanai and Allen Nicholson for giving me an opportunity to contribute to this wonderful special issue. I would like to apologize to all scientists whose research could not be properly discussed and cited here due to space limitations.

Competing Interests There are no competing interests to declare.

Funding This work was supported by Japan Society for the Promotion of Science (22K19239, 23K26864, 23H00358) to N.K.

References

Amrein H, Gorman M, Nothiger R (1988) The sex-determining gene tra-2 of Drosophila encodes a putative RNA binding protein. Cell 55:1025–1035

Belote JM, McKeown M, Boggs RT, Ohkawa R, Sosnowski BA (1989) Molecular genetics of transformer, a genetic switch controlling sexual differentiation in Drosophila. Dev Genet 10:143–154

Berget SM (1995) Exon recognition in vertebrate splicing. J Biol Chem 270:2411–2414

Cartegni L, Maconi M, Morandi E, Cobianchi F, Riva S, Biamonti G (1996) hnRNP A1 selectively interacts through its Gly-rich domain with different RNA-binding proteins. J Mol Biol 259:337–348

Cartegni L, Chew SL, Krainer AR (2002) Listening to silence and understanding nonsense: exonic mutations that affect splicing. Nat Rev Genet 3:285–298

Dominguez C, Schubert M, Duss O, Ravindranathan S, Allain FH (2011) Structure determination and dynamics of protein-RNA complexes by NMR spectroscopy. Prog Nucl Magn Reson Spectrosc 58:1–61

Dreyfuss G, Matunis MJ, Pinol-Roma S, Burd CG (1993) hnRNP proteins and the biogenesis of mRNA. Annu Rev Biochem 62:289–321

Dreyfuss G, Kim VN, Kataoka N (2002) Messenger-RNA-binding proteins and the messages they carry. Nat Rev Mol Cell Biol 3:195–205

Fu XD, Ares M Jr (2014) Context-dependent control of alternative splicing by RNA-binding proteins. Nat Rev Genet 15:689–701

Fu XD, Maniatis T (1990) Factor required for mammalian spliceosome assembly is localized to discrete regions in the nucleus. Nature 343:437–441

Fu XD, Maniatis T (1992) Isolation of a complementary DNA that encodes the mammalian splicing factor SC35. Science 256:535–538

Fu XD, Mayeda A, Maniatis T, Krainer AR (1992) General splicing factors SF2 and SC35 have equivalent activities in vitro, and both affect alternative 5′ and 3′ splice site selection. Proc Natl Acad Sci USA 89:11224–11228

Gao K, Masuda A, Matsuura T, Ohno K (2008) Human branch point consensus sequence is yUnAy. Nucleic Acids Res 36:2257–2267

Ge H, Manley JL (1990) A protein factor, ASF, controls cell-specific alternative splicing of SV40 early pre-mRNA in vitro. Cell 62:25–34

Ge H, Zuo P, Manley JL (1991) Primary structure of the human splicing factor ASF reveals similarities with Drosophila regulators. Cell 66:373–382

Geuens T, Bouhy D, Timmerman V (2016) The hnRNP family: insights into their role in health and disease. Hum Genet 135:851–867

Goralski TJ, Edstrom JE, Baker BS (1989) The sex determination locus transformer-2 of Drosophila encodes a polypeptide with similarity to RNA binding proteins. Cell 56:1011–1018

Gross T, Richert K, Mierke C, Lutzelberger M, Kaufer NF (1998) Identification and characterization of srp1, a gene of fission yeast encoding a RNA binding domain and a RS domain typical of SR splicing factors. Nucleic Acids Res 26:505–511

Hanamura A, Caceres JF, Mayeda A, Franza BR Jr, Krainer AR (1998) Regulated tissue-specific expression of antagonistic pre-mRNA splicing factors. RNA 4:430–444

Hansel-Hertsch R, Di Antonio M, Balasubramanian S (2017) DNA G-quadruplexes in the human genome: detection, functions and therapeutic potential. Nat Rev Mol Cell Biol 18:279–284

Hastings ML, Krainer AR (2001) Pre-mRNA splicing in the new millennium. Curr Opin Cell Biol 13:302–309

Howard JM, Sanford JR (2015) The RNAissance family: SR proteins as multifaceted regulators of gene expression. Wiley Interdiscip Rev RNA 6:93–110

Kataoka N (2017) Modulation of aberrant splicing in human RNA diseases by chemical compounds. Hum Genet 136:1237–1245

Kiledjian M, Dreyfuss G (1992) Primary structure and binding activity of the hnRNP U protein: binding RNA through RGG box. EMBO J 11:2655–2664
Krainer AR, Conway GC, Kozak D (1990) Purification and characterization of pre-mRNA splicing factor SF2 from HeLa cells. Genes Dev 4:1158–1171
Krainer AR, Mayeda A, Kozak D, Binns G (1991) Functional expression of cloned human splicing factor SF2: homology to RNA-binding proteins, U1 70K, and Drosophila splicing regulators. Cell 66:383–394
Kress TL, Krogan NJ, Guthrie C (2008) A single SR-like protein, Npl3, promotes pre-mRNA splicing in budding yeast. Mol Cell 32:727–734
Longman D, Johnstone IL, Caceres JF (2000) Functional characterization of SR and SR-related genes in Caenorhabditis elegans. EMBO J 19:1625–1637
Lopato S, Mayeda A, Krainer AR, Barta A (1996) Pre-mRNA splicing in plants: characterization of Ser/Arg splicing factors. Proc Natl Acad Sci USA 93:3074–3079
Lutzelberger M, Gross T, Kaufer NF (1999) Srp2, an SR protein family member of fission yeast: in vivo characterization of its modular domains. Nucleic Acids Res 27:2618–2626
Mancebo R, Lo PC, Mount SM (1990) Structure and expression of the Drosophila melanogaster gene for the U1 small nuclear ribonucleoprotein particle 70K protein. Mol Cell Biol 10:2492–2502
Manley JL, Krainer AR (2010) A rational nomenclature for serine/arginine-rich protein splicing factors (SR proteins). Genes Dev 24:1073–1074
Marcu A, Bassit B, Perez R, Pinol-Roma S (2001) Heterogeneous nuclear ribonucleoprotein complexes from Xenopus laevis oocytes and somatic cells. Int J Dev Biol 45:743–752
Matunis EL, Matunis MJ, Dreyfuss G (1992a) Characterization of the major hnRNP proteins from Drosophila melanogaster. J Cell Biol 116:257–269
Matunis MJ, Matunis EL, Dreyfuss G (1992b) Isolation of hnRNP complexes from Drosophila melanogaster. J Cell Biol 116:245–255
Ring HZ, Lis JT (1994) The SR protein B52/SRp55 is essential for Drosophila development. Mol Cell Biol 14:7499–7506
Roth MB, Murphy C, Gall JG (1990) A monoclonal antibody that recognizes a phosphorylated epitope stains lampbrush chromosome loops and small granules in the amphibian germinal vesicle. J Cell Biol 111:2217–2223
Samatanga B, Dominguez C, Jelesarov I, Allain FH (2013) The high kinetic stability of a G-quadruplex limits hnRNP F qRRM3 binding to G-tract RNA. Nucleic Acids Res 41:2505–2516
Scotti MM, Swanson MS (2016) RNA mis-splicing in disease. Nat Rev Genet 17:19–32
Shepard PJ, Hertel KJ (2009) The SR protein family. Genome Biol 10:242
Shkreta L, Delannoy A, Toutant J, Chabot B (2024) Regulatory interplay between SR proteins governs CLK1 kinase splice variants production. RNA 30:1596–1607
Siomi H, Matunis MJ, Michael WM, Dreyfuss G (1993a) The pre-mRNA binding K protein contains a novel evolutionarily conserved motif. Nucleic Acids Res 21:1193–1198
Siomi H, Siomi MC, Nussbaum RL, Dreyfuss G (1993b) The protein product of the fragile X gene, FMR1, has characteristics of an RNA-binding protein. Cell 74:291–298
Tacke R, Manley JL (1995) The human splicing factors ASF/SF2 and SC35 possess distinct, functionally significant RNA binding specificities. EMBO J 14:3540–3551
Wahl MC, Will CL, Luhrmann R (2009) The spliceosome: design principles of a dynamic RNP machine. Cell 136:701–718
Wang HY, Xu X, Ding JH, Bermingham JR Jr, Fu XD (2001) SC35 plays a role in T cell development and alternative splicing of CD45. Mol Cell 7:331–342
Xu X, Yang D, Ding JH, Wang W, Chu PH, Dalton ND, Wang HY, Bermingham JR Jr, Ye Z, Liu F et al (2005) ASF/SF2-regulated CaMKIIdelta alternative splicing temporally reprograms excitation-contraction coupling in cardiac muscle. Cell 120:59–72

Yomoda J, Muraki M, Kataoka N, Hosoya T, Suzuki M, Hagiwara M, Kimura H (2008) Combination of Clk family kinase and SRp75 modulates alternative splicing of Adenovirus E1A. Genes Cells 13:233–244

Zahler AM, Lane WS, Stolk JA, Roth MB (1992) SR proteins: a conserved family of pre-mRNA splicing factors. Genes Dev 6:837–847

Zhu J, Mayeda A, Krainer AR (2001) Exon identity established through differential antagonism between exonic splicing silencer-bound hnRNP A1 and enhancer-bound SR proteins. Mol Cell 8:1351–1361

Zhu L, Zhang Y, Zhang W, Yang S, Chen JQ, Tian D (2009) Patterns of exon-intron architecture variation of genes in eukaryotic genomes. BMC Genomics 10:47

Chapter 4
Evolution of Bacterial tRNA Nucleotidyltransferases

George H. Jones

Abstract The CCA end of tRNA molecules is required for their function in protein synthesis. As some tRNA genes do not encode the CCA end, that sequence must be added post-transcriptionally by tRNA nucleotidyltransferases. The existence of CCA-, CC- and A-adding tRNA nucleotidyltransferases raises interesting questions regarding the evolution of these enzymes. Which activity appeared first, for example, during bacterial evolution? This chapter discusses the activities and distribution of the tRNA nucleotidyltransferases, describes and evaluates existing models for the evolution of these enzymes and proposes a potentially useful and innovative approach to examine that evolution.

Keywords CCA-end · tRNA, tRNA nucleotidyltransferase · Bacteria · Evolution, phylogeny, ancestral reconstruction

4.1 Introduction

The 3′-CCA (cytidine-cytidine-adenine) ends of transfer RNA molecules are required for the participation of tRNAs in protein synthesis. Amino acids are esterified to the 3′-OH of the CCA terminus via a two-step reaction catalyzed by aminoacyl tRNA synthetases. The first step involves the formation of an aminoacyl-adenylate (aa-adenylate) via reaction with ATP. Subsequently, the amino acid is transferred from the aa-adenylate to the 3′-OH of the terminal adenosine in a tRNA molecule (Sprinzl and Cramer 1979). Genome sequencing studies indicate that the 3′terminal sequence is not encoded in all tRNA genes (Rammelt and Rossmanith 2016). Moreover, cellular endo- and exoribonucleases are known

G. H. Jones (✉)
Department of Biology, Emory University, Atlanta, GA, USA
e-mail: george.h.jones@emory.edu

A. Kanai, A. W. Nicholson (eds.), *Molecular Evolution of RNA Regulatory Enzymes and Their Systems*, Nucleic Acids and Molecular Biology 38,
https://doi.org/10.1007/978-3-032-03300-0_4

to remove the CCA ends from otherwise mature tRNA species (Wellner et al. 2018, 2025).

The addition of CCA ends to immature and damaged tRNA molecules is catalyzed by the enzyme ATP(CTP):tRNA nucleotidyltransferase (also referred to as CCA-adding enzyme or CCAase) (Deutscher 1990). tRNA nucleotidyltransferases (TNTs) add the CCA sequence one residue at a time to tRNA molecules that lack one or both C's and/or the A residue, producing mature tRNAs capable of functioning in protein synthesis. An intact CCA end is also required for the efficient maturation of tRNA 5′-ends by RNase P (Guerrier-Takada et al. 1984; Oh and Pace 1994), and the CCA end participates in tRNA quality control (Betat and Morl 2015; Wellner et al. 2018). TNTs are found in all domains of life and as some species do not encode the CCA end in any tRNA genes, some TNTs are essential enzymes (Yue et al. 1996). In humans, mutations in the nuclear TNT gene can cause pathologies of several types and severity (Slade et al. 2020).

In this chapter, the function, distribution and evolution of tRNA nucleotidyltransferases from bacteria will be discussed, with reference to TNTs from other organisms as appropriate.

4.2 Classification, Characteristics and Distribution of Bacterial tRNA Nucleotidyltranferases

Archaeal, eukaryotic and bacterial tRNA nucleotidyltransferases belong to a superfamily of nucleotidyltransferases that also includes DNA polymerase β, kanamycin nucleotidyltransferases and the poly(A) polymerases (Holm and Sander 1995). The members of the superfamily share sequence and structural motifs but there are significant differences between the enzymes from the three domains. Thus, Yue et al. proposed that the superfamily should be divided into two classes (Yue et al. 1996).

Class I was proposed to contain archaeal TNTs, eukaryotic poly(A) polymerases, eukaryotic terminal transferases, DNA polymerase β, several protein nucleotidyltransferases and a few other nucleotidyltransferases. Class II contains the eukaryotic and bacterial tRNA nucleotidyltransferases and the bacterial poly(A) polymerases (Yue et al. 1996).

More recent studies have shown that the Class II tRNA nucleotidyltransferases can be further divided into two subclasses. Many organisms contain a single enzyme that adds both the two C's and the A residue to immature and damaged tRNAs. However, some organisms contain two TNTs, one of which adds the two C's and another which adds the A residue (Tomita and Weiner 2001, 2002). This division of labor was identified initially in the Bacteria and CC- and A- adding enzymes have been characterized in the Aquificae (Tomita and Weiner 2001), Deinococcus and Synechocystis (Tomita and Weiner 2002), some Firmicutes (Bralley et al. 2005; Neuenfeldt et al. 2008) and some δ-Proteobateria (Bralley et al. 2009).

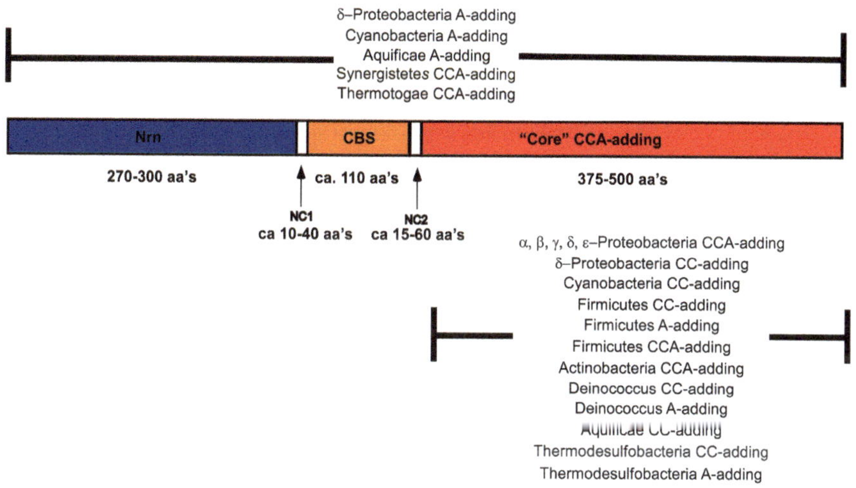

Fig. 4.1 Schematic representation of the domain organization of tRNA nucleotidyltransferases whose sequences and/or structures have been examined. The Nrn domain is shown in blue, the CBS domain in orange and the "core" CCA-adding domain is shown in red. Note that the "core CCA-adding" domain catalyzes CCA-addition in some species but A or CC-addition others. The regions labeled NC1 and NC2 are small regions whose sequences are not conserved in the various CCA- and A-adding enzymes. This figure is based on but is not identical to Fig. 4.2 from Jones (2019). (Reprinted with permission from the *Journal of Molecular Evolution*)

There are significant differences in the sizes of the CCA-, CC- and A-adding TNTs from various bacteria. The schematic representation of the domain organization of the bacterial TNTs shown in Fig. 4.1 explains this size variation.

The CCA-adding enzyme from *Thermotoga maritima*, and from the phylum Synergistetes and the A-adding enzymes from the Aquificae, Cyanobacteria (Tomita and Weiner 2001) and the δ-Proteobacteria (Bralley et al. 2009) have two accessory domains which are absent from other bacterial TNTs. The first of these is the Nrn domain, characteristic of the family of nanoRNases (the Nrn family), which degrade oligoribonucleotides (Liao et al. 2018). The second domain is the cystathionine β-synthetase domain (CBS), which regulates the activity of enzymes, e.g. inorganic pyrophosphatases, in various species (Anashkin et al. 2017). The role of these domains in facilitating the activity of the corresponding TNTs, if there is one, is unknown. There is no evidence that the Nrn and CBS domains are enzymatically active apart from any roles they may play in the CCA-adding activities of the TNTs in which they are found. Indeed, the Nrn and CBS domains can be deleted without affecting the activity of the remaining carboxy terminal end of the proteins, which contains the TNT catalytic site (Tomita and Weiner 2001). Figure 4.1 indicates that most modern bacterial TNTs lack the Nrn and CBS domains, whether their TNTs are CCA- or CC- and A- adding. It is likely that CCA- and A-adding enzymes that possess the Nrn and CBS domains will be identified in other bacterial species. To date no CC-adding enzymes have been shown to possess Nrn or CBS domains.

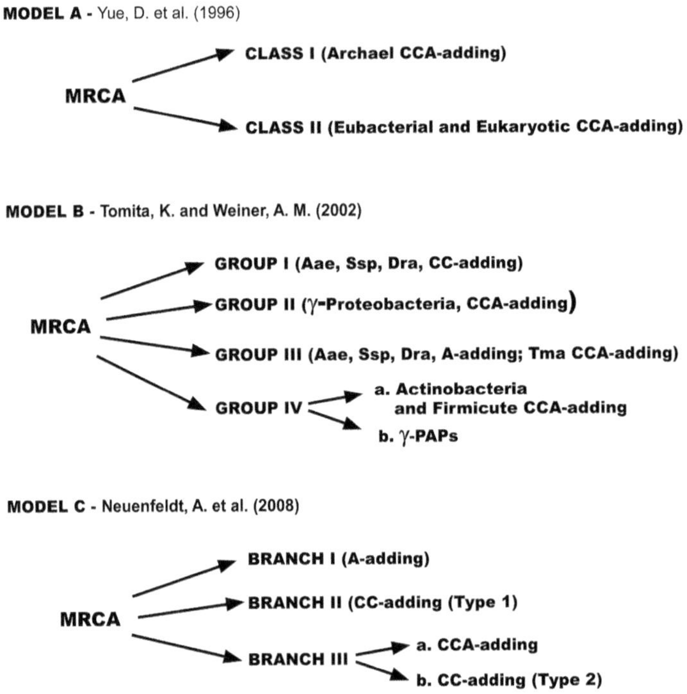

Fig. 4.2 Proposed models for the evolution of tRNA nucleotidyltransferases. Note that Models B and C are simplified versions of more sophisticated evolutionary models that were based on Neighbor Joining phylogenetic trees. Abbreviations used: MRCA Most Recent Common Ancestor, Aae *Aquifex aeolicus*, Ssp *Synechocystis sp*., Dra *Deinococcus radiodurans*, Tma *Thermotoga maritima*, γ γ-Proteobacteria, PAP poly(A) polymerase. The Firmicutes and Actinobacteria whose TNTs were analyzed by Tomita and Weiner (2002) were *Bacillus subtilis*; *Mycobacterium leprae* and *Mycobacterium tuberculosis*. In addition to the species listed above, additional species utilized in the analyses of Neuenfeldt et al. are listed in Figs. 4.4 and 4.5 of that publication. (Neuenfeldt et al. 2008)

The two-enzyme system has recently been shown to exist in some eukaryotes. The budding yeast, *Schizosaccharomyces pombe* was the first eukaryotic organism shown to contain separate CC- and A-adding TNTs (Reid et al. 2019). An interesting situation has recently been reported for the amoebozoan, *Acanthamoeba castellanii*. This organism contains two CCA-adding TNTs, one of which appears to be targeted to the mitochondrion. In addition, the organism contains both a CC- and an A-adding enzyme (Erber et al. 2020). The sequence of the A-adding enzyme bears significant resemblance to TNT sequences from the Proteobacteria, suggesting that the corresponding gene might not have been inherited via vertical transmission (Erber et al. 2020).

The crystal structures of the CCA-adding enzymes from *Geobacillus stearothermophilus* (formerly *Bacillus stearothermophilus*) (Li et al. 2002) and *Thermotoga maritima* (Toh et al. 2009) have been determined as have the structures of the CC- and A- adding enzymes from *Aquifex aeolicus* (Tomita et al. 2004; Yamashita et al.

Table 4.1 tRNA nucleotidyltransferases whose activity has been verified biochemically

Species	TNT activity	Reference
Escherichia coli	CCA-adding	Deutscher (1973)
Bacillus subtilis	CCA-adding	Raynal et al. (1998)
Mycobacterium leprae	CCA-adding	Raynal et al. (1998)
Geobacillus stearothermophilus	CCA-adding	Li et al. (2002)
Thermotoga maritima	CCA-adding	Tomita and Weiner (2001)
Streptomyces coelicolor	CCA-adding	Bralley et al. (2006)
Aquifex aeolicus	CC- and A-adding	Tomita and Weiner (2001)
Deinococcus radiodurans	CC- and A-adding	Tomita and Weiner (2002)
Synechocystis sp.	CC- and A-adding	Tomita and Weiner (2002)
Bacillus halodurans	CC- and A-adding	Bralley et al. (2005)
Geobacter sulfurreducens	CC- and A-adding	Bralley et al. (2009)
Thermus thermophilus	CC-adding	Neuenfeldt et al. (2008)
Bacillus clausii	CC-adding	Neuenfeldt et al. (2008)

2014). The amino acid sequences and crystal structures of the CCA-, CC- and A-adding enzymes show significant similarity suggesting that all three classes of TNT evolved from a common ancestor (Yue et al. 1996). It should be noted that the activities of only a few bacterial TNTs have been verified by biochemical assay. These TNTs are listed in Table 4.1.

TNTs are encoded by *cca* genes (Foulds et al. 1974). Reid et al. designated the gene for the CC-adding enzyme from *S. pombe* as *cca1* and that for the A-adding enzyme as *cca2* (Reid et al. 2019).

4.3 Encoding of the 3′-CCA End in tRNA Genes

One factor that may influence the evolution of and certainly relates to the evolution of tRNA nucleotidyltransferases is the frequency with which the CCA end is encoded in an organism's tRNA genes. In general terms, genome sequencing has revealed that all the tRNA genes of some bacteria encode the CCA end while only some tRNA genes in other bacteria encode CCA. Similarly, some Archaeal tRNA genes encode the CCA end while the tRNA genes of Eukaryotes do not (Klenk et al. 1997; Marck and Grosjean 2002).

More specifically, all the tRNA genes of *Escherichia coli* encode the 3′-CCA (Marck and Grosjean 2002) while ca. 40% of the tRNA genes from *Bacillus subtilis* do so (Vold 1985). Only 4 of 46 tRNA genes in the genome of the Archaeon *Archaeoglobus fulgidis* encode the 3′-CCA and it is noteworthy that *E. coli*, *B. subtilis* and *A. fulgidis* all contain a tRNA nucleotidyltransferase (Deutscher 1973; Raynal et al. 1998; Xiong et al. 2003), as do the Eukaryotes (Deutscher 1973). An interesting situation is presented by the Mycoplasmas. These organisms are considered to have the smallest genomes of any organism that can be grown in axenic culture, ca. 580 kb (Su and Baseman 1990). The genome of *Mycoplasma genitalium* contains 36 annotated tRNA genes, 17 of which encode the CCA end (Fraser et al. 1995). A BLAST search of the proteome of *M. genitalium,* using the CCA-adding TNT from *E. coli* or either the CC- or A-adding TNTs from *A. aeolicus* as queries, did not identify a TNT in that organism. As *M. genitalium* employs a parasitic lifestyle, CCA-addition may involve enzymes provided by its host.

In the context of the analysis of bacterial tRNA nucleotidyltransferases it is noteworthy that some bacteriophage genomes encode tRNA. Thus, the ca. 170 kb genome of bacteriophage T4 encodes tRNA for eight tRNAs, which recognize codons that are relatively minor in *E. coli* but are more frequent in T4 genes (Miller et al. 2003). Only one of those eight tRNA genes encodes the CCA end and the T4 genome does not encode a TNT. CCA-addition in T4 infected cells presumably utilizes the *E. coli* TNT.

It should again be noted that tRNA nucleotidyltransferase (*cca*) genes and, thus, tRNA nucleotidyltransferases are essential in some bacteria (Bralley et al. 2006; Reuss et al. 2016).

4.4 Evolution of Bacterial tRNA Nucleotidyltransferases: Questions That Must Be Addressed By Any Evolutionary Model

The foregoing analysis, coupled with existing structural, functional and phylogenetic data suggests several questions which must be addressed by any models proposed for the evolution of the bacterial tRNA nucleotidyltransferases.

1. If, as the available information suggests, CCA-, CC- and A-adding enzymes evolved from a common ancestor (MRCA, Most Recent Common Ancestor), what enzymatic activity did that ancestor possess?
2. How did the other activities evolve from that of the MRCA?
3. How did ancestral TNTs acquire the Nrn and CBS domains?
4. How were those domains lost in the evolution of modern TNTs?
5. What role did inheritance mechanisms other than vertical transmission play in the evolution of TNTs?
6. Are there evolutionary reasons for the existence of both CCA- and CC- and A-adding TNT systems?

4.5 Inferring the Activity of the MRCA from Phylogenies Based on Other Gene Products

One approach to an answer to the first question posed in the preceding section, whether the MRCA of modern bacterial tRNA nucleotidyltransferases was a CCA-, CC- or A-adding enzyme, would be to identify the most ancient extant bacterial phylum and assess the activity of the TNT in species from that phylum. Several approaches have been used to construct bacterial phylogenetic trees based on conserved genes and gene products. Those approaches include the use of protein synthesis elongation factors and RNA polymerase subunit sequences (Bocchetta et al. 2000), ribosomal protein sequences (Bocchetta et al. 2000; Hug et al. 2016; Cavalier-Smith and Chao 2020), conserved signature proteins and indels (Gupta 2016) and ribosomal RNA sequences (Burggraf et al. 1992; Yarza et al. 2010)

Unfortunately, even when generally robust phylogenetic methods such as Maximum Likelihood and Maximum Parsimony were used, the resulting analyses produced trees with different topologies and, of most importance to the discussion here, different phyla/species at the root of the resulting trees. Thus, the Thermotogae (Bocchetta et al. 2000), the Aquificae (Burggraf et al. 1992), the Chloroflexi (Cavalier-Smith and Chao 2020) and the Actinobacteria (Gupta 2016) have all been posited as the phylum at the root of the bacterial phylogenetic tree. If all these phyla contained only one of the two CCA-adding enzyme systems, it might still be possible to infer the activity of the MRCA from phylogenetic analyses. However, as noted above, the Thermotogae, for example, contain a CCA-adding TNT while the Aquificae contain the two-enzyme system. Thus, existing bacterial phylogenies do not allow the definitive assignment of the enzymatic activity of the bacterial TNT MRCA.

Apropos of bacterial phylogeny, Coleman et al. used a phylogenomic approach to model over 11,000 gene families to construct a rooted bacterial phylogenetic tree (Coleman et al. 2021). The root was placed between two major bacterial clades, the Gracilicutes and the Terrabacteria. Although a specific root phylum was not identified in the analyses of Coleman et al., the authors did predict a gene set for the MRCA (LUCA, Last Universal Common Ancestor in their parlance). That gene set included a tRNA nucleotidyltransferase, indicating, at least in the context of their analysis, that CCA-, CC- and/or A-addition is an ancient biochemical activity, present in the universal ancestor of modern bacteria (Coleman et al. 2021).

In a more recent, related study, Moody et al. constructed a rooted Maximum Likelihood phylogenetic tree using 27 vertically inherited genes from Archaea and Bacteria (Moody et al. 2022). Their analysis indicated that, of the bacterial species they examined, the one closest to the root of the bacterial phylogenetic tree was *Dictyoglomus turgidum*. *D. turgidum* is a Gram-negative, thermophilic bacterium which is a member of its own phylum, the Dictyoglomota (Saiki et al. 1985; Oren and Garrity 2021). The genome sequence of *D. turgidum* has been determined (Brumm et al. 2016) and the proteome was searched via BLAST analysis with the sequence of the CCA-adding TNT from Synergistetes as the query. *D. turgidum* was found to contain a protein of approximately the same length (839 aa) as the TNT from Synergistetes (877 aa) and with 35% identity and 55% similarity to the Synergistetes protein over 99% of its length. Of particular relevance to the present discussion is the observation that the protein from *D. turgidum* contained both the Nrn and CBS accessory domains, as do the CCA-adding enzymes from the Thermotogae and Synergistetes and the A-adding enzymes from the Cyanobacteria, the Aquificae and the δ-Proteobacteria (Fig. 4.1). Although no role has been adduced for these domains in the function or expression of the TNTs in which they occur, their association with the catalytic domain of the bacterial enzyme appears to be ancient.

4.6 Existing Models for TNT Evolution

Several research groups have posited evolutionary models for the tRNA nucleotidyltransferases. An early model was proposed by Yue et al. (1996) and is summarized in Fig. 4.2A. The model proposes an MRCA, which gave rise to the two classes of TNT discussed above. Class I contains the archaeal TNTs (and of course the other enzymes which are members of this class) while class II contains, the bacterial and eukaryotic TNTs as previously indicated.

A more detailed model was proposed by Tomita and Weiner (2002). This model derives from a phylogenetic analysis of several TNT sequences and is shown in Fig. 4.2B. The Neighbor Joining phylogeny posits four groups of TNTs. Group I contains the CC-adding enzymes from *Aquifex aeolicus*, *Synechocystis sp.* and *Deinococcus radiodurans*. Group II contains the CCA-adding enzymes from the γ-Proteobacteria and presumably from the closely related β-Proteobacteria as well. Group III contains the A-adding enzymes from *Aquifex*, *Synechocystis* and *Deinococcus* as well as the CCA-adding TNT from *T. maritima*, while Group IV contains the CCA-adding TNT from the Actinobacteria and the Firmicutes and the poly(A) polymerases from the γ-Proteobacteria.

The authors note that the CC- and A-adding enzymes do not group with each other in their analysis and argue that the division of CCA-adding activity between CC- and A-adding enzymes may have been the ancestral state. Thus, in their model the MRCA was either CC- or A-adding. Tomita and Weiner bolster their argument by noting that the bacterial species whose TNT sequences they analyzed were situated near the base of the 16S rRNA-based phylogenetic tree which was current at the time of their analysis (Tomita and Weiner 2002).

Neuenfeldt et al. also proposed an evolutionary model for TNTs based on a Neighbor Joining phylogenetic analysis (Neuenfeldt et al. 2008). Their model is shown in Fig. 4.2C. The model describes three branches to the tree, Branch I containing the A-adding enzymes, Branch II containing what they label as Type 1 CC-adding enzymes and Branch III which contains the CCA-adding enzymes and the Type 2 CC-adding enzymes. The Type 1 CC-adding enzymes were represented by proteins from Bacilli, while Type 2 enzymes were those from *Aquifex*, *Synechocystis* and *Deinococcus*, along with some other Cyanobacteria and Firmicutes. Neuenfeldt et al. argue that the appearance of the CC-adding enzymes in two different branches indicates that CC-addition evolved via two independent evolutionary events (Neuenfeldt et al. 2008). They argue further that the MRCA was a CCA-adding enzyme as the loss of A-adding ability would be consistent with their hypothesis that CC-adding activity was produced by a small deletion from a CCA-adding enzyme.

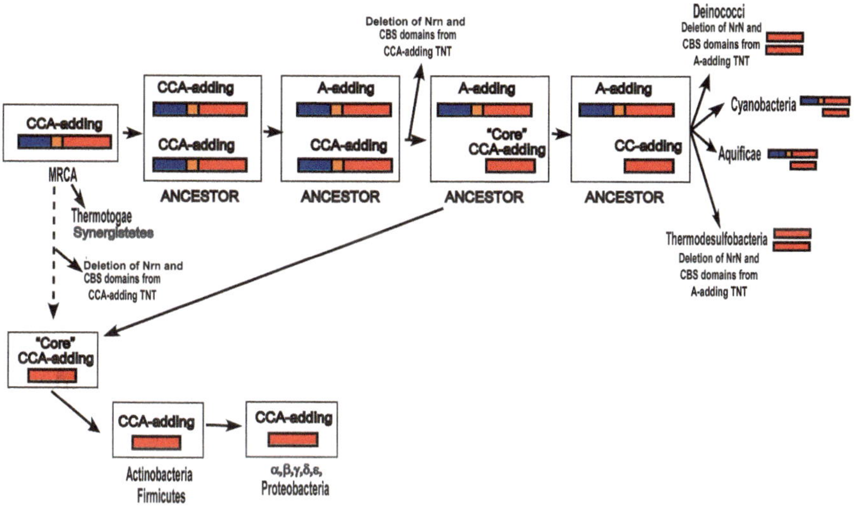

Fig. 4.3 Proposed model for the evolution of CCA-, CCA- and A-adding tRNA nucleotidyltransferases. Details of the model are given in the text. The color scheme used is identical to that of Fig. 4.1. For convenience NC1 and NC2 are not shown. This figure is derived from but is not identical to Fig. 4.4 from Jones (2019). Note that while changes to proteins are depicted in the figure, the mutations producing the ancestral states would have occurred in the relevant genes. (Reprinted with permission from the *Journal of Molecular Evolution*)

Jones (2019) proposed a comprehensive model for the evolution of bacterial TNTs and a summary of that model is depicted in Fig. 4.3. This model is also based on a Neighbor Joining phylogenetic analysis. The model posits a CCA-adding MRCA which gave rise to the CCA-adding activity in modern species that are placed near the root of the rRNA-based phylogenetic tree, e.g., the Thermotogae and Synergistetes. The MRCA contains the Nrn and CBS domains. A gene duplication gave rise to an ancestral species that initially contained two CCA-adding TNTs. Subsequent mutations converted one of these genes to that for an A-adding enzyme and the second to a CC-adding enzyme.

Over the course of the evolutionary history of the enzymes the Nrn and CBS domains were deleted in the genes for some CCA- and A-adding enzymes but retained in others. The model suggests two steps at which the accessory domains were lost to produce the "core" CCA-adding enzyme. In one scenario, those domains were lost from one of the duplicate genes during the separation of A- and CC-adding activities. The resulting ancestor would then have given rise to both the CC-adding TNT and an ancestor containing only the gene for the "core" CCA-adding enzyme. In the second scenario the Nrn and CBS domains were lost from the ancestral CCA-adding enzyme, producing an ancestor containing the "core" CCA-adding TNT. Subsequent evolution of that ancestor gave rise to modern CCA-adding enzymes in the phyla indicated in Fig. 4.1.

4.7 Critique of the Proposed Evolutionary Models

How do we evaluate the validity of the models for TNT evolution described above? The simple model proposed by Yue et al. does not provide a detailed evolutionary pathway and is likely to be accurate as far as it goes. An important omission from this model is the nature of the MRCA; Was it a CCA-, CC- or A-adding enzyme?

Although the models proposed by Tomita and Weiner, Neuenfeldt et al. and Jones were based on phylogenetic analysis, those analyses are subject to criticism. First, Tomita and Weiner and Neuenfeldt et al. used relatively few sequences to construct their phylogenetic trees. The robustness of their phylogenies may be compromised by this circumstance. Second, although Jones used ca. 120 protein sequences, some of those were poly(A) polymerases rather than TNTs. Third, in all three cases, the Neighbor Joining distance method was used to construct the phylogenetic trees. Distance methods can be useful and are computationally less demanding than other methods, but the literature suggests that a phylogenetic method such as Maximum Likelihood is more accurate than Neighbor Joining in inferring the phylogenies of conserved proteins, at least with sequences files of several hundred to one thousand (Som 2009; Yang and Rannala 2012).

Fourth, the *T. maritima* CCA-adding enzyme was used to root the phylogenetic tree in the model proposed by Jones. Use of the *T. maritima* enzyme as the outgroup was based on the placement of the species near the root of the rRNA-based phylogenetic tree, and as discussed above other phylogenies also place the Thermotogae at the root of the bacterial tree (Bocchetta et al. 2000). However, the most recent release from the Living Tree Project depicting the rRNA-based bacterial phylogenetic tree (https://imedea.uib-csic.es/mmg/ltp/#LTP) places several species closer to the root than *T. maritima*, including *Coprothermobacteria* and *Synergistetes*. The previous analysis suggested that Synergistetes contain a CCA-adding enzyme (Jones 2019), and a BLAST search of *Coprothermobacteria* genomes suggests that this phylum also contains a CCA-adding TNT. However, it must be noted that the most recent release from the Living Tree Project places *Aquifex* closest to the root of the bacterial phylogenetic tree (https://imedea.uib-csic.es/mmg/ltp/#LTP), and Aquifex contains separate CC- and A-adding enzymes. Moreover, and as noted above, other approaches place different species at the root of the bacterial phylogenetic tree.

4.8 The Role of Horizontal Gene Transfer in the Evolution of Bacterial Nucleotidyltransferases

Abundant evidence exists for the acquisition of genetic information by horizontal transfer (Horizontal Gene Transfer, HGT) between species (Brown 2003; Soucy et al. 2015; Koonin 2016). The "gold standard" for the detection of HGT has traditionally been phylogenetic incongruence, that is, the appearance of genes or proteins in clades other than those in which their sister species cluster in phylogenetic trees (Ravenhall et al. 2015; Koonin 2016). Phylogenetic analysis of the bacterial TNTs has identified several examples of apparent HGT.

In the phylogeny described by Jones, the TNT from the α-proteobacterium, *Magnetofaba australis* appeared in the clade containing the β, γ-TNTs (Jones 2019). Additionally, the TNT from *Helicobacter pylori*, an ε-proteobacterium, appeared in a clade containing the CC-adding enzymes from species such as *D. radiodurans*, while the A-adding enzyme from *B. halodurans* appeared in the clade containing A-adding TNTs from organisms such as *A. aeolicus* (Jones 2019). Although the placement of the CC- and A-adding enzymes from *Geobacter sulfurreducens* was not incongruent in relation to other CC- and A-adding enzymes, it was incongruent based on the placement of the species, a δ-proteobacterium, in phylogenies derived using rRNA sequences and other gene products (Gupta 2016; Hug et al. 2016; Cavalier-Smith and Chao 2020; Coleman et al. 2021).

Calculation of alien indices (Rancurel et al. 2017), which are indicative of likely HGT, indicated with high probability that the genes for the *G. sulfurreducens* CC- and A-adding enzymes were inherited by horizontal gene transfer from a member of the Nitrospirota and a member of the Aquificae, respectively (Jones 2019). It should be noted that the δ-Proteobacteria are shown in Fig. 4.1 among phyla that contain a single CCA-adding TNT and among those that contain separate CC- and A-adding enzymes. This dual placement reflects the likelihood that the two enzyme system in the δ-Proteobacteria arose by HGT and the observation that some δ-Proteobacteria contain a single TNT, which is likely to be a CCA-adding enzyme (Jones 2019).

Alien index calculations also indicated that the *M. australis*, *H. pylori* and *B. halodurans* (A-adding) TNT genes were acquired by HGT. It is likely that other examples of HGT will be discovered as the evolutionary history of the bacterial TNTs is explored further.

It should finally be noted that the horizontal transfer of the gene for a putative CCA-adding TNT from an α-proteobacterium to the eukaryotic Holozoa has been reported (Betat et al. 2015), and as noted above, there is reason to believe that the gene for at least one of the TNTs found in *A. castellanii* may have been inherited horizontally (Erber et al. 2020).

4.9 Ancestral Sequence Reconstruction as an Approach to the Evolution of Bacterial tRNA Nucleotidyltransferases

Ancestral Sequence Reconstruction (ASR) uses extant gene or protein sequences to predict the sequence of ancestral genes or proteins (Cai et al. 2004; Mascotti 2022). ASR programs use a phylogenetic tree and generate ancestral sequences corresponding to an MRCA and sequences corresponding to each node in the tree.

ASR has already been employed to answer an intriguing evolutionary question regarding the catalytic activity of the TNT from the γ-Proteobacteria. Hager et al.

observed that the class II TNT from *E. coli* catalyzes CCA addition with high efficiency even though the tRNA substrate binds only weakly to the enzyme during that reaction (Hager et al. 2022). Indeed, the binding efficiency is sufficiently weak that binding constants for the interaction between the substrate and the *E. coli* TNT cannot be measured (Shi et al. 1998; Hager et al. 2022).

To explore this apparent paradox, Hager et al. reconstructed ancestral TNTs using sequences of ca. 100 TNTs from various γ-proteobacterial genomes (Hager et al. 2022). They synthesized and overexpressed the corresponding gene for one of the ASR-generated TNTs, purified the protein and analyzed its function in comparison with the *E. coli* TNT in CCA addition. Hager et al. observed that the ancestral TNT bound tRNA substrates much more efficiently than the *E. coli* enzyme, such that it was possible to calculate a binding constant for the interaction between the ancestral enzyme and the substrate. This increased binding affinity had two consequences affecting the catalytic efficiency of the ancestral enzyme. First, that increased affinity enhanced the efficiency of the reverse reaction, in which CCA was removed from the end of mature tRNAs. Second, the increased affinity led to a processive rather than a distributive catalytic mechanism for the ancestral enzyme, resulting in substrate inhibition of the ancestral TNT. The authors conclude that the reduced efficiency of substrate binding to the modern *E. coli* TNT, paradoxically, leads to an increase rather than a decrease in catalytic efficiency, since both CCA removal and substrate inhibition are significantly reduced with the modern enzyme (Hager et al. 2022).

To complement the existing analyses of the evolutionary history of bacterial TNTs, ASR was used here to obtain a preliminary look at the problem. Approximately 150 sequences of bacterial CCA-, CC- and A-adding enzymes were used to construct a Maximum Likelihood, unrooted phylogenetic tree. This tree was then used, along with the sequence alignment that generated the tree, as inputs to the FireProt ASR program (Khan et al. 2021). FireProt produced an MRCA from the tree and the associated sequences and predicted the sequences at each node of the phylogenetic tree.

To gain a preliminary understanding of the activity of the putative MRCA, SWISSDOCK ligand docking software (Eberhardt et al. 2021; Bugnon et al. 2024) was used to model the interaction of the protein with ATP and CTP. As a control, the docking software was also used to predict the interaction of the *T. maritima* CCA-adding enzyme with its substrates. This control allowed comparison of the docking models with the enzyme-substrate interactions determined by x-ray crystallography (Toh et al. 2009).

The results of the docking analysis are shown in Fig. 4.4.The docking interactions shown for the *T. maritima* enzyme are essentially identical to those indicated by the structural studies for both ATP and CTP (Toh et al. 2009) and it is noteworthy that the docking analysis predicts the involvement of the same amino acids in forming H-bonds with ATP as with CTP. It should also be noted that a corresponding set of residues has been shown to be involved in substrate binding in studies of the TNT from *G. stearothermophilus* (Fig. 4.5) (Li et al. 2002).

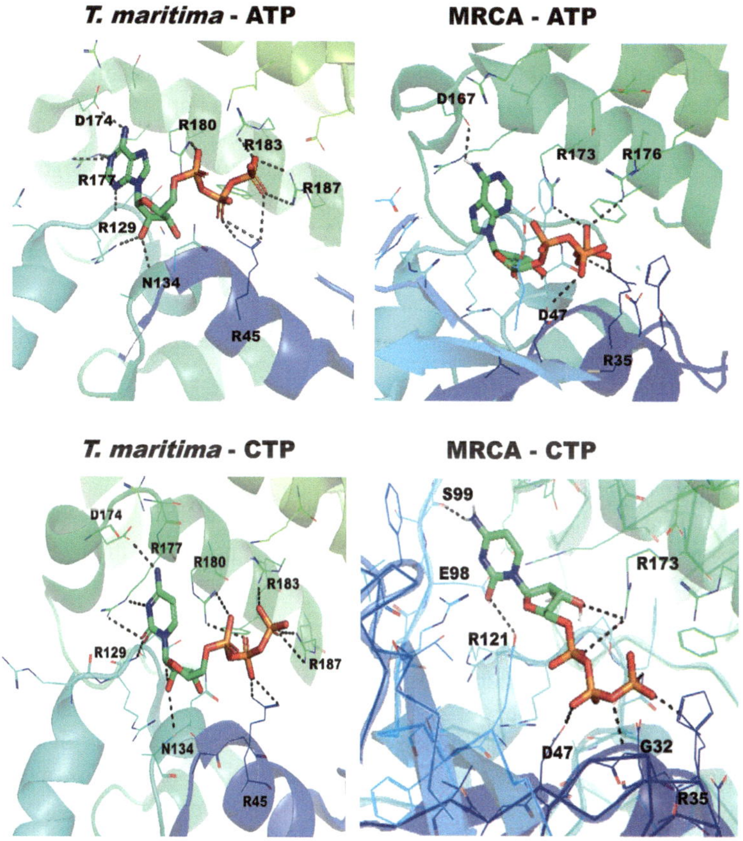

Fig. 4.4 Models depicting the binding of ATP and CTP at the active sites of the *T. maritima* CCA-adding enzyme and the MRCA predicted by FireProt. Docking was performed using the Autodock Vina program in SWISSDOCK (Eberhardt et al. 2021; Bugnon et al. 2024). The models were visualized using PyMOL 3.1.3.1

T. maritima	173	DDPTRILRAIRFEQR
G. stearothermophilus	153	EDALRMMRAVRFVSQ
MRCA	166	EDPLRMLRAVRFAAR

Fig. 4.5 Alignment of residues involved in ATP and CTP binding by the TNTs from *T. maritima* and *G. stearothermophilus* and putatively involved in ATP binding in the MRCA. Residues highlighted in black are identical in the sequences of at least two of the indicated enzymes, and similar residues are highlighted in gray. Residues in red for *G. stearothermophilus* and *T. maritima* were shown by x-ray crystallography to form H-bonds with ATP and CTP and the residues predicted to interact with ATP in the MRCA are also shown in red. (Li et al. 2002; Toh et al. 2009)

A similarly positioned, although a smaller number, of residues are predicted to be involved in ATP binding by the MRCA (Figs. 4.4 and 4.5). However, the docking software did not predict the involvement of this group of residues in the binding of CTP to the MRCA (Fig. 4.4). This result would seem to imply that either (a) if the MRCA is a CCA-adding enzyme, the CTP binding pocket is comprised of different amino acid residues in that protein as compared with extant TNTs, or (b) the MRCA does not bind CTP and is, therefore, an A-adding enzyme. Note that one of the models described above speculates that the MRCA was not a CCA-adding TNT (Tomita and Weiner 2002).

4.10 Concluding Remarks

The foregoing analysis identifies important questions that must be addressed in any model for the evolution of bacterial TNTs and suggests some approaches to explore their evolutionary history. One important issue which was not considered in this discussion is the evolution of another member of the nucleotidyltransferase superfamily, viz. the bacterial poly(A) polymerases (Holm and Sander 1995; Yue et al. 1996; Jones 2021). These enzymes are related to the bacterial TNTs in both sequence and structure and presumably arose from the MRCA that gave rise to the TNTs. A scheme for the evolution of bacterial poly(A) polymerases from a TNT ancestor has recently been proposed (Jones 2021).

A second issue which evolutionary modeling might address is the disparate explanations for the substrate specificities of the A- and CC-adding enzymes. Based on crystallographic studies, the Tomita group argues that A-adding enzymes are restricted to A-addition because the head region of the protein, containing the catalytic site, is less flexible than is the case in CCA-adding TNTs (Tomita et al. 2004). Thus, they argue that the head region cannot extend to the residue that precedes the first added C and, therefore, can only add the terminal A-residue once the Cs are already in place. They argue further that CC-adding enzymes are restricted to C-addition because tRNA-CC dissociates from the enzyme after the second C is added and can then only bind to the A-adding enzyme (Yamashita et al. 2014).

Tretbar et al. propose a different explanation for the specificity of CCA- and A-adding enzymes. They presented data indicating that a deletion of the C-terminal region of the *D. radiodurans* and *B. halodurans* A-adding enzymes converted them to CCA-adding enzymes. They argue that the C-terminal sequences inhibit CC-addition in the A-adding enzymes and propose that mutations in this region led to the evolution of CCA-adding from A-adding TNTs (Tretbar et al. 2011). It should be noted, however, that Tomita and coworkers were unable to convert the *A. aeolicus* A-adding enzyme to CCA-addition by deletion of the corresponding C-terminal fragment (Yamashita et al. 2015).

As regards CC-addition, Neuenfeldt et al. identified a flexible loop in the sequence of CCA-adding TNTs which when deleted converted the enzymes from CCA-addition to CC-addition only (Neuenfeldt et al. 2008). The flexible loop appears to be present in all bacterial CCA-adding TNTs but absent from CC-adding enzymes (Neuenfeldt et al. 2008). Moreover, when the loop from the *B. subtilis* CCA-adding enzyme was inserted in the appropriate position in the CC-adding TNT from *B. halodurans*, the resulting mutant protein displayed CCA-adding activity (Neuenfeldt et al. 2008).

It should again be noted that Tomita and coworkers obtained a different result in their analyses, albeit with different enzymes. Yamashita et al. reported that insertion of the flexible loop from the *A. aeolicus* A-adding or the *T. maritima* CCA-adding enzyme into the appropriate position failed to convert the *A. aeolicus* CC-adding enzyme to CCA-addition (Yamashita et al. 2014). It should also be noted that the *S. pombe* CC-adding enzyme contains a flexible loop with conserved sequence motifs that are similar to those found in bacterial CCA-adding enzymes (Reid et al. 2019; Jones 2019). It is hoped that detailed explication of the evolutionary history of the bacterial TNTs will help to resolve these apparently conflicting results.

The most intriguing question that arises from the fact that modern organisms can contain a CCA-adding enzyme, CC- and A-adding enzymes or both, is what is the evolutionary advantage, if there is one, of having one TNT rather than two or two rather than one. Approaches such as Ancestral Reconstruction, coupled with the expression and biochemical analysis of the predicted MRCA and other ancestral species, may provide answers to this and the other intriguing questions posed in this discussion.

References

Anashkin VA, Baykov AA, Lahti R (2017) Enzymes regulated via cystathionine beta-synthase domains. Biochemistry (Mosc) 82:1079–1087

Betat H, Mede T, Tretbar S, Steiner L, Stadler PF, Morl M, Prohaska SJ (2015) The ancestor of modern Holozoa acquired the CCA-adding enzyme from Alphaproteobacteria by horizontal gene transfer. Nucleic Acids Res 43:6739–6746

Betat H, Morl M (2015) The CCA-adding enzyme: a central scrutinizer in tRNA quality control. BioEssays 37:975–982

Bocchetta M, Gribaldo S, Sanangelantoni A, Cammarano P (2000) Phylogenetic depth of the bacterial genera *Aquifex* and *Thermotoga* inferred from analysis of ribosomal protein, elongation factor, and RNA polymerase subunit sequences. J Mol Evol 50:366–380

Bralley P, Chang SA, Jones GH (2005) A phylogeny of bacterial RNA nucleotidyltransferases: *Bacillus halodurans* contains two tRNA nucleotidyltransferases. J Bacteriol 187:5927–5936

Bralley P, Cozad M, Jones GH (2009) *Geobacter sulfurreducens* contains separate C- and A-adding tRNA nucleotidyltransferases and a poly(A) polymerase. J Bacteriol 191:109–114

Bralley P, Gust B, Chang S, Chater KF, Jones GH (2006) RNA 3′-tail synthesis in *Streptomyces*: *in vitro* and *in vivo* activities of RNase PH, the *SCO3896* gene product and polynucleotide phosphorylase. Microbiology 152:627–636

Brown JR (2003) Ancient horizontal gene transfer. Nat Rev Genet 4:121–132

Brumm PJ, Gowda K, Robb FT, Mead DA (2016) The complete genome sequence of hyperthermophile *Dictyoglomus turgidum* DSM 6724 reveals a specialized carbohydrate fermentor. Front Microbiol 7:1979

Bugnon M, Rohrig UF, Goullieux M, Perez MAS, Daina A, Michielin O, Zoete V (2024) SwissDock 2024: major enhancements for small-molecule docking with attracting cavities and AutoDock Vina. Nucleic Acids Res 52:W324–W332

Burggraf S, Olsen GJ, Stetter KO, Woese CR (1992) A phylogenetic analysis of *Aquifex pyrophilus*. Syst Appl Microbiol 15:352–356

Cai W, Pei J, Grishin NV (2004) Reconstruction of ancestral protein sequences and its applications. BMC Evol Biol 4:33

Cavalier-Smith T, Chao EE (2020) Multidomain ribosomal protein trees and the planctobacterial origin of neomura (eukaryotes, archaebacteria). Protoplasma 257:621–753

Coleman GA, Davin AA, Mahendrarajah TA, Szantho LL, Spang A, Hugenholtz P, Szollosi GJ, Williams TA (2021) A rooted phylogeny resolves early bacterial evolution. Science 372:eabe0511

Deutscher MP (1973) Synthesis and functions of the -C-C-A terminus of transfer RNA. Prog Nucleic Acid Res Mol Biol 13:51–92

Deutscher MP (1990) Ribonucleases, tRNA nucleotidyltransferase, and the 3′ processing of tRNA. Prog Nucleic Acid Res Mol Biol 39:209–240

Eberhardt J, Santos-Martins D, Tillack AF, Forli S (2021) AutoDock Vina 1.2.0: new docking methods, expanded force field, and python bindings. J Chem Inf Model 61:3891–3898

Erber L, Betat H, Mörl M (2020) CCA-addition gone wild: unusual occurrence and phylogeny of four different tRNA nucleotidyltransferases in *Acanthamoeba castellanii*. Mol Biol Evol 38:1006–1017

Foulds J, Hilderman RH, Deutscher MP (1974) Mapping of the locus for *Escherichia coli* transfer ribonucleic acid nucleotidyltransferase. J Bacteriol 118:628–632

Fraser CM, Gocayne JD, White O, Adams MD, Clayton RA, Fleischmann RD, Bult CJ, Kerlavage AR, Sutton G, Kelley JM, Fritchman RD, Weidman JF, Small KV, Sandusky M, Fuhrmann J, Nguyen D, Utterback TR, Saudek DM, Phillips CA, Merrick JM, Tomb JF, Dougherty BA, Bott KF, Hu PC, Lucier TS, Peterson SN, Smith HO, Hutchison CA 3rd, Venter JC (1995) The minimal gene complement of *Mycoplasma genitalium*. Science 270:397–403

Guerrier-Takada C, Mcclain WH, Altman S (1984) Cleavage of tRNA precursors by the RNA subunit of *E. coli* ribonuclease P (M1 RNA) is influenced by 3′-proximal CCA in the substrates. Cell 38:219–224

Gupta RS (2016) Impact of genomics on the understanding of microbial evolution and classification: the importance of Darwin's views on classification. FEMS Microbiol Rev 40:520–553

Hager M, Pohler MT, Reinhardt F, Wellner K, Hubner J, Betat H, Prohaska S, Morl M (2022) Substrate affinity versus catalytic efficiency: ancestral sequence reconstruction of tRNA nucleotidyltransferases solves an enzyme puzzle. Mol Biol Evol 39:msac250

Holm L, Sander C (1995) DNA polymerase beta belongs to an ancient nucleotidyltransferase superfamily. Trends Biochem Sci 20:345–347

Hug LA, Baker BJ, Anantharaman K, Brown CT, Probst AJ, Castelle CJ, Butterfield CN, Hernsdorf AW, Amano Y, Ise K, Suzuki Y, Dudek N, Relman DA, Finstad KM, Amundson R, Thomas BC, Banfield JF (2016) A new view of the tree of life. Nat Microbiol 1:16048

Jones GH (2019) Phylogeny and evolution of RNA 3′-nucleotidyltransferases in bacteria. J Mol Evol 87:254–270

Jones GH (2021) Acquisition of *pcnB* [poly(A) polymerase I] genes via horizontal transfer from the beta, gamma-Proteobacteria. Microb Genom 7:000508

Khan RT, Musil M, Stourac J, Damborsky J, Bednar D (2021) Fully automated ancestral sequence reconstruction using FireProt(ASR). Curr Protoc 1:e30

Klenk HP, Clayton RA, Tomb JF, White O, Nelson KE, Ketchum KA, Dodson RJ, Gwinn M, Hickey EK, Peterson JD, Richardson DL, Kerlavage AR, Graham DE, Kyrpides NC, Fleischmann RD, Quackenbush J, Lee NH, Sutton GG, Gill S, Kirkness EF, Dougherty BA, Mckenney K, Adams MD, Loftus B, Peterson S, Reich CI, Mcneil LK, Badger JH, Glodek A, Zhou L, Overbeek R, Gocayne JD, Weidman JF, Mcdonald L, Utterback T, Cotton MD, Spriggs T, Artiach P, Kaine BP, Sykes SM, Sadow PW, D'Andrea KP, Bowman C, Fujii C, Garland SA, Mason TM, Olsen

GJ, Fraser CM, Smith HO, Woese CR, Venter JC (1997) The complete genome sequence of the hyperthermophilic, sulphate-reducing archaeon *Archaeoglobus fulgidus*. Nature 390:364–370

Koonin EV (2016) Horizontal gene transfer: essentiality and evolvability in prokaryotes, and roles in evolutionary transitions. F1000Res 5:F1000 Faculty Rev-1805

Li F, Xiong Y, Wang J, Cho HD, Tomita K, Weiner AM, Steitz TA (2002) Crystal structures of the *Bacillus stearothermophilus* CCA-adding enzyme and its complexes with ATP or CTP. Cell 111:815–824

Liao H, Liu M, Guo X (2018) The special existences: nanoRNA and nanoRNase. Microbiol Res 207:134–139

Marck C, Grosjean H (2002) tRNomics: analysis of tRNA genes from 50 genomes of Eukarya, Archaea, and Bacteria reveals anticodon-sparing strategies and domain-specific features. RNA 8:1189–1232

Mascotti ML (2022) Resurrecting enzymes by ancestral sequence reconstruction. Methods Mol Biol 2397:111–136

Miller ES, Kutter E, Mosig G, Arisaka F, Kunisawa T, Ruger W (2003) Bacteriophage T4 genome. Microbiol Mol Biol Rev 67:86–156. table of contents

Moody ERR, Mahendrarajah TA, Dombrowski N, Clark JW, Petitjean C, Offre P, Szollosi GJ, Spang A, Williams TA (2022) An estimate of the deepest branches of the tree of life from ancient vertically evolving genes. Elife 11:e66695

Neuenfeldt A, Just A, Betat H, Mörl M (2008) Evolution of tRNA nucleotidyltransferases: a small deletion generated CC-adding enzymes. Proc Natl Acad Sci USA 105:7953–7958

Oh BK, Pace NR (1994) Interaction of the 3′-end of tRNA with ribonuclease P RNA. Nucleic Acids Res 22:4087–4094

Oren A, Garrity GM (2021) Valid publication of the names of forty-two phyla of prokaryotes. Int J Syst Evol Microbiol 71. https://doi.org/10.1099/ijsem.0.005056

Rammelt C, Rossmanith W (2016) Repairing tRNA termini: news from the 3′ end. RNA Biol 13:1182–1188

Rancurel C, Legrand L, Danchin EGJ (2017) Alienness: rapid detection of candidate horizontal gene transfers across the tree of life. Genes (Basel) 8:248–261

Ravenhall M, Skunca N, Lassalle F, Dessimoz C (2015) Inferring horizontal gene transfer. PLoS Comput Biol 11:e1004095

Raynal LC, Krisch HM, Carpousis AJ (1998) The *Bacillus subtilis* nucleotidyltransferase is a tRNA CCA-adding enzyme. J Bacteriol 180:6276–6282

Reid NE, Ngou JS, Joyce PBM (2019) *Schizosaccharomyces pombe* contains separate CC- and A-adding tRNA nucleotidyltransferases. Biochem Biophys Res Commun 508:785–790

Reuss DR, Commichau FM, Gundlach J, Zhu B, Stulke J (2016) The blueprint of a minimal cell: MiniBacillus. Microbiol Mol Biol Rev 80:955–987

Saiki T, Kobayashi Y, Kawagoe K, Beppu T (1985) *Dictyoglomus thermophilum* gen. nov., sp. nov., a Chemoorganotrophic, Anaerobic, Thermophilic Bacterium. Int J Syst Evol Microbiol 35:253–259

Shi PY, Maizels N, Weiner AM (1998) CCA addition by tRNA nucleotidyltransferase: polymerization without translocation? EMBO J 17:3197–3206

Slade A, Kattini R, Campbell C, Holcik M (2020) Diseases associated with defects in tRNA CCA addition. Int J Mol Sci 21:3780

Som A (2009) ML or NJ-MCL? A comparison between two robust phylogenetic methods. Comput Biol Chem 33:373–378

Soucy SM, Huang J, Gogarten JP (2015) Horizontal gene transfer: building the web of life. Nat Rev Genet 16:472–482

Sprinzl M, Cramer F (1979) The -C-C-A end of tRNA and its role in protein biosynthesis. Prog Nucleic Acid Res Mol Biol 22:1–69

Su CJ, Baseman JB (1990) Genome size of *Mycoplasma genitalium*. J Bacteriol 172:4705–4707

Toh Y, Takeshita D, Numata T, Fukai S, Nureki O, Tomita K (2009) Mechanism for the definition of elongation and termination by the class II CCA-adding enzyme. EMBO J 28:3353–3365

Tomita K, Fukai S, Ishitani R, Ueda T, Takeuchi N, Vassylyev DG, Nureki O (2004) Structural basis for template-independent RNA polymerization. Nature 430:700–704

Tomita K, Weiner AM (2001) Collaboration between CC- and A-adding enzymes to build and repair the 3′-terminal CCA of tRNA in *Aquifex aeolicus*. Science 294:1334–1336

Tomita K, Weiner AM (2002) Closely related CC- and A-adding enzymes collaborate to construct and repair the 3′-terminal CCA of tRNA in *Synechocystis* sp. and *Deinococcus radiodurans*. J Biol Chem 277:48192–48198

Tretbar S, Neuenfeldt A, Betat H, Mörl M (2011) An inhibitory C-terminal region dictates the specificity of A-adding enzymes. Proc Natl Acad Sci USA 108:21040–21045

Vold BS (1985) Structure and organization of genes for transfer ribonucleic acid in *Bacillus subtilis*. Microbiol Rev 49:71–80

Wellner K, Betat H, Morl M (2018) A tRNA's fate is decided at its 3′ end: collaborative actions of CCA-adding enzyme and RNases involved in tRNA processing and degradation. Biochim Biophys Acta Gene Regul Mech 1861:433–441

Wellner K, Gnauck J, Bernier D, Bernhart SH, Betat H, Morl M (2025) Two complementing in vivo selection systems based on CCA-trimming exonucleases as a tool to monitor, select and evaluate enzymatic features of tRNA nucleotidyltransferases. RNA Biol 22:1–14

Xiong Y, Li F, Wang J, Weiner AM, Steitz TA (2003) Crystal structures of an archaeal class I CCA-adding enzyme and its nucleotide complexes. Mol Cell 12:1165–1172

Yamashita S, Martinez A, Tomita K (2015) Measurement of acceptor-TPsiC helix length of tRNA for terminal A76-addition by A-adding enzyme. Structure 23:830–842

Yamashita S, Takeshita D, Tomita K (2014) Translocation and rotation of tRNA during template-independent RNA polymerization by tRNA nucleotidyltransferase. Structure 22:315–325

Yang Z, Rannala B (2012) Molecular phylogenetics: principles and practice. Nat Rev Genet 13:303–314

Yarza P, Ludwig W, Euzeby J, Amann R, Schleifer KH, Glockner FO, Rossello-Mora R (2010) Update of the all-species living tree project based on 16S and 23S rRNA sequence analyses. Syst Appl Microbiol 33:291–299

Yue D, Maizels N, Weiner AM (1996) CCA-adding enzymes and poly(A) polymerases are all members of the same nucleotidyltransferase superfamily: characterization of the CCA-adding enzyme from the archaeal hyperthermophile *Sulfolobus shibatae*. RNA 2:895–908

Chapter 5
Polyribonucleotide Phosphorylase: An Ancient Enzyme Conserved in Bacteria and Eukarya

Federica Briani

Abstract Polyribonucleotide phosphorylase (PNPase) is a conserved phosphorolytic exonuclease involved in RNA metabolism in bacteria and eukaryotic organelles. Discovered in the mid-twentieth century, PNPase has since been extensively studied for its role in RNA stability control. PNPase belongs to the PDX protein family, sharing structural similarities with RNase PH and the archaeal and eukaryotic exosomes, and functions as a homotrimer with the three subunits forming a central channel where phosphorolysis occurs. In bacteria, PNPase is a key component of a multiprotein complex—the RNA degradosome—facilitating RNA turnover and quality control. In human mitochondria, it forms the mitochondrial degradosome and contributes to RNA trafficking and decay. Structural analyses reveal a dynamic interplay between its catalytic and RNA-binding domains, allowing PNPase to switch between an open and a closed conformation. Additionally, its activity is modulated, at least in *vitro*, by small molecules. Despite decades of research, questions remain regarding its precise mechanistic regulation, interactions, and functional diversification.

Keywords PNPase · RNase PH · Exosome · RNA degradation · Post-transcriptional regulation · Mitochondrial proteins · *PNPT1* · RNA degradosome

5.1 Introduction

Polyribonucleotide phosphorylase (PNPase) is an evolutionary conserved enzyme catalyzing the phosphorolytic cleavage of RNA, an activity integral to the control of gene expression in diverse organisms. The discovery of PNPase in the mid-twentieth century thanks to the work of Severo Ochoa and Marianne Grunberg-Manago was

F. Briani (✉)
Università degli Studi di Milano, Dipartimento di Bioscienze, Milan, Italy
e-mail: federica.briani@unimi.it

A. Kanai, A. W. Nicholson (eds.), *Molecular Evolution of RNA Regulatory Enzymes and Their Systems*, Nucleic Acids and Molecular Biology 38,
https://doi.org/10.1007/978-3-032-03300-0_5

recognized by a Nobel Prize awarded to Ochoa in 1959 and marked the inception of decades of research about PNPase activity and its multifaceted roles in different organisms.

The initial studies on PNPase focused on the biochemical characterization of the bacterial PNPase catalytic activity and on the elegant auto-regulation system of the PNPase *pnp* gene in *Escherichia coli* (reviewed by Littauer and Grunberg-Manago 2002; Briani et al. 2016). The resolution of the crystal structure of PNPase from *Streptomyces antibioticus* by Martyn Symmons and collaborators paved the way for a deeper comprehension of PNPase activity (Symmons et al. 2000). Equally relevant was the discovery, first in *E. coli* and then in other bacteria, that PNPase partners with other proteins forming the RNA degradosome complex to execute coordinated RNA degradation (Carpousis et al. 1994; Py et al. 1994, 1996; Tejada-Arranz et al. 2020). Studies on the human PNPase began when the cDNA of the nuclear *PNPT1* PNPase gene was cloned and PNPase was demonstrated to be mainly located in mitochondria (Leszczyniecka et al. 2002; Piwowarski et al. 2003). More recently, it was found that dysregulated expression of human PNPase can affect the progression of some tumors, whereas PNPase mutations cause neurological disorders ranging from hereditary hearing loss to multisystem oxidative phosphorylation (OXPHOS) deficiency disorders, spinocerebellar ataxia type 25, Aicardi-Goutiéres and Leigh syndromes (Sarkar et al. 2006; Eaton et al. 2018; Rius et al. 2019; Agarwal et al. 2020; Bamborschke et al. 2021; Falchi et al. 2022; Barbier et al. 2022; Revert-Ros et al. 2024). These discoveries emphasize the importance of efforts aimed at understanding PNPase function in different organisms and highlight its potential as a therapeutic target.

5.2 The PDX Family of Phosphorolytic Exonucleases

PNPase belongs to the PDX family of phosphorolytic exonucleases, which also comprises RNase PH (Zuo and Deutscher 2001). These enzymes use inorganic phosphate (Pi) as a nucleophile to cleave the phosphodiester bond between the terminal nucleotide at the 3′-end of an RNA molecule and the upstream chain. Consequently, RNA degradation by PDX-family exonucleases releases nucleoside 5′-diphosphates (NDPs), whereas hydrolytic exonucleases release nucleoside monophosphates (NMPs) (Fig. 5.1).

PNPase and RNase PH share a conserved domain, known as the RPH domain (Mian 1997). The RPH domain is often the only domain present in RNase PH homologues in different species. Accordingly, the monomeric unit of bacterial RNase PH is usually about 250 amino acids long. In contrast, the RPH domain is duplicated in the more complex PNPase, which is typically around 750 amino acids long and contains additional domains, unrelated to the RPH domain (Fig. 5.2).

The catalytically active RNase PH is composed of six identical subunits arranged around a central channel. Such doughnut-like structure formed by six RPH domains is conserved in PNPase, in which three subunits associate, and in the archaeal and

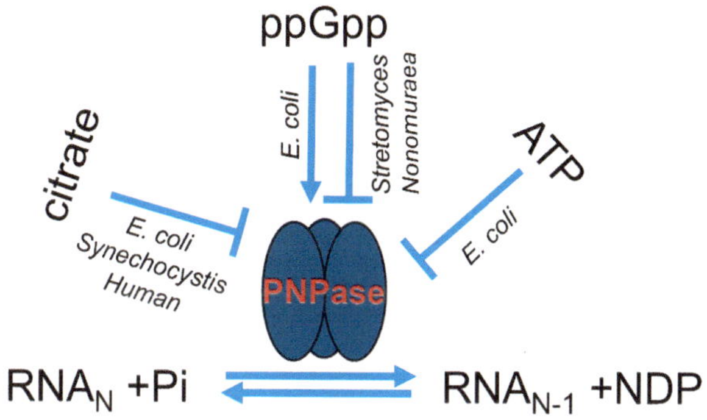

Fig. 5.1 PNPase activity and effectors. PNPase processively degrades RNA molecules by removing nucleotides one by one from the 3′-end and generating NDPs. It catalyzes also NDPs polymerization and RNA tailing. Small molecules can act as inhibitors (T-bar arrows) or activators (directional arrow) in different organisms. Only molecules directly affecting the *in vitro* activity of purified PNPase are mentioned

Fig. 5.2 Domain structure of PDX exonucleases. LS, localization signal: PNP1 and PNP2, PNPase RPH domains; ALL- α, all- α helical domain (AAHD); KH and S1, RNA binding domains. LS is present only in mitochondrial and chloroplast PNPases and absent in bacterial enzymes. RPH, RNase PH domain

eukaryotic exosome, a multi-protein intracellular complex formed by three heterodimers composed by two divergent RNase PH-like subunits (Fig. 5.3). RNase PH orthologues are found in all domains of life, whereas PNPase is restricted to Bacteria and Eukarya, being absent in Archaea (Fig. 5.4). This distribution, coupled with the high conservation of the RPH domain between RNase PH and the two RPH domains of PNPase (referred to as PNP1 and PNP2; Fig. 5.2), supports the hypothesis that both proteins derived from an ancestral RPH domain present in the Most Recent Common Ancestor (MRCA) of life (Fig. 5.4) (Leszczyniecka et al. 2004). Phylogenetic analysis of RNase PH/PNPase sequences from vertebrate taxa suggests that PNP2 and RPH originated from a duplication event of an ancestral PNP1 domain before the divergence of the Bacteria and Archaea/Eukarya lineages. A subsequent duplication and divergence of the RPH-PNP2 common ancestor may have led to the co-existence of PNPase and RNase PH in the MRCA, whereas a gene loss event may have resulted in the elimination of PNPase in the Archaea/Eukarya lineage (Sokhi et al. 2014). PNPase would have then been re-acquired by eukaryotes through the endosymbiotic events that led to the emergence of their ancestor, consistent with its localization in the mitochondria and chloroplasts of eukaryotic cells (Fig. 5.5). In this scenario, the absence of PNPase in the mitochondria of some unicellular eukaryotes would be attributed to successive gene loss events in specific

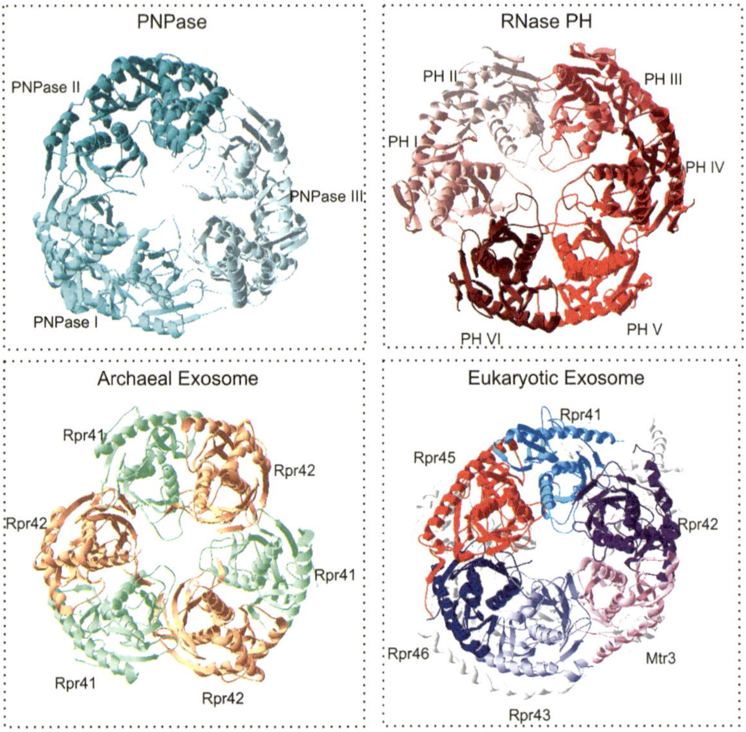

Fig. 5.3 Structure of the *E. coli* PNPase and RNase PH and of the archaeal and eukaryotic exosomes. Only domains constituting the rings are shown. The subunits are named. (Source: Briani et al. (2016))

taxonomic groups. It remains unclear whether these events conferred a selective advantage or became fixed through genetic drift (Fig. 5.5). In this regard, it is intriguing that while RNase PH-like proteins are conserved in all domains of life, their phosphorolytic activity is not. In fact, with the exception of the Rrp41 subunit in *Arabidopsis thaliana,* the RPH subunits composing the eukaryotic exosome are catalytically inactive in both yeast and higher eukaryotes, in which the ribonucleolytic activity of the exosome is driven by the association with a hydrolytic subunit unrelated to RNase PH (Chekanova et al. 2000; reviewed by Lykke-Andersen et al. 2009). It was speculated that during the evolution of the eukaryotic cell, RNA phosphorolysis was replaced by hydrolysis because phosphorolysis is an equilibrium reaction with NDP polymerization (Fig. 5.1). Under standard conditions, NDP polymerization should be slightly favoured over RNA degradation, but in cells, the equilibrium is likely shifted toward degradation due to high phosphate (Pi) and low NDP concentrations. The evolution of complex eukaryotic cells may have favoured RNA hydrolysis, which is irreversible, to prevent fluctuations in local Pi or NDP concentrations from impacting RNA metabolism (Wahle 2007).

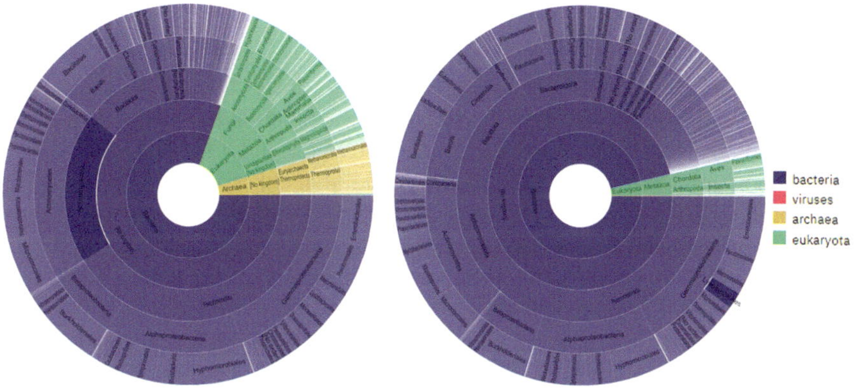

Fig. 5.4 Spread of RNase PH and PNPase in different taxonomic groups. InterPro sunburst charts for RNase PH (IPR050080, on the left) and PNPase (PIRSF005499, on the right) based on homologous entries retrieved in 22,551 and 26,247 species, respectively. The taxonomic ranks represented from the innermost to the outermost circle are: domain, kingdom, phylum, class, and order.

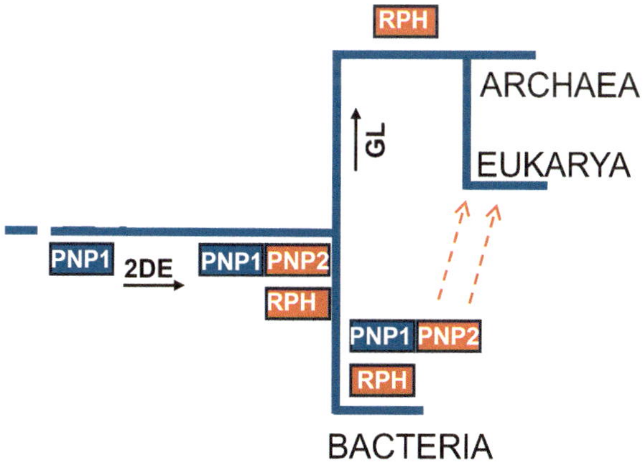

Fig. 5.5 Possible evolutionary scenario of PDX nucleases. An initial duplication event may have led to the emergence of a PNPase ancestor with duplicated PNP domains. RNase PH could have arisen from a second duplication event involving either the PNP2 domain alone or the entire PNPase ancestor gene, followed by the loss of the PNP1 encoding region. The common ancestor of Archaea and Eukarya may have subsequently lost the PNPase gene while undergoing RPH duplication events, ultimately leading to the evolution of the exosome. DE, duplication event; GL, gene loss event; dotted arrows, PNPase acquisition through the endosymbiotic events leading to the emergence and evolution of the eukaryotic cell

Whatever the reason for the loss of phosphorolytic activity in the eukaryotic exosome, the addition of a hydrolytic subunit, by relieving the selective pressure to maintain catalytic activity, allowed the catalytic sites of the exosome's RPH-like subunits to mutate, which reflects what occurred in the bacterial PNPase PNP1 domain (see below).

5.3 Structure and Activity of PNPase

All PNPases share a common domain structure, in which the three so called “core domains”, namely the two PNP1 and PNP2 RNase PH-like domains and the all α-helical domain (AAHD) connecting them, are followed by the KH and S1 RNA binding domains (Symmons et al. 2002). In addition, eukaryotic PNPases have an organelle localization signal at the very N-terminal end. The arrangement of three monomers composing the active enzyme, with the core domains delimiting a central channel in which the catalytic site is located, is conserved among PNPases of different species (Figs. 5.2 and 5.3).

PNPase was the first enzyme shown able to synthesize RNA from nucleotide precursors, and its activity was exploited in the early stages of molecular biology to synthesize RNA *in vitro* (Fry 2016). However, research has primarily linked PNPase to RNA degradation *in vivo*, though in bacterial cells and in chloroplasts the enzyme also performs 3′-end RNA tailing (Mohanty and Kushner 2000, 2003; Yehudai-Resheff et al. 2001; Pobre and Arraiano 2015).

5.3.1 *PNPase Catalytic Activity and Structure of the Active Site*

In vitro, the reactions catalysed by PNPase are the phosphorolytic degradation of the RNA, in presence of Mg^{2+} and Pi, and the reverse template-independent NDPs polymerization reaction (Fig. 5.1), as summarized by the equation.

$$RNA_n + Pi \leftrightharpoons RNA_{n-1} + NDP$$

The enzyme also catalyses the exchange between inorganic phosphate and the β-phosphate of NDPs (Littauer and Grunberg-Manago 2002).

In presence of Mn^{2+}, PNPases of different organisms degrade (or trim) single stranded DNA (ssDNA) molecules and catalyse the reverse reaction of dNDPs polymerization. Although PNPase has been implicated in DNA repair in distantly related organisms as *E. coli* and *Bacillus subtilis*, the biological relevance of its activity on DNA molecules is unknown (Chou and Singer 1971; Cardenas et al. 2011; Briani et al. 2016; Carzaniga et al. 2017; Pizzoccheri et al. 2025).

RNA molecules are degraded by PNPase provided they have single stranded tails at the 3′end with a minimal estimated length of 12 nt and 6–9 nt for human and *E. coli* PNPase, respectively (Table 5.1). This characteristic of PNPase is due to the diameter of the channel housing the active site, which is capable of accommodating only single-stranded RNAs (Spickler and Mackie 2000; Symmons et al. 2000; Andrade et al. 2009; Lin et al. 2012; Viegas et al. 2020).

Pi concentration deeply impacts PNPase phosphorolytic activity, with PNPases of different organisms having optimal Pi concentration ranging from 0.1 to 10 mM or higher. In particular, the phosphorolytic activity of human PNPase is inhibited *in*

Table 5.1 Comparison between the *in vitro* activity of *E. coli* and human PNPase

	Organism	
	E. coli	*H. sapiens*
Optimal [Pi][a]	10–20 mM	0.1 mM
Length of phosphorolysis end-product[b]	2 nt	4 nt
ssRNA tail required for RNA binding[c]	6–9 nt	12 nt
Preferred RNA homopolymers for binding[d]	poly(U); poly(A)	poly(U); poly(G)

[a]Littauer and Grunberg-Manago (2002), Portnoy et al. (2008)
[b]Cheng and Deutscher (2005)
[c]Spickler and Mackie (2000), Andrade et al. (2009), Lin et al. (2012)
[d]Lisitsky and Schuster (1999), Yehudai-Resheff et al. (2003), and Portnoy et al. (2008)

vitro at Pi concentrations above 2.5 mM and is optimal at low Pi levels. In contrast, bacterial and chloroplast PNPases have optimal Pi concentration around 10 mM or higher (Table 5.1) (Yehudai-Resheff et al. 2001; Littauer and Grunberg-Manago 2002; Portnoy et al. 2008; Pizzoccheri et al. 2025). This is a relevant aspect because the Pi concentration in the mitochondrial matrix and cytosol of mammalian cells is estimated between 2 and 10 mM, similar to that in *E. coli* (Soboll et al. 1978; Sarpel et al. 1982; Amin and Peterkofsky 1995; Nesmeyanova 2000), potentially affecting PNPase activity.

Crystallographic analysis identified the active site of PNPase in the PNP2 domain by mapping the binding sites of either analogues of the small-molecule involved in phosphorolysis, such as Mn^{2+} (replacing Mg^{2+}) and tungstate (substituting for Pi), or citrate, an inhibitor of both bacterial and human PNPases (Symmons et al. 2000; Shi et al. 2008; Nurmohamed et al. 2009; Lin et al. 2012; Stone et al. 2017). The relevance of amino acids identified as important for catalysis, which are highly conserved in PNPases of distantly related organisms, was confirmed by studying the effect of mutations changing such residues. In particular, mutational studies of human PNPase have assigned a critical role in catalysis to four residues, namely R446, S484, D538 and D544, which surround the citrate binding site. These four residues are all conserved in the *E. coli* enzyme (R399, S439, D486 and D492, respectively; Fig. 5.6), whose active site has similar geometry (Portnoy et al. 2008; Lin et al. 2012; Falchi et al. 2022).

In *E. coli* PNPase, S439 is located in close proximity to the so-called Pi loop, i.e. the region implicated in Pi binding (residues 433–437 in the *E. coli* enzyme), whereas D486 and D492 coordinate Mg^{2+} together with K494 and support metal-assisted catalysis (Symmons et al. 2002; Nurmohamed et al. 2009). The four residues are conserved not only in PNPases of different organisms, but also in RNase PH and the archaeal exosome, which is a phosphorolytic machine, but not in the catalytically inactive subunits of the *Saccharomyces cerevisiae* exosome.

Although the catalytic site is located in the PNP2 domain, the PNP1 and AAHD domains contribute to PNPase enzymatic activity (Jarrige et al. 2002; Yehudai-Resheff et al. 2003; Briani et al. 2007; Portnoy et al. 2008). While the precise role of the AAHD domain in catalysis remains elusive, evidence suggests that the PNP1 domain may assist in binding the RNA substrate and maintaining its proper

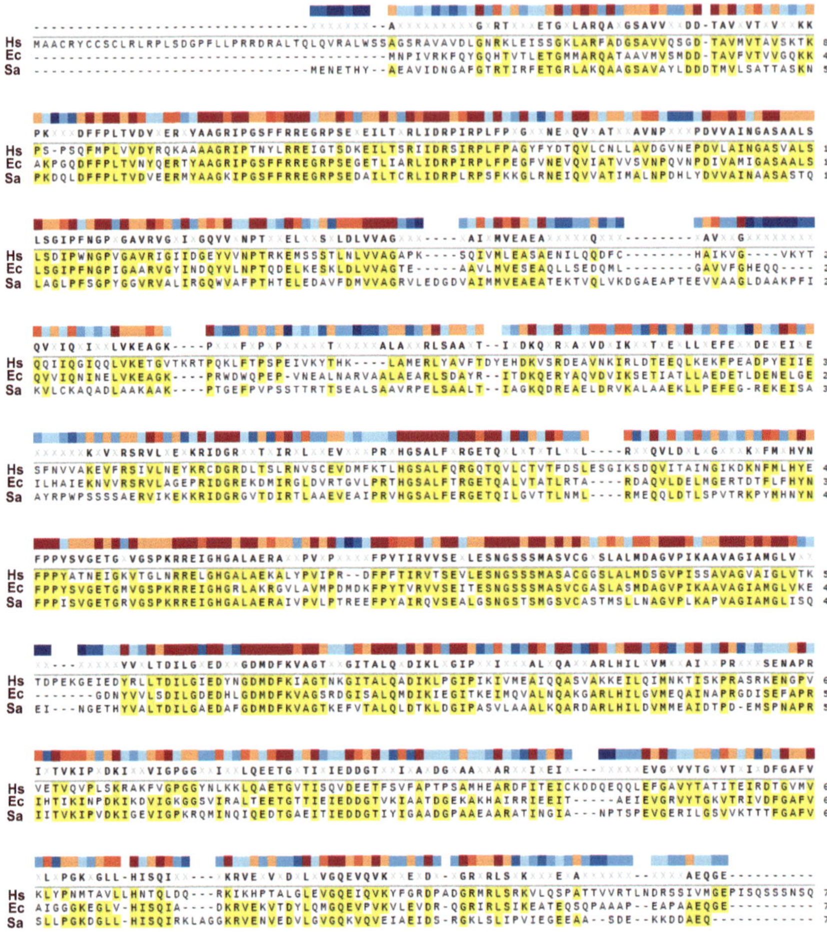

Fig. 5.6 Alignment of *E. coli*, *S. antibioticus* and *H. sapiens* PNPase. *E. coli* (Ec; UniProtKB Accession number: P05055), *S. antibioticus* (Sa; UniProtKB Accession number: Q53597) and *H. sapiens* (Hs; UniProtKB Accession number: Q8TCS8) PNPase aligned with ClustalW and visualized with SnapGene viewer. Amino acids conserved in at least two sequences are highlighted in yellow and reported in bold above the alignment. Colored blocks refer to sequence conservation. Red boxes indicate conserved positions, with darker shades of red corresponding to higher conservation, and blue indicates less conserved regions, with darker shades of blue representing lower conservation

orientation for catalysis. Structural analysis and mutagenesis studies have identified a sequence of four conserved residues (F77-F78-R79-R80 in *E. coli* PNPase) as crucial for enzymatic activity. These residues form the so-called FFRR-loop, which creates a constriction point near the channel entrance (Symmons et al. 2000, 2002; Jarrige et al. 2002; Briani et al. 2007). The FFRR loop is not clearly resolved in some PNPase structures, probably because it is highly mobile in the absence of the

RNA substrate. However, this sequence is highly conserved among PNPases from different organisms, following an FX(R/K)RE consensus, where X represents any amino acid (Nurmohamed et al. 2009; Carzaniga et al. 2014). The FFRR loop has been predicted to be important for RNA binding and in fact, stacking contacts between F77 and an RNA base were detected in crystallographic analysis of *E. coli* PNPase bound to RNA. On the other hand, R79A and R80A mutations alter long-range H-bonding interactions with the catalytic site in the PNP2 domain and affect both RNA and ADP/Pi binding (Carzaniga et al. 2014). In human PNPase, F77 is replaced by a tyrosine (Y116), which can maintain stacking with RNA, and the amino acids involved in long-range hydrogen bonding interactions with R80 (corresponding to R119 in the human PNPase) are conserved (Symmons et al. 2002; Nurmohamed et al. 2009; Falchi et al. 2022). It has been proposed that the evolution of the mobile FFRR loop is linked to the origin of the induced-fit mechanism in the interaction between PNPase and RNA. In this model, RNA entry shapes the enzyme central channel, promoting a reciprocal orientation between the RNA and the catalytic site that enhances enzymatic activity (Nurmohamed et al. 2009; Carzaniga et al. 2014). Notably, the FFRR loop is not conserved in RPH/PNP2 domains suggesting that it was absent in their common ancestor.

In vitro, PNPase catalytic activity is modulated by small molecules that directly bind the enzyme (Fig. 5.1). Notably, citrate, a Krebs cycle intermediate, binds and inhibits PNPase in distantly related organisms, including humans, linking PNPase activity to the cellular energy state. Another connection is evidenced by findings that ATP and ppGpp inhibit and activate *E. coli* PNPase, respectively. ATP and ppGpp levels are respectively low and high when bacteria are in stationary phase (Murray et al. 2003; Buckstein et al. 2008; Del Favero et al. 2008; Siculella et al. 2010; Nurmohamed et al. 2011; Stone et al. 2017).

Overall, the picture emerging from these—still fragmentary—observations is that PNPase should be maximally active when the cells are in conditions of energy deprivation, and this could be an evolutionary conserved feature. This can be rationalized considering that phosphorolysis, which generates NDPs, and the reverse NDP polymerization involved in the synthesis of heteropolymeric RNA tails are less costly processes than RNA hydrolysis, generating NMPs, and the polymerization of ATP by polyadenyl polymerase (Deutscher and Reuven 1991; Briani et al. 2016). Curiously, ppGpp acts as an inhibitor for PNPase of *Nonomuraea*, which is insensitive to ATP, and *Streptomyces*. In these Gram-positive bacteria, a plethora of secondary metabolites are synthesized when the cells stop dividing and enter the stationary phase, suggesting that down-regulating PNPase may help stabilize cellular RNAs when transcription is not efficient (Siculella et al. 2010; Gatewood and Jones 2010).

5.3.2 *Role of the KH and S1 Domains in RNA Binding by PNPase*

Efficient and processive RNA degradation by PNPase requires the simultaneous interaction of the RNA substrate with both the catalytic site and the KH-S1 RNA-binding platform located at the C-terminus of each PNPase monomer (García-Mena et al. 1999; Jarrige et al. 2002; Amblar et al. 2007; Briani et al. 2007; Matus-Ortega et al. 2007; Shi et al. 2008; Wong et al. 2013; Viegas et al. 2020). Accordingly, based only on kinetic analysis of *E. coli* PNPase activity, T. Godefroy identified two RNA binding sites, I and II, which correspond, respectively, to the FFRR loop and possibly other basic residues in the central channel, and the KH-S1 platform (Godefroy 1970). In addition to mediating RNA substrate binding, the KH and S1 domains have also been implicated in maintaining the structural integrity of the trimeric assembly of both *E. coli* and human PNPase (García-Mena et al. 1999; Amblar et al. 2007; Matus-Ortega et al. 2007; Shi et al. 2008; Lin et al. 2012; Pizzoccheri et al. 2025).

The KH-S1 platform is absent or disordered in the available structures of PNPases from different organisms, suggesting that it can be highly mobile (Symmons et al. 2000; Shi et al. 2008; Nurmohamed et al. 2009; Hardwick et al. 2012; Lin et al. 2012; Golzarroshan et al. 2018). This prediction was confirmed by a cryo-electron microscopy (cryo-EM) analysis in which the *E. coli* PNPase bound to the RNA substrate was compared with the unbound (apo) form (Dendooven et al. 2021). In the apo-form, the KH-S1 platform is found in different states due to molecular motions relative to the enzyme core, whereas in the PNPase-RNA complex, the KH-S1 platforms of the three protomers wrap around the RNA substrate. This results in a prolongation of the central channel of the core, into which the RNA inserts. In particular, the RNA interacts with the KH domain of one protomer, which conveys it into the catalytic channel where it forms the stacking interaction with F77 of the FFRR loop mentioned before (Fig. 5.7) (Dendooven et al. 2021). In contrast, in the presence of a small RNA (sRNA) bound by the RNA chaperone Hfq, a ternary complex is formed involving PNPase, the sRNA and Hfq. No direct PNPase-Hfq interaction occurs in the complex but extensive interactions of the RNA with Hfq and with both the KH and S1 domains are established and reroute the RNA away from the active site. In this ternary Hfq-sRNA-PNPase complex, PNPase KH and S1 domains are in an open conformation resembling more the apo state than the "close" conformation typical of the phosphorolytic mode (Fig. 5.7) (Bandyra et al. 2016; Dendooven et al. 2021).

The results of this cryo-EM analysis help explain how PNPase can shift from a catalytic/phosphorolytic mode to a protective/carrier mode depending on the RNA substrate. In fact, while many *E. coli* transcripts are stabilized in the absence of PNPase, which is an expected consequence for the absence of an exonuclease, others exhibit decreased stability, suggesting a PNPase protective role against their degradation by other nucleases (De Lay and Gottesman 2011; Pobre and Arraiano 2015; Falchi et al. 2023). The ability to switch between a catalytic/phosphorolytic

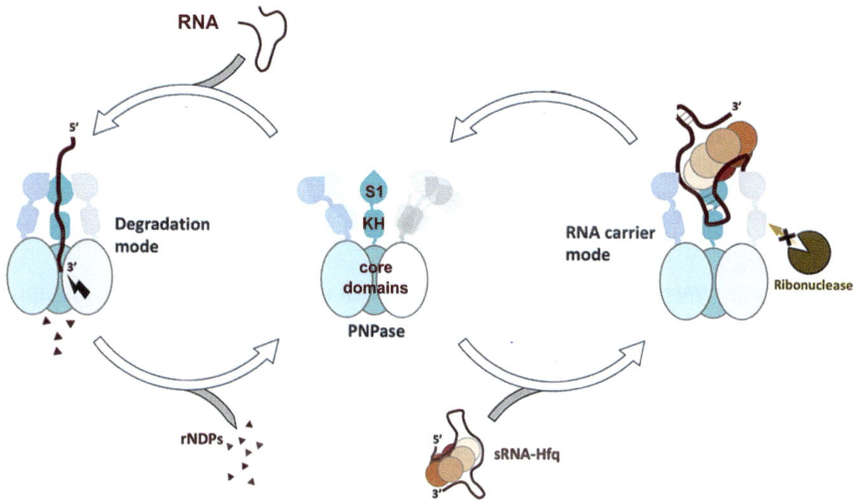

Fig. 5.7 *E. coli* PNPase in phosphorolytic and protective/carrier modes. Explanations are in the text. (Source: Modified from Dendooven et al. (2021); published under Creative Commons CC-BY license (Rightslink®))

and a protective/carrier mode is conserved in human PNPase, which has been implicated both in the transport of several non-coding RNAs from the cytoplasm into mitochondria and in the degradation of mitochondrial transcripts (see Briani et al. 2016 and references therein). Indeed, recent cryo-EM data have shown that, like *E. coli* PNPase, human PNPase transitions between a closed conformation, where the S1 domains interact with each other, and an open conformation, where each S1 domain is directed towards the RPH2 domain of the flanking protomer (Li et al. 2025). Interestingly, at least for some imported RNAs, PNPase-dependent transport seems to be mediated by secondary structures recognised by the S1 domain, reminiscent of the recognition of RNA hairpins by the *E. coli* orthologue (Nurmohamed et al. 2009; Wang et al. 2010, 2012b; Cheng et al. 2018; Golzarroshan et al. 2018).

5.4 PNPase Is a Component of Bacterial and Mitochondrial Degradosomes

In bacteria, exo- and endonucleases involved in RNA decay are associated in multiprotein complexes that include non-nucleolytic proteins (Carpousis 2007; Tejada-Arranz et al. 2020). This concept emerged from the discovery of the *E. coli* RNA degradosome, a high-molecular-weight complex composed of the essential endonuclease RNase E, PNPase, the DEAD-box RNA helicase RhlB, and the glycolytic enzyme enolase (Carpousis et al. 1994; Py et al. 1994, 1996). Other factors have been detected in substoichiometric amount and the composition of the complex can vary in response to environmental conditions (Raynal and Carpousis 1999; Khemici et al. 2004; Prud'homme-Généreux et al. 2004; Regonesi et al. 2006). The assembly of an endonuclease, an exonuclease, and an RNA helicase into a single complex likely facilitates the coordination of RNA decay activities. This organization is particularly advantageous given that both RNase E and PNPase are hindered by RNA secondary structures, which the helicase can resolve. In contrast, the functional contribution of enolase within the complex remains less well understood. However, its presence suggests a functional link between RNA degradation and central metabolism.

The term "*RNA degradosome*" has since been used to describe other multiprotein complexes involved in RNA decay across bacterial species, even when they are not composed of proteins homologous to those found in the *E. coli* degradosome. For instance, in the Gram-positive bacteria *Bacillus subtilis* and *Staphylococcus aureus*, the degradosome components are PNPase, the endonuclease RNase Y, the helicase CshA, and the metabolic enzyme phosphofructokinase (Tejada-Arranz et al. 2020). The common trait among degradosomes described in different species is the presence of at least an RNase and an RNA helicase of the DEAD-box family. According to this broad definition, in *E. coli*, PNPase forms two distinct degradosomes: the RNase E-based complex previously described, and a minimal degradosome composed of a dimer of RhlB that directly interacts with an unidentified region of PNPase, efficiently degrading double-stranded RNA (dsRNA) substrates in the 3′ to 5′ direction (Portier 1975; Liou et al. 2002; Lin and Lin-Chao 2005). It is worth noting that the assembly of PNPase into different types of complexes influences its intracellular localization. Specifically, the RNase E-based RNA degradosome is associated with the inner membrane, whereas the "minimal degradosome", composed of PNPase and RhlB, and unbound PNPase reside in the cytoplasm (Liou et al. 2001).

Interestingly, the physical association between PNPase and an RNA helicase has been conserved throughout evolution. In fact, in the mitochondrial matrix, human PNPase forms the mitochondrial degradosome (mtDEG), namely a complex with SUV3, a DExH-box RNA helicase, able to efficiently degrade double stranded RNAs (dsRNAs). This activity is relevant because the mitochondrial genome is transcribed in both sense and antisense direction, leading to the formation of dsRNA molecules. The S1 domain of PNPase mediates the interaction with SUV3 (Murphy et al. 1975; Szczesny et al. 2009; Wang et al. 2009; Venø et al. 2011; Borowski et al. 2013; Rhee et al. 2013; Li et al. 2025).

Both bacterial and mitochondrial RNA degradosomes have variable compositions, as they can transiently include factors that adapt them to environmental conditions or specific substrates, modulating PNPase phosphorolytic activity (Prud'homme-Géńreux et al. 2004; Chujo et al. 2012; Pietras et al.,2018b; Falchi et al. 2022). One such factor is the poly(A) polymerase (PAP1 in *E. coli* and mtPAP in human mitochondria), which has been identified as a component of both the *E. coli* and mitochondrial degradosomes. In both cases, degradosomes containing poly(A) polymerase have been implicated in the regulation of the length of poly(A) tails (Wang et al. 2014; Mildenhall et al. 2016). The evolutionary conservation of this interaction highlights a functional role of coordinating exonucleolytic trimming and RNA tailing. Tailing would provide single-stranded "entry points" for exonucleases, thereby facilitating the degradation of highly structured RNAs.

5.5 Activity of PNPase in Bacterial and Human Cells

5.5.1 *PNPase Role in the Bacterial RNA Metabolism*

PNPase has a major role in *E. coli* mRNA decay, as shown by the increase in the levels and stability of hundreds of mRNAs observed in bacteria lacking the PNPase *pnp* gene (Pobre and Arraiano 2015; Falchi et al. 2023). It is also responsible for the degradation of stable RNAs such as incorrectly processed ribosomal RNAs and tRNAs and sRNAs not protected by the RNA chaperone Hfq (Li et al. 2002; Cheng and Deutscher 2003; Andrade and Arraiano 2008; Andrade et al. 2012). However, PNPase is not essential in *E. coli*, indicating that other enzymes can compensate for its absence. Indeed, the simultaneous deletion of *pnp* and one between the *rnb* or *rnr* genes encoding the hydrolytic exonucleases RNase II and RNase R, respectively, is not viable. On the contrary, the double *rnb rnr* deletion is tolerated, suggesting that PNPase can compensate for essential functions not shared by the two hydrolytic enzymes (Donovan and Kushner 1986; Cheng et al. 1998). Only RNase R and PNPase-based degradosomes have the ability to digest structured RNAs, whereas RNase II is blocked by RNA hairpins such as the intrinsic transcription terminators frequently found at the end of transcriptional units. Moreover, only PNPase and RNase R, and not RNase II, are able to degrade RNA fragments deriving from the incorrect maturation of the 16S and 23S rRNAs. The lack of this quality control pathway in the double *pnp rnr* mutant may cause viability loss linked to ribosome malfunctioning. On the other hand, it is unlikely that RNase R could sustain the whole burden of RNA degradation, because it seems relatively scarce, at least in optimal growth conditions, and mostly associated with ribosomes (Py et al. 1994; Coburn and Mackie 1996; Cheng and Deutscher 2002, 2003, 2005; Chen and Deutscher 2005; Lin and Lin-Chao 2005; Ishihama et al. 2008; Dos Santos et al. 2020).

Consistent with a pivotal role of PNPase in the degradation of structured RNA, *pnp* mutants in distantly related bacteria are cold sensitive, i.e. unable to grow at temperatures below 18–16 °C. In these conditions, stabilization of RNA secondary structures occurs, which results in higher transcript stability. A dedicated set of chaperones and RNA helicases are expressed at low temperature to allow translation and recycling of transcripts, and the expression of the *pnp* gene is induced (Luttinger et al. 1996; Goverde et al. 1998; Phadtare et al. 1999; Zangrossi et al. 2000; Zhang and Gross 2021).

PNPase has been implicated in the response to oxidative stress in different bacteria. *E. coli* PNPase preferentially binds oxidized RNAs containing

8-oxo-7,8-dihydroguanosine (8-oxo-rG) and stalls when a damaged nucleotide reaches the catalytic site. Residues with key roles in the catalysis, and in particular the highly conserved R399, are critical in damaged RNA recognition. It has been speculated that PNPase stalling may be instrumental to recruit RNA degradative machines, like the RNA degradosome, onto damaged RNAs and recycle them (Hayakawa et al. 2001; Jarrige et al. 2002; Wu et al. 2009; Han et al. 2022; Buchser et al. 2023; Miller et al. 2024).

PNPase has been associated in different bacteria with the maturation of non-coding RNAs derived from processing of longer transcripts. Moreover, in mutants lacking PNPase, a number of mRNAs are less abundant and less stable (Piazza et al. 1996; Briani et al. 2002; Mohanty and Kushner 2003; Sesto et al. 2014; Pobre and Arraiano 2015; Chou-Zheng and Hatoum-Aslan 2019; Spanka and Klug 2021; Falchi et al. 2023; König et al. 2024). Although for many transcripts, this is probably an indirect effect of the enzyme absence, for some, it might result from a direct interaction between PNPase and the mRNA, protecting the transcript from degradation by other nucleases. It would be interesting to investigate whether mRNAs are capable of forming complexes akin to those previously described for PNPase and Hfq-bound RNAs, particularly given that Rho-independent terminators, which are commonly found at the ends of bacterial transcriptional units, have been identified as Hfq-binding sites. (Otaka et al. 2011; Sauer and Weichenrieder 2011; Chao et al. 2012; Cameron and De Lay 2016; Bandyra et al. 2016; Dendooven et al. 2021; Sinha and De Lay 2022).

PNPase protective activity towards sRNAs is not limited to Hfq-dependent sRNAs. Notably, the full-length CsrB and CsrC sRNAs, which interact with the global regulator CsrA to modulate its activity, are also stabilized by PNPase. Conversely, degradation fragments of CsrB accumulate in *pnp* mutants, suggesting that PNPase may exert opposing effects on the stability of the intact CsrB transcript versus its decay intermediates (Viegas et al. 2007; Vakulskas et al. 2016; Carzaniga et al. 2021). Overall, these findings indicate that *in vivo*, PNPase is not exclusively dedicated to RNA degradation and that its exonucleolytic activity can be modulated based on the intrinsic properties of the RNA substrate and/or the interaction with other proteins.

In *E. coli*, PNPase is responsible for the synthesis of heteropolymeric tails at the 3′-end of some RNAs. It was proposed that RNA tailing by PNPase could be important to target structured regions through cycles of polymerization and phosphorolysis, influenced by local concentrations of Pi and NDPs (Mohanty and Kushner 2000, 2011). While the physiological relevance of this activity in *E. coli* is still unclear, RNA tailing has an important role in *Synechocystis* cyanobacteria and in *Streptomyces* spp., which lack the poly(A)polymerase PAPI and rely only on PNPase—essential in these bacteria—for RNA tailing. In fact, only *Streptomyces,* and not *E. coli,* PNPase is able to perform RNA tailing *in vitro* in conditions that support phosphorolysis (Rott et al. 2003; Sohlberg et al. 2003; Bralley et al. 2006; Gatewood and Jones 2010; Jones and Mackie 2013; Jones 2018). This ability may be pivotal for degrading RNAs with stable secondary structures resulting from the transcription of *Streptomyces* spp. genomes, which have a very high GC-content, through cycles of polymerization and phosphorolysis.

5.5.2 PNPase Role in the RNA Metabolism of Eukaryotic Organelles

PNPase plays a major role in controlling RNA metabolism in mitochondria, from RNA degradation and processing to RNA trafficking and stabilization. Accordingly, *PNPT1* is a human core-fitness gene (i.e., a gene invariantly essential across cell types and conditions), and its orthologues are essential in *Drosophila melanogaster* and in murine embryonic fibroblasts (Wang et al. 2010; Hart et al. 2015; Pajak et al. 2019). In human mitochondria, PNPase has a dual function, being involved in both the transport of some non- coding RNAs, such as RNase P H1 RNA, MRP RNA and 5S rRNA, from the cytoplasm and in RNA degradation (reviewed by (Falchi et al. 2022)). These functions are performed by physically separated PNPase pools, as the PNPase fraction in the mitochondrial intermembrane space should be mainly, or exclusively, dedicated to the import of non-coding RNAs from the cytoplasm, a process involving RNA binding by PNPase without phosphorolysis. The recognition of RNAs destined for import by PNPase requires specific features of the target RNAs e.g., the presence of a stem-loop structure, which could per se have an inhibitory effect on PNPase activity (Wang et al. 2010, 2012a, b; Cheng et al. 2018).

Conversely, a mainly degradative function has been ascribed to PNPase located in the mitochondrial matrix, where it forms the mtDEG with the SUV3 helicase (Szczesny et al. 2009; Wang et al. 2010, 2012a; Borowski et al. 2013). A relevant function of the mtDEG is to degrade the antisense RNA deriving from the transcription of the light strand of the mitochondrial genome, thus preventing the accumulation of dsRNA. Both SUV3 and PNPase are required to degrade the dsRNA, but only the second one is necessary to prevent dsRNA accumulation in the cytosol, suggesting that PNPase may be involved in both the degradation and the trafficking of dsRNA. In at least some of the neurological diseases due to *PNPT1* mutations, dsRNA accumulation in the cytosol could actually be the main cause of the pathology by triggering an aberrant activation of interferon type I signaling leading to dysregulated inflammation and tissue damage (Dhir et al. 2018; Pennisi et al. 2022; Barbier et al. 2022).

The preferential binding of G-rich RNA, which is a peculiar feature of the human PNPase not shared with the bacterial enzyme (Table 5.1), may be an adaptation to optimize degradation of the G-rich antisense RNA deriving from the transcription of the light strand of the mitochondrial genome (Portnoy et al. 2008; Jedynak-Slyvka et al. 2021). On this line, it has been found that in the mitochondrial matrix, the mtDEG co-localizes with GRSF1, a protein that specifically recognizes G-rich RNA stretches and melts G-quadruplexes (G4) stimulating their degradation by the mtDEG. G4 sequences are rare in bacteria, which do not have a GRSF1 orthologue.

In fact, GRSF1 is unique to vertebrates, whose mitochondrial genomes are rich in sequences potentially able to form G4s (Pietras et al. 2018a, b).

The ability of mtDEG to target RNA molecules engaged in double-stranded structures is most likely pivotal for another relevant function of the complex, namely the removal of R-loops, which are RNA-DNA hybrids formed by the persistent annealing of a nascent transcript with the template DNA helix. These structures are potentially dangerous because they can interfere with DNA replication, leading to mitochondrial genome instability. Interestingly, ectopic expression of human PNPase in *E. coli* leads to R-loop accumulation, suggesting that the human enzyme may recognize such structures even in the bacterial context, but it could be unable to process them, thereby interfering with their resolution by the bacterial factors responsible for this function (Aguilera and García-Muse 2012; Silva et al. 2018; Falchi et al. 2023; Pizzoccheri et al. 2025).

The human PNPase has high affinity for oxidized RNA like its *E. coli* orthologue; however, its role in the protection against reactive oxygen species (ROS) is unclear. In fact, PNPase depletion in Hela cells increases sensitivity to H_2O_2, but its overproduction in the same cells has been associated with either H_2O_2 resistance or ROS accumulation (Sarkar et al. 2004; Hayakawa and Sekiguchi 2006; Wu and Li 2008; Taylor et al. 2024).

Even less explored is whether the NDP polymerization activity of human PNPase, detected *in vitro*, has any relevance *in vivo*. In fact, the enzyme responsible for RNA tailing in mitochondria is mtPAP, the product of the *PAPD1* gene, whose poly(A) tailing activity is actually counteracted by PNPase (Tomecki et al. 2004; Nagaike et al. 2005; Portnoy et al. 2008; Borowski et al. 2013). Tails produced by PNPase are expected to be heteropolymeric, whereas mtPAP synthesizes homopolymeric poly(A) tails; thus, mitochondrial RNA sequencing could, in principle, distinguish between PNPase- and mtPAP-dependent tailing.

Beside controlling the import and degradation of mtRNAs, the human PNPase has an important role in the massive degradation of polyadenylated RNAs induced by a pro-apoptotic stimulus. In these conditions, PNPase is released from the mitochondrial intermembrane space and digests cytosolic transcripts with poly(A) tails (or other single stranded 3′-ends). Silencing PNPase expression reduces apoptosis in different types of cells. Also, a *PNPT1* pathological mutation causing the E475G amino acid change in PNPase strongly impacts the transcriptome of apoptotic 293 T cells (Sarkar et al. 2005; Chen et al. 2006; Liu et al. 2018; Pizzoccheri et al. 2025).

In contrast with the essential role of mitochondrial PNPase in mammals, chloroplast PNPase is not essential in the alga *Chlamydomonas reinhardtii* and the plant *Arabidopsis thaliana*. In both these organisms, PNPase regulates the adaptation to phosphate availability (Yehudai-Resheff et al. 2007; Marchive et al. 2009). However, relevant differences exist in the role played by PNPase in RNA metabolism. In *C. reinhardtii,* poly(A)-rich heteropolymeric RNA tails found in presence of PNPase were almost missing in the absence of the enzyme (Zimmer et al. 2009), suggesting that like in cyanobacteria, PNPase may be responsible for RNA tailing in *Chlamydomonas* chloroplasts. In contrast, in *Arabidopsis* chloroplasts, polyadenylated RNAs accumulated upon PNPase depletion. In this organism, PNPase was involved in the formation of mature 23S rRNA and mRNAs from longer transcripts and in tRNA degradation, all mechanisms relying on the enzyme phosphorolytic activity (Walter et al. 2002). Expanding the analysis to include chloroplasts from other plants and algae would be important to determine how representative the results obtained from *Chlamydomonas* and *Arabidopsis* are.

5.6 Conclusions

PNPase demonstrates remarkable evolutionary adaptability, participating in a wide range of cellular processes and intramolecular interactions to fulfill diverse roles across different bacterial species and eukaryotic organelles. In spite of 70 years of research on bacterial PNPases, a number of questions are still open. In fact, not only the mechanistic details of PNPase catalytic mechanism are unclear but also general aspects related to its physiological role in bacterial and eukaryotic cells await clarification. How is the assembly of bacterial PNPase into different complexes regulated? What is the relevance of PNPase RNA tailing activity in bacteria and eukaryotic organelles? How and to what extent is PNPase modulation by small molecules evolutionarily conserved, and how does it impact PNPase function *in vivo*? Through which mechanisms do different mutations in PNPase lead to distinct neurological diseases in terms of symptoms, severity, and progression throughout the patients' lives? Furthermore, PNPase ability to act on oxidized RNAs and on DNA, regulated by divalent cations, broadens its functional spectrum and opens new avenues for research into its role in RNA and DNA damage response. Addressing these open questions will be crucial not only for uncovering the full spectrum of PNPase functions but also for understanding how its evolutionary diversification has shaped its roles across different species.

Acknowledgments I am grateful to Gianni Dehò and Thomas Carzaniga for critically reading the manuscript and providing insightful comments that improved its clarity and quality.

Bibliography

Agarwal S, Behring M, Kim H et al (2020) TRIP13 promotes metastasis of colorectal cancer regardless of p53 and microsatellite instability status. Mol Oncol 14:3007–3029. https://doi.org/10.1002/1878-0261.12821

Aguilera A, García-Muse T (2012) R loops: from transcription byproducts to threats to genome stability. Mol Cell 46:115–124. https://doi.org/10.1016/j.molcel.2012.04.009

Amblar M, Barbas A, Gomez-Puertas P, Arraiano CM (2007) The role of the S1 domain in exoribonucleolytic activity: substrate specificity and multimerization. RNA 13:317–327. https://doi.org/10.1261/RNA.220407

Amin N, Peterkofsky A (1995) A dual mechanism for regulating cAMP levels in *Escherichia coli*. J Biol Chem 270:11803–11805. https://doi.org/10.1074/jbc.270.20.11803

Andrade JM, Arraiano CM (2008) PNPase is a key player in the regulation of small RNAs that control the expression of outer membrane proteins. RNA 14:543–551. https://doi.org/10.1261/rna.683308

Andrade JM, Pobre V, Matos AM, Arraiano CM (2012) The crucial role of PNPase in the degradation of small RNAs that are not associated with Hfq. RNA 18:844–855. https://doi.org/10.1261/rna.029413.111

Andrade JM, Pobre V, Silva IJ et al (2009) The role of 3′–5′ exoribonucleases in RNA degradation. Prog Mol Biol Transl Sci 85:187–229. https://doi.org/10.1016/S0079-6603(08)00805-2

Bamborschke D, Kreutzer M, Koy A et al (2021) PNPT1 mutations may cause Aicardi-Goutières-Syndrome. Brain Dev 43:320–324. https://doi.org/10.1016/j.braindev.2020.10.005

Bandyra KJ, Sinha D, Syrjanen J et al (2016) The ribonuclease polynucleotide phosphorylase can interact with small regulatory RNAs in both protective and degradative modes. RNA 22:360–372. https://doi.org/10.1261/RNA.052886.115

Barbier M, Bahlo M, Pennisi A et al (2022) Heterozygous PNPT1 variants cause spinocerebellar ataxia Type 25. Ann Neurol 92:122–137. https://doi.org/10.1002/ANA.26366

Borowski LS, Dziembowski A, Hejnowicz MS et al (2013) Human mitochondrial RNA decay mediated by PNPase-hSuv3 complex takes place in distinct foci. Nucleic Acids Res 41:1223–1240. https://doi.org/10.1093/nar/gks1130

Bralley P, Gust B, Chang S et al (2006) RNA 3′-tail synthesis in Streptomyces: in vitro and in vivo activities of RNase PH, the SCO3896 gene product and polynucleotide phosphorylase. Microbiology (N Y) 152:627–636. https://doi.org/10.1099/MIC.0.28363-0

Briani F, Carzaniga T, Dehò G (2016) Regulation and functions of bacterial PNPase. Wiley Interdiscip Rev RNA 7:241–258. https://doi.org/10.1002/wrna.1328

Briani F, Del Favero M, Capizzuto R et al (2007) Genetic analysis of polynucleotide phosphorylase structure and functions. Biochimie 89:145–157. https://doi.org/10.1016/j.biochi.2006.09.020

Briani F, Del Vecchio E, Migliorini D et al (2002) RNase E and polyadenyl polymerase I are involved in maturation of CI RNA, the P4 phage immunity factor. J Mol Biol 318:321–331. https://doi.org/10.1016/S0022-2836(02)00085-2

Buchser R, Sweet P, Anantharaman A, Contreras L (2023) RNAs as sensors of oxidative stress in bacteria. Annu Rev Chem Biomol Eng 14:265–281. https://doi.org/10.1146/ANNUREV-CHEMBIOENG-101121-070250

Buckstein MH, He J, Rubin H (2008) Characterization of nucleotide pools as a function of physiological state in Escherichia coli. J Bacteriol 190:718–726. https://doi.org/10.1128/JB.01020-07

Cameron TA, De Lay NR (2016) The phosphorolytic exoribonucleases polynucleotide phosphorylase and RNase PH stabilize sRNAs and facilitate regulation of their mRNA targets. J Bacteriol 198:3309–3317. https://doi.org/10.1128/JB.00624-16

Cardenas PP, Carzaniga T, Zangrossi S et al (2011) Polynucleotide phosphorylase exonuclease and polymerase activities on single-stranded DNA ends are modulated by RecN, SsbA and RecA proteins. Nucleic Acids Res 39:9250–9261. https://doi.org/10.1093/nar/gkr635

Carpousis AJ (2007) The RNA degradosome of *Escherichia coli*: an mRNA-degrading machine assembled on RNase E. Ann Rev Microbiol 61:71–87. https://doi.org/10.1146/annurev.micro.61.080706.093440

Carpousis AJ, Van Houwe G, Ehretsmann C, Krisch HM (1994) Copurification of E. coli RNAase E and PNPase: evidence for a specific association between two enzymes important in RNA processing and degradation. Cell 76:889–900. https://doi.org/10.1016/0092-8674(94)90363-8

Carzaniga T, Falchi FA, Forti F et al (2021) Different csra expression levels in c versus k-12 E. coli strains affect biofilm formation and impact the regulatory mechanism presided by the Csrb and Csrc small RNAs. Microorganisms 9:1010. https://doi.org/10.3390/microorganisms9051010

Carzaniga T, Mazzantini E, Nardini M et al (2014) A conserved loop in polynucleotide phosphorylase (PNPase) essential for both RNA and ADP/phosphate binding. Biochimie 97:49–59. https://doi.org/10.1016/j.biochi.2013.09.018

Carzaniga T, Sbarufatti G, Briani F, Dehò G (2017) Polynucleotide phosphorylase is implicated in homologous recombination and DNA repair in Escherichia coli. BMC Microbiol 17:81. https://doi.org/10.1186/s12866-017-0980-z

Chao Y, Papenfort K, Reinhardt R et al (2012) An atlas of Hfq-bound transcripts reveals 3′ UTRs as a genomic reservoir of regulatory small RNAs. EMBO J 31:4005. https://doi.org/10.1038/EMBOJ.2012.229

Chen C, Deutscher MP (2005) Elevation of RNase R in response to multiple stress conditions. J Biol Chem 280:34393–34396. https://doi.org/10.1074/JBC.C500333200

Chen H-W, Rainey RN, Balatoni CE et al (2006) Mammalian polynucleotide phosphorylase is an intermembrane space RNase that maintains mitochondrial homeostasis. Mol Cell Biol 26:8475–8487. https://doi.org/10.1128/MCB.01002-06

Cheng Y, Liu P, Zheng Q et al (2018) Mitochondrial trafficking and processing of telomerase RNA TERC. Cell Rep 24:2589–2595. https://doi.org/10.1016/J.CELREP.2018.08.003

Cheng ZF, Deutscher MP (2003) Quality control of ribosomal RNA mediated by polynucleotide phosphorylase and RNase R. Proc Natl Acad Sci USA 100:6388–6393. https://doi.org/10.1073/pnas.1231041100

Cheng ZF, Deutscher MP (2002) Purification and characterization of the Escherichia coli exoribonuclease RNase R. comparison with RNase II. J Biol Chem 277:21624–21629. https://doi.org/10.1074/JBC.M202942200

Cheng ZF, Deutscher MP (2005) An important role for RNase R in mRNA decay. Mol Cell 17:313–318. https://doi.org/10.1016/J.MOLCEL.2004.11.048

Cheng ZF, Zuo Y, Li Z et al (1998) The vacB gene required for virulence in Shigella flexneri and Escherichia coli encodes the exoribonuclease RNase R. J Biol Chem 273:14077–14080. https://doi.org/10.1074/JBC.273.23.14077

Chou JY, Singer MF (1971) Deoxyadenosine diphosphate as a substrate and inhibitor of polynucleotide phosphorylase of Micrococcus luteus. I. Deoxyadenosine diphosphate as a substrate for polymerization and the exchange reaction with inorganic 32 P. J Biol Chem 246:7486–7496

Chou-Zheng L, Hatoum-Aslan A (2019) A type iii-a crispr-cas system employs degradosome nucleases to ensure robust immunity. Elife 8. https://doi.org/10.7554/eLife.45393

Chujo T, Ohira T, Sakaguchi Y et al (2012) LRPPRC/SLIRP suppresses PNPase-mediated mRNA decay and promotes polyadenylation in human mitochondria. Nucleic Acids Res 40:8033–8047. https://doi.org/10.1093/NAR/GKS506

Coburn GA, Mackie GA (1996) Differential sensitivities of portions of the mRNA for ribosomal protein S20 to 3′-exonucleases dependent on oligoadenylation and RNA secondary structure. J Biol Chem 271:15776–15781. https://doi.org/10.1074/JBC.271.26.15776

Chekanova JA, Shaw RJ, Wills MA, Belostotsky DA. Poly(A) tail-dependent exonuclease AtRrp41p from Arabidopsis thaliana rescues 5.8 S rRNA processing and mRNA decay defects of the yeast ski6 mutant and is found in an exosomesized complex in plant and yeast cells. J Biol Chem. (2000) 275(42):33158-66. https://doi.org/10.1074/jbc.M005493200. PMID: 10930416

De Lay N, Gottesman S (2011) Role of polynucleotide phosphorylase in sRNA function in Escherichia coli. RNA 17:1172–1189. https://doi.org/10.1261/RNA.2531211

Del Favero M, Mazzantini E, Briani F et al (2008) Regulation of Escherichia coli polynucleotide phosphorylase by ATP. J Biol Chem 283:27355–27359. https://doi.org/10.1074/jbc.C800113200

Dendooven T, Sinha D, Roeselová A et al (2021) A cooperative PNPase-Hfq-RNA carrier complex facilitates bacterial riboregulation. Mol Cell 81:2901–2913.e5. https://doi.org/10.1016/j.molcel.2021.05.032

Deutscher MP, Reuven NB (1991) Enzymatic basis for hydrolytic versus phosphorolytic mRNA degradation in Escherichia coli and Bacillus subtilis. Proc Natl Acad Sci USA 88:3277–3280. https://doi.org/10.1073/pnas.88.8.3277

Dhir A, Dhir S, Borowski LS et al (2018) Mitochondrial double-stranded RNA triggers antiviral signalling in humans. Nature 560:238–242

Donovan WP, Kushner SR (1986) Polynucleotide phosphorylase and ribonuclease II are required for cell viability and mRNA turnover in Escherichia coli K-12. Proc Natl Acad Sci USA 83:120–124. https://doi.org/10.1073/pnas.83.1.120

Dos Santos RF, Andrade JM, Pissarra J et al (2020) Hfq and RNase R Mediate rRNA processing and degradation in a Novel RNA quality control process. MBio 11:1–14. https://doi.org/10.1128/mBio.02398-20

Eaton A, Bernier FP, Goedhart C et al (2018) Is PNPT1-related hearing loss ever non-syndromic? Whole exome sequencing of adult siblings expands the natural history of PNPT1-related disorders. Am J Med Genet A 176:2487–2493. https://doi.org/10.1002/ajmg.a.40516

Falchi FA, Forti F, Carnelli C et al (2023) Human PNPase causes RNA stabilization and accumulation of R-loops in the Escherichia coli model system. Sci Rep 13:11771. https://doi.org/10.1038/s41598-023-38924-x

Falchi FA, Pizzoccheri R, Briani F (2022) Activity and function in human cells of the evolutionary conserved exonuclease polynucleotide phosphorylase. Int J Mol Sci 23:1652. https://doi.org/10.3390/ijms23031652

Fry M (2016) The deciphering of the genetic code. In: Landmark experiments in molecular biology, pp 421–480. https://doi.org/10.1016/B978-0-12-802074-6.00010-2

García-Mena J, Das A, Sánchez-Trujillo A et al (1999) A novel mutation in the KH domain of polynucleotide phosphorylase affects autoregulation and mRNA decay in Escherichia coli. Mol Microbiol 33:235–248. https://doi.org/10.1046/J.1365-2958.1999.01451.X

Gatewood ML, Jones GH (2010) (p)ppGpp inhibits polynucleotide phosphorylase from Streptomyces but not from Escherichia coli and increases the stability of bulk mRNA in Streptomyces coelicolor. J Bacteriol 192:4275–4280. https://doi.org/10.1128/JB.00367-10

Godefroy T (1970) Kinetics of polymerization and phosphorolysis reactions of Escherichia coli polynucleotide phosphorylase. Evidence for multiple binding of polynucleotide in phosphorolysis. Eur J Biochem 14:222–231. https://doi.org/10.1111/J.1432-1033.1970.TB00281.X

Golzarroshan B, Lin C-L, Li C-L et al (2018) Crystal structure of dimeric human PNPase reveals why disease-linked mutants suffer from low RNA import and degradation activities. Nucleic Acids Res 46:8630–8640. https://doi.org/10.1093/nar/gky642

Goverde RLJ, In't Veld JHJ H, Kusters JG, Mooi FR (1998) The psychrotrophic bacterium Yersinia enterocolitica requires expression of pnp, the gene for polynucleotide phosphorylase, for growth at low temperature (5°C). Mol Microbiol 28:555–569. https://doi.org/10.1046/j.1365-2958.1998.00816.x

Han R, Jiang J, Fang J, Contreras LM (2022) PNPase and RhlB interact and reduce the cellular availability of oxidized RNA in Deinococcus radiodurans. Microbiol Spectr 10. https://doi.org/10.1128/SPECTRUM.02140-22

Hardwick SW, Gubbey T, Hug I et al (2012) Crystal structure of Caulobacter crescentus polynucleotide phosphorylase reveals a mechanism of RNA substrate channelling and RNA degradosome assembly. Open Biol 2:120028. https://doi.org/10.1098/RSOB.120028

Hart T, Chandrashekhar M, Aregger M et al (2015) High-resolution CRISPR screens reveal fitness genes and genotype-specific cancer liabilities. Cell 163:1515–1526. https://doi.org/10.1016/j.cell.2015.11.015

Hayakawa H, Kuwano M, Sekiguchi M (2001) Specific binding of 8-oxoguanine-containing RNA to polynucleotide phosphorylase protein. Biochemistry 40:9977–9982. https://doi.org/10.1021/BI010595Q

Hayakawa H, Sekiguchi M (2006) Human polynucleotide phosphorylase protein in response to oxidative stress. Biochemistry 45:6749–6755. https://doi.org/10.1021/BI052585L

Ishihama Y, Schmidt T, Rappsilber J et al (2008) Protein abundance profiling of the Escherichia coli cytosol. BMC Genomics 9:102. https://doi.org/10.1186/1471-2164-9-102

Jarrige A-C, Bréchemier-Baey D, Mathy N et al (2002) Mutational analysis of polynucleotide phosphorylase from Escherichia coli. J Mol Biol 321:397–409. https://doi.org/10.1016/S0022-2836(02)00645-9

Jedynak-Slyvka M, Jabczynska A, Szczesny RJ (2021) Human mitochondrial RNA processing and modifications: overview. Int J Mol Sci 22. https://doi.org/10.3390/IJMS22157999

Jones GH (2018) Novel aspects of polynucleotide phosphorylase function in Streptomyces. Antibiotics 7:25. https://doi.org/10.3390/ANTIBIOTICS7010025

Jones GH, Mackie GA (2013) Streptomyces coelicolor polynucleotide phosphorylase can polymerize nucleoside diphosphates under phosphorolysis conditions, with implications for the degradation of structured RNAs. J Bacteriol 195:5151–5159. https://doi.org/10.1128/JB.00936-13

Khemici V, Toesca I, Poljak L et al (2004) The RNase E of Escherichia coli has at least two binding sites for DEAD-box RNA helicases: functional replacement of RhlB by RhlE. Mol Microbiol 54:1422–1430. https://doi.org/10.1111/J.1365-2958.2004.04361.X

König F, Svensson SL, Sharma CM (2024) Interplay of two small RNAs fine-tunes hierarchical flagella gene expression in Campylobacter jejuni. Nat Commun 15. https://doi.org/10.1038/s41467-024-48986-8

Leszczyniecka M, DeSalle R, Kang DC, Fisher PB (2004) The origin of polynucleotide phosphorylase domains. Mol Phylogenet Evol 31:123–130. https://doi.org/10.1016/j.ympev.2003.07.012

Leszczyniecka M, Kang DC, Sarkar D et al (2002) Identification and cloning of human polynucleotide phosphorylase, hPNPase old-35, in the context of terminal differentiation and cellular senescence. Proc Natl Acad Sci USA 99:16636–16641. https://doi.org/10.1073/pnas.252643699

Li Y-C, Wang C-H, Patra M et al (2025) Structural insights into human PNPase in health and disease. Nucleic Acids Res 53:13–14. https://doi.org/10.1093/NAR/GKAF119

Li Z, Reimers S, Pandit S, Deutscher MP (2002) RNA quality control: degradation of defective transfer RNA. EMBO J 21:1132–1138. https://doi.org/10.1093/EMBOJ/21.5.1132

Lin CL, Wang YT, Yang WZ et al (2012) Crystal structure of human polynucleotide phosphorylase: insights into its domain function in RNA binding and degradation. Nucleic Acids Res 40:4146–4157. https://doi.org/10.1093/NAR/GKR1281

Lin PH, Lin-Chao S (2005) RhlB helicase rather than enolase is the β-subunit of the Escherichia coli polynucleotide phosphorylase (PNPase)-exoribonucleolytic complex. Proc Natl Acad Sci USA 102:16590–16595. https://doi.org/10.1073/pnas.0500994102

Liou GG, Chang HY, Lin CS, Lin-Chao S (2002) Dead box RhlB RNA helicase physically associates with exoribonuclease PNPase to degrade double-stranded RNA independent of the degradosome-assembling region of RNase E. J Biol Chem 277:41157–41162. https://doi.org/10.1074/jbc.M206618200

Liou GG, Jane WN, Cohen SN et al (2001) RNA degradosomes exist in vivo in Escherichia coli as multicomponent complexes associated with the cytoplasmic membrane via the N-terminal region of ribonuclease E. Proc Natl Acad Sci USA 98:63–68. https://doi.org/10.1073/pnas.98.1.63

Lisitsky I, Schuster G (1999) Preferential degradation of polyadenylated and polyuridinylated RNAs by the bacterial exoribonuclease polynucleotide phosphorylase. Eur J Biochem 261:468–474. https://doi.org/10.1046/J.1432-1327.1999.00285.X

Littauer UZ, Grunberg-Manago M (2002) Polynucleotide phosphorylase. In: Creighton TE (ed) Encyclopedia of molecular biology. Wiley, New York, pp 1911–1918

Liu X, Fu R, Pan Y et al (2018) PNPT1 release from mitochondria during apoptosis triggers decay of poly(A) RNAs. Cell 174:187–201.e12. https://doi.org/10.1016/j.cell.2018.04.017

Luttinger A, Hahn J, Dubnau D (1996) Polynucleotide phosphorylase is necessary for competence development in Bacillus subtilis. Mol Microbiol 19:343–356. https://doi.org/10.1046/j.1365-2958.1996.380907.x

Lykke-Andersen S, Brodersen DE, Jensen TH (2009) Origins and activities of the eukaryotic exosome. J Cell Sci 122:1487–1494. https://doi.org/10.1242/JCS.047399

Marchive C, Yehudai-Resheff S, Germain A et al (2009) Abnormal physiological and molecular mutant phenotypes link chloroplast polynucleotide phosphorylase to the phosphorus deprivation response in Arabidopsis. Plant Physiol 151:905–924. https://doi.org/10.1104/PP.109.145144

Matus-Ortega ME, Regonesi ME, Piña-Escobedo A et al (2007) The KH and S1 domains of Escherichia coli polynucleotide phosphorylase are necessary for autoregulation and growth at low temperature. Biochim Biophys Acta 1769:194–203. https://doi.org/10.1016/J.BBAEXP.2007.01.008
Mian I (1997) Comparative sequence analysis of ribonucleases HII, III, II PH and D. Nucleic Acids Res 25:3187–3195. https://doi.org/10.1093/nar/25.16.3187
Mildenhall KB, Wiese N, Chung D et al (2016) RNase E-based degradosome modulates polyadenylation of mRNAs after Rho-independent transcription terminators in Escherichia coli. Mol Microbiol 101:645–655. https://doi.org/10.1111/mmi.13413
Miller LG, Kim W, Schowe S et al (2024) Selective 8-oxo-rG stalling occurs in the catalytic core of polynucleotide phosphorylase (PNPase) during degradation. Proc Natl Acad Sci USA 121:e2317865121. https://doi.org/10.1073/PNAS.2317865121
Mohanty BK, Kushner SR (2000) Polynucleotide phosphorylase functions both as a 3′ → 5′ exonuclease and a poly(A) polymerase in Escherichia coli. Proc Natl Acad Sci USA 97:11966–11971. https://doi.org/10.1073/pnas.220295997
Mohanty BK, Kushner SR (2003) Genomic analysis in Escherichia coli demonstrates differential roles for polynucleotide phosphorylase and RNase II in mRNA abundance and decay. Mol Microbiol 50:645–658. https://doi.org/10.1046/j.1365-2958.2003.03724.x
Mohanty BK, Kushner SR (2011) Bacterial/archaeal/organellar polyadenylation. Wiley Interdiscip Rev RNA 2:256–276. https://doi.org/10.1002/wrna.51
Murphy WI, Attardi B, Tu C, Attardi G (1975) Evidence for complete symmetrical transcription in vivo of mitochondrial DNA in HeLa cells. J Mol Biol 99:809–814. https://doi.org/10.1016/s0022-2836(75)80187-2
Murray HD, Schneider DA, Gourse RL (2003) Control of rRNA expression by small molecules is dynamic and nonredundant. Mol Cell 12:125–134. https://doi.org/10.1016/S1097-2765(03)00266-1
Nagaike T, Suzuki T, Katoh T, Ueda T (2005) Human mitochondrial mRNAs are stabilized with polyadenylation regulated by mitochondria-specific poly(A) polymerase and polynucleotide phosphorylase. J Biol Chem 280:19721–19727. https://doi.org/10.1074/JBC.M500804200
Nesmeyanova MA (2000) Polyphosphates and enzymes of polyphosphate metabolism in Escherichia coli. Biochemistry (Mosc) 65:309–314
Nurmohamed S, Vaidialingam B, Callaghan AJ, Luisi BF (2009) Crystal structure of Escherichia coli Polynucleotide Phosphorylase core bound to RNase E, RNA and manganese: implications for catalytic mechanism and RNA degradosome assembly. J Mol Biol 389:17–33. https://doi.org/10.1016/j.jmb.2009.03.051
Nurmohamed S, Vincent HA, Titman CM et al (2011) Polynucleotide phosphorylase activity may be modulated by metabolites in Escherichia coli. J Biol Chem 286:14315–14323. https://doi.org/10.1074/JBC.M110.200741
Otaka H, Ishikawa H, Morita T, Aiba H (2011) PolyU tail of rho-independent terminator of bacterial small RNAs is essential for Hfq action. Proc Natl Acad Sci USA 108:13059–13064. https://doi.org/10.1073/PNAS.1107050108
Pajak A, Laine I, Clemente P et al (2019) Defects of mitochondrial RNA turnover lead to the accumulation of double-stranded RNA in vivo. PLoS Genet 15:e1008240. https://doi.org/10.1371/JOURNAL.PGEN.1008240
Pennisi A, Rötig A, Roux CJ et al (2022) Heterogeneity of PNPT1 neuroimaging: Mitochondriopathy, interferonopathy or both? J Med Genet 59:204–208. https://doi.org/10.1136/jmedgenet-2020-107367
Phadtare S, Alsina J, Inouye M (1999) Cold-shock response and cold-shock proteins. Curr Opin Microbiol 2:175–180. https://doi.org/10.1016/S1369-5274(99)80031-9
Piazza F, Zappone M, Sana M et al (1996) Polynucleotide phosphorylase of Escherichia coli is required for the establishment of bacteriophage P4 immunity. J Bacteriol 178:5513–5521. https://doi.org/10.1128/jb.178.18.5513-5521.1996

Pietras Z, Wojcik MA, Borowski LS et al (2018a) Dedicated surveillance mechanism controls G-quadruplex forming non-coding RNAs in human mitochondria. Nat Commun 9:2558

Pietras Z, Wojcik MA, Borowski LS et al (2018b) Controlling the mitochondrial antisense–role of the SUV3-PNPase complex and its co-factor GRSF1 in mitochondrial RNA surveillance. Mol Cell Oncol 5:e156452

Piwowarski J, Grzechnik P, Dziembowski A et al (2003) Human polynucleotide phosphorylase, hPNPase, is localized in mitochondria. J Mol Biol 329:853–857. https://doi.org/10.1016/S0022-2836(03)00528-X

Pizzoccheri R, Falchi FA, Alloni A et al (2025) Pathological PNPase variants with altered RNA binding and degradation activity affect the phenotype of bacterial and human cell models. NAR Molecular Medicine 2:ugae028. https://doi.org/10.1093/NARMME/UGAE028

Pobre V, Arraiano CM (2015) Next generation sequencing analysis reveals that the ribonucleases RNase II, RNase R and PNPase affect bacterial motility and biofilm formation in E. coli. BMC Genomics 16:72. https://doi.org/10.1186/s12864-015-1237-6

Portier C (1975) Quaternary structure of polynucleotide phosphorylase from Escherichia coli: evidence of a complex between two types of polypeptide chains. Eur J Biochem 55:573–582. https://doi.org/10.1111/J.1432-1033.1975.TB02194.X

Portnoy V, Palnizky G, Yehudai-Resheff S et al (2008) Analysis of the human polynucleotide phosphorylase (PNPase) reveals differences in RNA binding and response to phosphate compared to its bacterial and chloroplast counterparts. RNA 14:297–309. https://doi.org/10.1261/rna.698108

Prud'homme-Généreux A, Beran RK, Iost I et al (2004) Physical and functional interactions among RNase E, polynucleotide phosphorylase and the cold-shock protein, CsdA: evidence for a "cold shock degradosome.". Mol Microbiol 54:1409–1421. https://doi.org/10.1111/J.1365-2958.2004.04360.X

Py B, Causton H, Mudd EA, Higgins CF (1994) A protein complex mediating mRNA degradation in Escherichia coli. Mol Microbiol 14:717–729. https://doi.org/10.1111/j.1365-2958.1994.tb01309.x

Py B, Higgins CF, Krisch HM, Carpousis AJ (1996) A DEAD-box RNA helicase in the Escherichia coli RNA degradosome. Nature 381:169–172. https://doi.org/10.1038/381169a0

Raynal LC, Carpousis AJ (1999) Poly(A) polymerase I of Escherichia coli: characterization of the catalytic domain, an RNA binding site and regions for the interaction with proteins involved in mRNA degradation. Mol Microbiol 32:765–775. https://doi.org/10.1046/J.1365-2958.1999.01394.X

Regonesi ME, Del Favero M, Basilico F et al (2006) Analysis of the Escherichia coli RNA degradosome composition by a proteomic approach. Biochimie 88:151–161. https://doi.org/10.1016/j.biochi.2005.07.012

Revert-Ros F, Ventura I, Prieto-Ruiz JA et al (2024) Polyribonucleotide phosphorylase is overexpressed in hepatocellular cancer, promoting epithelial phenotype maintenance and tumor progression. Pathol Res Pract 264:155713. https://doi.org/10.1016/j.prp.2024.155713

Rhee HW, Zou P, Udeshi ND et al (2013) Proteomic mapping of mitochondria in living cells via spatially restricted enzymatic tagging. Science 339:1328–1331

Rius R, Van Bergen NJ, Compton AG et al (2019) Clinical spectrum and functional consequences associated with bi-allelic pathogenic Pnpt1 variants. J Clin Med 8:2020. https://doi.org/10.3390/jcm8112020

Rott R, Zipor G, Portnoy V et al (2003) RNA polyadenylation and degradation in cyanobacteria are similar to the chloroplast but different from Escherichia coli. J Biol Chem 278:15771–15777. https://doi.org/10.1074/JBC.M211571200

Sarkar D, Lebedeva IV, Emdad L et al (2004) Human polynucleotide phosphorylase (hPNPaseold-35): a potential link between aging and inflammation. Cancer Res 64:7473–7478. https://doi.org/10.1158/0008-5472.CAN-04-1772

Sarkar D, Park ES, Emdad L et al (2005) Defining the domains of human Polynucleotide Phosphorylase (hPNPase OLD-35) mediating cellular senescence. Mol Cell Biol 25:7333–7343. https://doi.org/10.1128/mcb.25.16.7333-7343.2005

Sarkar D, Park ES, Fisher PB (2006) Defining the mechanism by which IFN-β downregulates c-myc expression in human melanoma cells: Pivotal role for human polynucleotide phosphorylase (hPNPaseold-35). Cell Death Differ 13:1541–1553. https://doi.org/10.1038/sj.cdd.4401829

Sarpel G, Barp AN, Lubansky HJ, Omachi A (1982) Erythocyte phosphate content in Huntington's disease. Neurosci Lett 31:91–96. https://doi.org/10.1016/0304-3940(82)90060-X

Sauer E, Weichenrieder O (2011) Structural basis for RNA 3′-end recognition by Hfq. Proc Natl Acad Sci USA 108:13065–13070. https://doi.org/10.1073/PNAS.1103420108

Sesto N, Touchon M, Andrade JM et al (2014) A PNPase dependent CRISPR system in Listeria. PLoS Genet 10. https://doi.org/10.1371/journal.pgen.1004065

Shi Z, Yang WZ, Lin-Chao S et al (2008) Crystal structure of Escherichia coli PNPase: central channel residues are involved in processive RNA degradation. RNA 14:2361–2371. https://doi.org/10.1261/rna.1244308

Siculella L, Damiano F, Di Summa R et al (2010) Guanosine 5′-diphosphate 3′-diphosphate (ppGpp) as a negative modulator of polynucleotide phosphorylase activity in a "rare" actinomycete. Mol Microbiol 77:716–729. https://doi.org/10.1111/J.1365-2958.2010.07240.X

Silva S, Camino LP, Aguilera A (2018) Human mitochondrial degradosome prevents harmful mitochondrial R loops and mitochondrial genome instability. Proc Natl Acad Sci 115:11024–11029. https://doi.org/10.1073/pnas.1807258115

Sinha D, De Lay NR (2022) Target recognition by RNase E RNA-binding domain AR2 drives sRNA decay in the absence of PNPase. Proc Natl Acad Sci USA 119. https://doi.org/10.1073/pnas.2208022119

Soboll S, Sholz R, Heldt HW (1978) Subcellular metabolite concentrations: dependence of Mitochondrial and Cytosolic ATP systems on the metabolic state of perfuse rat liver. Eur J Biochem 87:377–390. https://doi.org/10.1111/j.1432-1033.1978.tb12387.x

Sohlberg B, Huang J, Cohen SN (2003) The Streptomyces coelicolor polynucleotide phosphorylase homologue, and not the putative poly(A) polymerase, can polyadenylate RNA. J Bacteriol 185:7273–7278. https://doi.org/10.1128/JB.185.24.7273-7278.2003

Sokhi UK, DeSalle R, Bacolod MD et al (2014) Evolutionary dynamics of polynucelotide phosphorylases. Mol Phylogenet Evol 73:77–86. https://doi.org/10.1016/j.ympev.2014.01.020

Spanka DT, Klug G (2021) Maturation of utr-derived srnas is modulated during adaptation to different growth conditions. Int J Mol Sci 22. https://doi.org/10.3390/ijms222212260

Spickler C, Mackie GA (2000) Action of RNase II and polynucleotide phosphorylase against RNAs containing stem-loops of defined structure. J Bacteriol 182:2422–2427. https://doi.org/10.1128/JB.182.9.2422-2427.2000

Stone CM, Butt LE, Bufton JC et al (2017) Inhibition of homologous phosphorolytic ribonucleases by citrate may represent an evolutionarily conserved communicative link between RNA degradation and central metabolism. Nucleic Acids Res 45:4655–4666. https://doi.org/10.1093/nar/gkx114

Symmons MF, Jones GH, Luisi BF (2000) A duplicated fold is the structural basis for polynucleotide phosphorylase catalytic activity, processivity, and regulation. Structure 8:1215–1226. https://doi.org/10.1016/S0969-2126(00)00521-9

Symmons MF, Williams MG, Luisi BF et al (2002) Running rings around RNA: a superfamily of phosphate-dependent RNases. Trends Biochem Sci 27:11–18. https://doi.org/10.1016/S0968-0004(01)01999-5

Szczesny RJ, Borowski LS, Brzezniak LK et al (2009) Human mitochondrial RNA turnover caught in flagranti: Involvement of hSuv3p helicase in RNA surveillance. Nucleic Acids Res 38:279–298. https://doi.org/10.1093/nar/gkp903

Taylor KE, Miller LG, Contreras LM (2024) RNA-binding proteins that preferentially interact with 8-oxoG-modified RNAs: our current understanding. Biochem Soc Trans 52:111–122. https://doi.org/10.1042/BST20230254

Tejada-Arranz A, de Crécy-Lagard V, de Reuse H (2020) Bacterial RNA degradosomes: molecular machines under tight control. Trends Biochem Sci 45:42. https://doi.org/10.1016/J.TIBS.2019.10.002

Tomecki R, Dmochowska A, Gewartowski K et al (2004) Identification of a novel human nuclear-encoded mitochondrial poly(A) polymerase. Nucleic Acids Res 32:6001–6014. https://doi.org/10.1093/NAR/GKH923

Vakulskas CA, Leng Y, Abe H et al (2016) Antagonistic control of the turnover pathway for the global regulatory sRNA CsrB by the CsrA and CsrD proteins. Nucleic Acids Res 44:7896–7910. https://doi.org/10.1093/nar/gkw484

Venø ST, Kulikowicz T, Pestana C et al (2011) The human Suv3 helicase interacts with replication protein A and flap endonuclease 1 in the nucleus. Biochem J 440:293. https://doi.org/10.1042/BJ20100991

Viegas SC, Matos RG, Arraiano CM (2020) The bacterial counterparts of the eukaryotic exosome: an evolutionary perspective. Methods Mol Biol 2062:37–46. https://doi.org/10.1007/978-1-4939-9822-7_2

Viegas SC, Pfeiffer V, Sittka A et al (2007) Characterization of the role of ribonucleases in Salmonella small RNA decay. Nucleic Acids Res 35:7651–7664. https://doi.org/10.1093/NAR/GKM916

Wahle E (2007) Wrong PH for RNA degradation. Nat Struct Mol Biol 14:5–7

Walter M, Kilian J, Kudla J (2002) PNPase activity determines the efficiency of mRNA 3′-end processing, the degradation of tRNA and the extent of polyadenylation in chloroplasts. EMBO J 21:6905–6914. https://doi.org/10.1093/EMBOJ/CDF686

Wang DDH, Guo XE, Modrek AS et al (2014) Helicase SUV3, polynucleotide phosphorylase, and mitochondrial polyadenylation polymerase form a transient complex to modulate mitochondrial mRNA polyadenylated tail lengths in response to energetic changes. J Biol Chem 289:16727–16735. https://doi.org/10.1074/jbc.M113.536540

Wang DDH, Shu Z, Lieser SA et al (2009) Human mitochondrial SUV3 and polynucleotide phosphorylase form a 330-kDa heteropentamer to cooperatively degrade double-stranded RNA with a 3′-to-5′ directionality. J Biol Chem 284:20812–20821. https://doi.org/10.1074/jbc.M109.009605

Wang G, Chen HW, Oktay Y et al (2010) PNPASE regulates RNA import into mitochondria. Cell 142:456–467. https://doi.org/10.1016/j.cell.2010.06.035

Wang G, Shimada E, Koehler CM, Teitell MA (2012a) PNPASE and RNA trafficking into mitochondria. Biochim Biophys Acta Gene Regul Mech 1819:998–1007

Wang G, Shimada E, Zhang J et al (2012b) Correcting human mitochondrial mutations with targeted RNA import. Proc Natl Acad Sci USA 109:4840–4845. https://doi.org/10.1073/PNAS.1116792109/-/DCSUPPLEMENTAL/PNAS.201116792SI.PDF

Wong AG, McBurney KL, Thompson KJ et al (2013) S1 and KH domains of polynucleotide phosphorylase determine the efficiency of RNA binding and autoregulation. J Bacteriol 195:2021–2031. https://doi.org/10.1128/JB.00062-13

Wu J, Jiang Z, Liu M et al (2009) Polynucleotide phosphorylase protects Escherichia coli against oxidative stress. Biochemistry 48:2012–2020. https://doi.org/10.1021/BI801752P

Wu J, Li Z (2008) Human polynucleotide phosphorylase reduces oxidative RNA damage and protects HeLa cell against oxidative stress. Biochem Biophys Res Commun 372:288–292. https://doi.org/10.1016/J.BBRC.2008.05.058

Yehudai-Resheff S, Hirsh M, Schuster G (2001) Polynucleotide phosphorylase functions as both an exonuclease and a poly(A) polymerase in spinach chloroplasts. Mol Cell Biol 21:5408. https://doi.org/10.1128/MCB.21.16.5408-5416.2001

Yehudai-Resheff S, Portnoy V, Yogev S et al (2003) Domain analysis of the chloroplast polynucleotide phosphorylase reveals discrete functions in RNA degradation, polyadenylation, and sequence homology with exosome proteins. Plant Cell 15:2003–2019. https://doi.org/10.1105/TPC.013326

Yehudai-Resheff S, Zimmer SL, Komine Y, Stern DB (2007) Integration of chloroplast nucleic acid metabolism into the phosphate deprivation response in Chlamydomonas reinhardtii. Plant Cell 19:1023–1038. https://doi.org/10.1105/tpc.106.045427

Zangrossi S, Briani F, Ghisotti D et al (2000) Transcriptional and post-transcriptional control of polynucleotide phosphorylase during cold acclimation in Escherichia coli. Mol Microbiol 36:1470–1480. https://doi.org/10.1046/j.1365-2958.2000.01971.x

Zhang Y, Gross CA (2021) Cold shock response in bacteria. Annu Rev Genet 55:377–400. https://doi.org/10.1146/annurev-genet-071819-031654

Zimmer SL, Schein A, Zipor G et al (2009) Polyadenylation in Arabidopsis and Chlamydomonas organelles: the input of nucleotidyltransferases, poly(A) polymerases and polynucleotide phosphorylase. Plant J 59:88–99. https://doi.org/10.1111/J.1365-313X.2009.03853.X

Zuo Y, Deutscher MP (2001) Exoribonuclease superfamilies: structural analysis and phylogenetic distribution. Nucleic Acids Res 29:1017–1026. https://doi.org/10.1093/nar/29.5.1017

Chapter 6
RNA Processing and Decay Pathways in Archaea: Emerging Insights on the Function and Evolution of the Archaeal Specific Ski2-Like RNA Helicases

Charles Cirio, Maël Ansart, Emmanuelle Schmitt, Marie Bouvier, and Béatrice Clouet-d'Orval

Abstract In this chapter, we examine the current understanding of gene expression regulation by post-transcriptional RNA processing and decay pathways in Archaea, addressing critical bottlenecks and scientific advances. We highlight the RNA degrading machines and the associated RNA helicases engaged in these processes. Our focus is on how our understanding of these pathways contributes to a more comprehensive understanding of the evolutionary relationship between Archaea and Eukarya. Integrating insights from biochemical and evolutionary research has enabled us to identify Ski2-like RNA helicases as major players in RNA decay pathways in Archaea and Eukarya. In conclusion, we propose that this RNA helicase family has evolved to connect the RNA-degrading enzymes and ribosomes, in order to regulate gene expression across Eukarya and Archaea. Moreover, the presence of the ASH-Ski2 helicase in the Asgard clade, which is more closely related to Eukarya, supports the hypothesis that a common ancestor of the Ski2-like RNA helicase existed prior the emergence of the Eukarya.

Keywords Archaea · RNA exosome · aRNase J · Ski2-like RNA Helicase

Highlights
This study examines the molecular evolution of archaeal RNA-degrading enzymes and RNA helicase families, with a particular focus on the Ski2-like family.

C. Cirio · E. Schmitt
Laboratoire de Biologie Structurale de la Cellule (BIOC), CNRS, Ecole polytechnique, Institut Polytechnique de Paris, Palaiseau, France

M. Ansart · M. Bouvier · B. Clouet-d'Orval (✉)
MCD, Centre de Biologie Intégrative (CBI), CNRS, Université de Toulouse, Toulouse, France
e-mail: Beatrice.Clouet-dOrval@univ-tlse3.fr

A. Kanai, A. W. Nicholson (eds.), *Molecular Evolution of RNA Regulatory Enzymes and Their Systems*, Nucleic Acids and Molecular Biology 38,
https://doi.org/10.1007/978-3-032-03300-0_6

6.1 Introduction

Cells are capable of adapting their growth and metabolism to a changing environment by regulating the expression of their genes. As RNA is the sole direct product of the genome, ensuring its stability and integrity is crucial for gene expression. Post-transcriptional RNA maturation and degradation pathways regulate the stability of messenger RNAs and regulatory non-coding RNAs. It also gives rise to the mature forms of ribosomal RNAs and transfer RNAs. The maintenance of cellular homeostasis is contingent upon the presence of specific RNAs in adequate quantities and with optimal quality (Wolin and Maquat 2019). A detailed analysis of the differences and similarities that exist between the key RNA surveillance and decay machines in the three domains of life can provide a more comprehensive understanding of their functional characteristics (Dominski et al. 2013; Stoecklin and Mühlemann 2013; Clouet-d'Orval et al. 2018). The objective of this chapter is to provide a comprehensive overview of the recent advances in RNA decay processes in archaea, with a focus on RNA helicases of the Ski2-like family. In the cell, RNA helicases are intricately linked to RNA-related processes and the life cycle of RNA molecules by continually reorganising RNA and ribonucleoprotein (RNP) complexes (Anantharaman et al. 2002; Singleton et al. 2007; Jankowsky and Fairman 2007; Pyle 2008; Linder and Jankowsky 2011).

Archaea represent a unique domain of life and studying them is of critical importance for understanding the evolution of living organisms (Fig. 6.1) (Woese and Fox 1977; Woese et al. 1990; Raymann et al. 2015). Demonstrating that the phylogenetic signal carried by DNA replication is highly consistent with that of the other two key information processing machineries (translation and transcription) strengthened the existence of a robust organismal tree for the Archaea (Raymann et al. 2014). In addition, studying archaea has become essential to depict the evolutionary history of eukaryotes. Indeed, in-depth phylogenomic studies have provided compelling evidence that Eukarya and Archaea are sister groups that exhibit closely related mechanisms for propagating and decoding genetic information (Williams et al. 2013; Raymann et al. 2015; Eme et al. 2017; Betts et al. 2018; Williams et al. 2020). Therefore, Archaea are now viewed as models of importance to study the molecular mechanisms that are conserved between Eukarya and Archaea. Recent findings have also revealed novel traits that transcend the behaviours of archaeal extremophiles and their supposed 'primitiveness', transforming our understanding of the biology of Archaea. It has been established that Archaea are present in a significant proportion of the Earth's ecosystems, and they are regarded as key species in plant, animal and human microbiomes (Adam et al. 2017; Borrel et al. 2020). Furthermore, it is widely acknowledged that they play pivotal roles in the context of significant geochemical cycles (Offre et al. 2013). Nonetheless, Archaea are poorly known and understudied.

Archaea represent a very diverse phylogenetic group that includes free-living, extremophiles, mesophiles, symbionts, and opportunistic organisms (Tahon et al. 2021). An archaeal tree, obtained with a comprehensive set of more than 3000

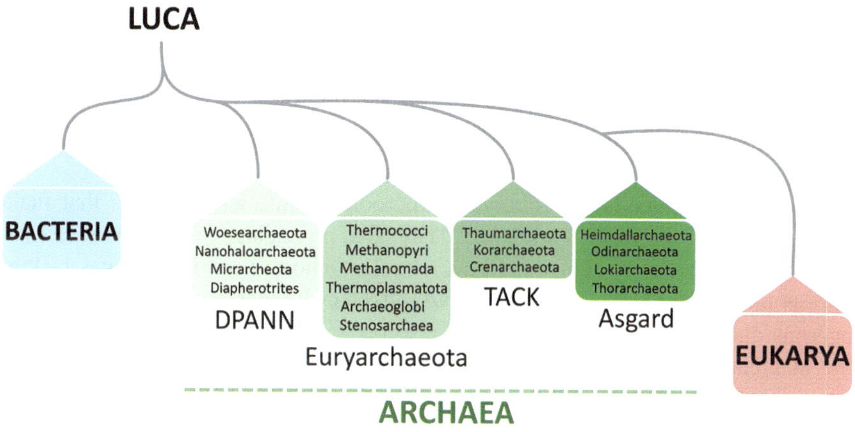

Fig. 6.1 Schematic representation of tree of life showing the eukaryal domain emerging from the archaeal domain. Recent hypothesis implicates that the eukaryotic emerged from Lokiarchaea clade and are nested within the Heimdallarchaeota (Eme et al. 2023; Zhang et al. 2025). The four clades of the archaeal domain, along with their major phylogenetic groups, are indicated. The. LUCA stand for "Last Universal Common Ancestor"

uncultured and cultured archaeal genomes, identified four clades: the Euryarchaea, and three superphyla, namely Asgard, DPANN and TACK (Baker et al. 2020) (Fig. 6.1). Recently, a standardized archaeal taxonomy for the Genome Taxonomy Database (GTDB) reclassified three major monophyletic units from the former Euryarchaeota and one phylum that unites the Thaumarchaeota–Aigarchaeota–Crenarchaeota–Korarchaeota (TACK) superphylum into a single phylum (Rinke et al. 2021). This chapter, which reviews earlier work, employs the original nomenclature (Baker et al. 2020). The Euryarchaea contain the greatest number and diversity of cultured lineages (Baker et al. 2020). The methanogens are divided into several named classes: these include Methanobacteria, Methanococci, Methanomicrobia, Methanonatronarchaeia and Methanopyri. Cultivation of other members has been achieved in halophilic (i.e. Halobacteria), acidophilic (i.e. Thermoplasmata) or thermophilic environments (i.e. Archaeoglobi, Methanopyri and Thermococci). The discovery of extremophiles belonging to the classes Thermoplasmata, Thermococci and Archaeoglobi has provided numerous cultured representatives for physiological studies and genomic analyses. This has advanced our understanding of the biological adaptations of life in extreme conditions. The nomenclature of DPANN and TACK Archaea was initially designated on the basis of the phyla present in them. The acronym 'DPANN' is an abbreviation for the taxonomic groups Diapherotrites, Parvarchaeota, Aenigmarchaeota, Nanohaloarchaeota and Nanoarchaeota (Dombrowski et al. 2019). The acronym 'TACK', on the other hand, is an abbreviation for the taxonomic groups Thaumarchaeota, Aigarchaeota, Crenarchaeota and Korarchaeota (Guy and Ettema 2011). Finally, the recent discovery of the Asgard archaea superphylum has provided substantial support for the hypothesis that eukaryotic cellular organisms evolved from archaea (Spang et al.

2015; Seitz et al. 2016; Zaremba-Niedzwiedzka et al. 2017; Spang et al. 2018). The study of the morphology, physiology and genomics of Asgard archaea has provided valuable insights into the evolutionary history of eukaryotes, positioning them as the most eukaryotic (Williams et al. 2020; Liu et al. 2021; Eme et al. 2023; Zhang et al. 2025). This finding indicates that many components specific to eukaryotes were already present in their archaeal ancestors.

Archaea are unicellular prokaryotic microorganisms, with features that make them resemble bacteria at first glance. They have no nuclear compartment or complex endomembrane systems, and have circular genomes encoding genes lacking introns (except for those encoding tRNA). Archaeal genes are often organized in operons and co-expressed. Translation also occurs co-transcriptionally. However, the machineries that archaea use to replicate and express their genetic information—DNA replication, transcription and translation—are more closely related to their eukaryotic counterparts, as explained above. Concomitantly, archaea exhibit a set of shared cellular hallmarks that differentiate them from both eukaryotes and bacteria. Over the past few decades, biochemical and structural studies of diverse clades of archaeal organisms have helped to clarify the regulatory mechanisms and macromolecular machines that are responsible for major functions of the cell, such as genome maintenance, cell division, metabolism and the transmission of genetic information. RNA processing pathways occupy a pivotal role in the regulation of gene expression. In order to ensure their functionality within the cell, all RNA transcripts undergo multiple maturation steps in addition to covalent chemical modifications. The life cycle of an RNA molecule is intricately linked to RNA helicases which use nucleoside triphosphates to bind or remodel RNA and RNA–protein complexes (RNP). RNA helicases are the largest class of enzymes involved in cellular RNA metabolism, including all aspects of gene expression and regulation. They are also highly conserved proteins, present in all domains of life (Anantharaman et al. 2002). To date, the majority of archaeal actors and pathways have been identified based on their similarity to those in bacteria and eukarya. Therefore, it is of critical importance to address the knowledge gap concerning the fundamental post-transcriptional pathways that regulate gene expression, and to decipher the key regulators at the nexus of multiple RNA metabolic pathways in Archaea.

6.2 RNA-Degrading Complexes in Archaea: How They Compare to Their Eukaryal Counterparts

6.2.1 mRNA Decay Pathways

A deep understanding of RNA metabolism—including all events in the life cycle of RNA molecules, including their synthesis, folding/unfolding, modification, processing and degradation—is critical. RNA biology has become an important and rapidly evolving field at the centre of post-transcriptional gene expression regulation. The continuous breakdown and synthesis of mRNAs allow cells to adapt rapidly to a changing environment. In essence, mRNA decay pathways are critical by enabling transient adaptations and rapid responses to external stimuli. The archaea, which inhabit harsh environments, offer compelling model systems for identifying RNA quality control and decay-associated pathways. Many archaea are exposed to extreme conditions, including temperatures above 100 °C, high pressure, and oxidative/radiation conditions, all of which challenge RNA stability.

Most mechanistic studies have focused on bacterial and eukaryal systems (Stoecklin and Mühlemann 2013; Durand et al. 2015; Wagner and Romby 2015). Identifying the main drivers of RNA decay and processing pathways in Archaea has proven challenging. Only fragmented information has been reported on the enzymes and RNA-binding proteins that carry out RNA maturation and degradation within archaeal cells, and numerous open questions remain (Clouet-d'Orval et al. 2018; Li et al. 2025). The stability of an mRNA molecule depends on: (i) the cell-specific set of RNases; (ii) the sequence and structural features of the mRNA molecule itself; and (iii) RNA-binding proteins, or regulatory RNAs. More precisely, the quantity and quality of RNAs are regulated by endoribonucleases, and by 5′–3′ or 3′–5′ exoribonucleases that are associated with RNA helicases. Most ribonucleases (RNases) that function in Archaea have been identified on the basis of their homologies with bacterial and eukaryotic orthologs. To date, with the exception of RNase W (Vayssières et al. 2024), no archaeal-specific ribonucleases have been identified. Moreover, no double-stranded endoribonuclease activity has been identified in archaea. These studies suggest that the processes that finely regulate the quality and stability of archaeal RNA molecules, mostly possess mosaic eukaryotic and bacterial characteristics (Brochier-Armanet et al. 2011; Eme et al. 2017; Clouet-d'Orval et al. 2018; Phung et al. 2020). Therefore, studying these mechanisms in Archaea also provides an opportunity to improve our understanding of how RNA biology has evolved across the three domains of life.

Briefly summarized, eukaryal mRNA is usually monocistronic and must pass through the nuclear membrane before translation can occur. These mRNAs are characterised by their extensive half-lives varying from 3 minutes to over 90 minutes (Wang et al. 2002). The synthesis of eukaryal mRNA is a complex, hierarchical, and multi-step process involving RNA polymerase II mediated transcription, 5′-end m_7G capping, 3′-end polyadenylation, splicing, nuclear-export and chemical modifications, such as base editing. In this setting, the turnover of mRNA is initiated by the stepwise shortening of the 3′ poly(A) tail (deadenylation), and then proceeds via the action of decapping enzymes that remove the protective cap structure from the mRNA. The transcript is then degraded from either end by processive 5′–3′ and 3′–5′ exoribonucleases. Different RNase complexes (Pan2-Pan3, Ccr4-Not, Xrn1, RNA exosome) and decapping enzymes (Dcp2, DcpS) are involved in these pathways (Stoecklin and Mühlemann 2013) (Fig. 6.2a). These factors are embedded within interaction networks that regulate their enzymatic activities and ensure the correct substrates are recruited. These intricate steps are orchestrated and controlled by RNA surveillance and decay pathways (Inada 2017) (Fig. 6.2a). Both nuclear RNA quality control pathways and cytoplasmic mRNA decay pathways

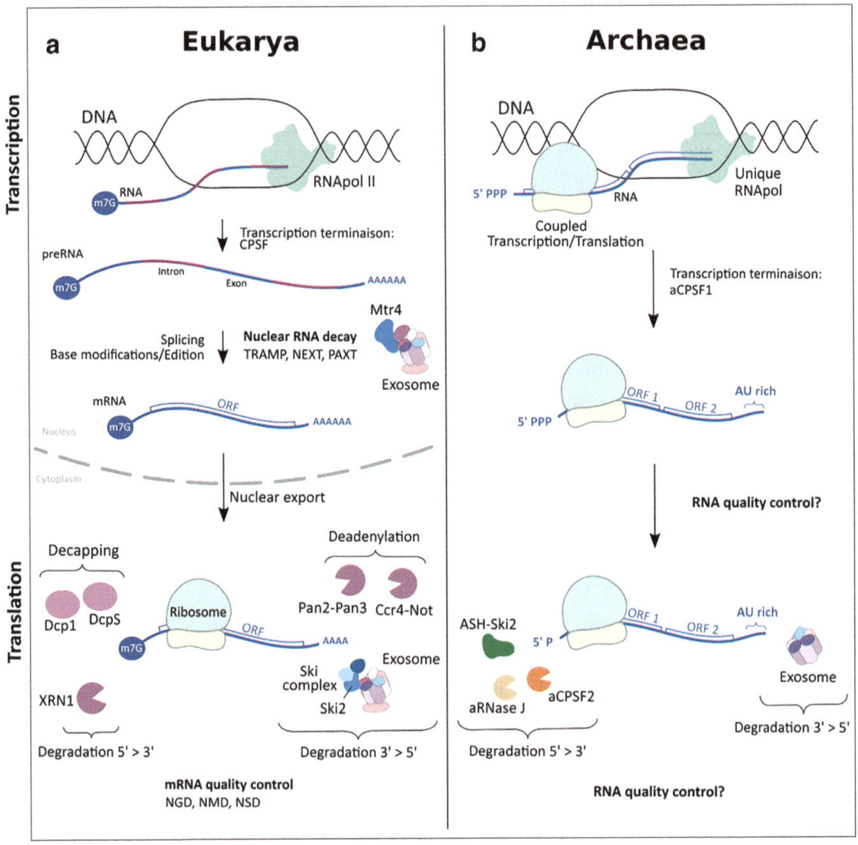

Fig. 6.2 Overview of gene expression in eukaryotes and archaea. Schematic view and key actors of post-transcriptional regulation in eukaryotes (**a**) and archaea (**b**). CPSF cleavage and polyadenylation specificity factor, TRAMP Trf4/Air2/Mtr4 polyadenylation complex, NEXT nuclear exosome targeting complex, PAXT PolyA exosome targeting, NGD No-Go decay, NMD nonsense-mediated decay, NSD non-STOP decay

rely on RNA exosome machineries with dual endo- and 3′–5′ exoribonucleolytic activities, as well as ribonucleases from the XRN family with 5′–3′ exoribonucleolytic activity (Chlebowski et al. 2013; Nagarajan et al. 2013). Within the nucleus, the RNA exosome works alongside the ribonucleases Rrp6 and Dis3, as well as associated factors to form the TRAMP, NEXT or PAXT complexes. The complexes are responsible for the specific degradation of non-coding RNAs, aberrant or cryptic RNAs, introns, or intermediates in the maturation of rRNAs, tRNAs or small regulatory RNAs (Fig. 6.2a) (Chlebowski et al. 2013; Puno et al. 2019). Within the cytoplasm, the RNA exosome, which is associated with the Dis3 ribonuclease, is involved in co-translational quality control mechanisms that allow the elimination of various aberrant mRNAs (e.g. those resulting from errors in transcription or processing, or errors induced by stresses) (Fig. 6.2a). In addition, numerous quality control systems are in place to differentiate between paused and terminally stalled ribosomes that require rescue. Three cytoplasmic mRNA quality control pathways have been reported to monitor mRNAs for translational errors: the nonsense-mediated decay (NMD) pathway, the non-stop decay (NSD) pathway, and the no-go decay (NGD) pathway (Inada 2013, 2017) (Fig. 6.2a). In the NMD pathway, aberrant or faulty mRNA molecules harbouring premature termination codons (PTCs) are recognised and rapidly degraded to prevent the expression of potentially harmful truncated polypeptides (Karousis and Mühlemann 2019).

Recent data on eukaryal RNA surveillance pathways has revealed an intimate coupling between translation and mRNA degradation (Schmidt et al. 2016; Tesina et al. 2019; Zinoviev et al. 2020; Kögel et al. 2022). This suggests that RNA decay plays a major role in coordinating post-transcriptional regulation (Heck and Wilusz 2018). Indeed, unproductive translation, characterized by stalled ribosomes, leads to ribosome rescue and mRNA decay (Tuck et al. 2020). Recent studies have demonstrated the association of RNases with the ribosome, via collaborative interactions. The structural basis of mRNA decay is gradually being elucidated in eukaryotic systems. Evidence of this comes from the identification of a direct physical coupling mechanism that occurs through the formation of a cytoplasmic RNA exosome-ribosome supercomplex (Kögel et al. 2022; Kögel et al. 2024).

6.2.2 Little Knowledge of mRNA Decay Pathways in Archaea

In archaea, the synthesis of all cellular RNAs is performed by a unique RNA polymerase (RNAP) that shares several features with eukaryal RNA polymerase II, including sequence conservation and structural similarities (Blombach et al. 2019). Similarly, it has recently been proposed that archaeal transcription termination is assisted by the archaeal termination factor aCPSF1, which is orthologous to the eukaryal CPSF73 termination factor (Phung et al. 2013; Sanders et al. 2020; Yue et al. 2020; Li et al. 2021). Despite homologies between the transcription machineries of archaeal and eukaryal organisms, archaeal mRNA molecules possess bacterial features. These include the absence of introns, 5′ methyl caps, and long 3′ poly A tails (Fig. 6.2b). They can also be polycistronic and contain Shine-Dalgarno sequences (Schmitt et al. 2020) and their 5′ untranslated region (5′-UTR) can contribute to their stability (Cao et al. 2014).

In Archaea, mRNA half-lives range from a few seconds to over 80 minutes (min): in *Methanococcus vannielii,* from 7 to 57 min (Hennigan and Reeve 1994); in *Sulfolobus solfataricus* and *S. acidocaldarius,* from 2 to 20 min (Bini et al. 2002; Andersson et al. 2006); and *in Haloferax mediterranei,* from 4 to 80 min (Jäger et al. 2002). These measurements must be taken cautiously, bearing in mind that different methods and growth conditions were used. Nevertheless, mRNA turnover plays a major regulatory role in determining protein content of cells. The analysis of differential expression and half-lives in *Methanosarcina acetivorans* provided the evidence of drastic genome-wide shifts in RNA half-lives that are modulated by growth substrates. This led to the conclusion that the expression of more than half of the metabolic genes are controlled by the degradation of their transcripts (Peterson et al. 2016). In *Methanosarcina mazei*, large 5′ UTRs with stem-loop structures have been shown to enhance the stability of the mRNA responsible for methanol-derived methanogenesis (Cao et al. 2014). In contrast, the mRNAs from *H. volcanii* (Euryarchaeota) and *S. solfataricus* (Creanrchaeota) are predominantly leaderless transcripts (Londei 2005; Babski et al. 2016). Genome-wide sequencing of 5′-monophosphate RNA transcripts from a methanogenic archaeon, allowing the mapping of processing sites, revealed that mRNA processing regulates ribosomal protein synthesis and stoichiometry (Qi et al. 2017).

A study of *S. solfataricus* suggests that the triphosphorylated 5′ ends of mRNA can play the role of a cap-like structure. Indeed, the archaeal translation initiation factor aIF2-γ seems to protect transcripts from a 5′ end-dependent degradation by binding to the triphosphoryl moiety of the 5′ end (Hasenöhrl et al. 2008; Arkhipova et al. 2015). In this context, the aIF2-γ subunit would play a similar role as that of the eukaryotic cap-complex. In parallel, NAD-capped RNAs were detected in *Sulfolobus acidocaldarius* and *H. volcanii* using liquid chromatography coupled with tandem mass spectrometry (LC-MS/MS) and NAD captureSeq (Gomes-Filho et al. 2022). It was shown that NAD-capping influenced RNA turnover by promoting degradation.

Several studies provide substantial evidence for a directional 5′ to 3′ degradation pathway in archaeal cells (Clouet-d'Orval et al. 2018; Phung et al. 2020) (Fig. 6.2b). However, no homologues of XRN1, the primary cytosolic 5′–3′ exoribonuclease responsible for deadenylation-dependent turnover of decapped mRNA in eukaryotes, have yet been identified in archaeal genomes. Nevertheless, the widespread occurrence of 5′–3′ exoribonucleolytic activity in members of the archaeal β-CASP family, such as aRNase J and aCPSF2, suggests the existence of a strictly conserved 5′–3′ degradation pathway in Archaea (Fig. 6.2b). Over the past decade, the key importance of β-CASP nucleases in archaeal RNA metabolism has emerged in various biological contexts (Clouet-d'Orval et al. 2010; Phung et al. 2013; Dominski et al. 2013).

6.2.3 Archaeal 5′–3′ Exoribonucleases of the β-CASP Family

The ribonucleases of the β-CASP family are part of an ubiquitous enzyme family that operate in critical RNA processing and/or maturation pathways. Members of this family harbour a metallo-β-lactamase domain (MβL), composed of a four layered αβ/βα architecture, into which a β-CASP domain is inserted (Fig. 6.3a). Several crystal structures of bacterial, eukaryal and archaeal members have been reported to have an identical structural topology consisting of two domains, with a catalytic site formed at their interface involving the coordination of two Zn^{2+} ions by conserved histidine and aspartic acid residues (Dominski et al. 2013). These enzymes, catalysing both endo- and 5′–3′ exoribonucleolytic degradation, are unique among all

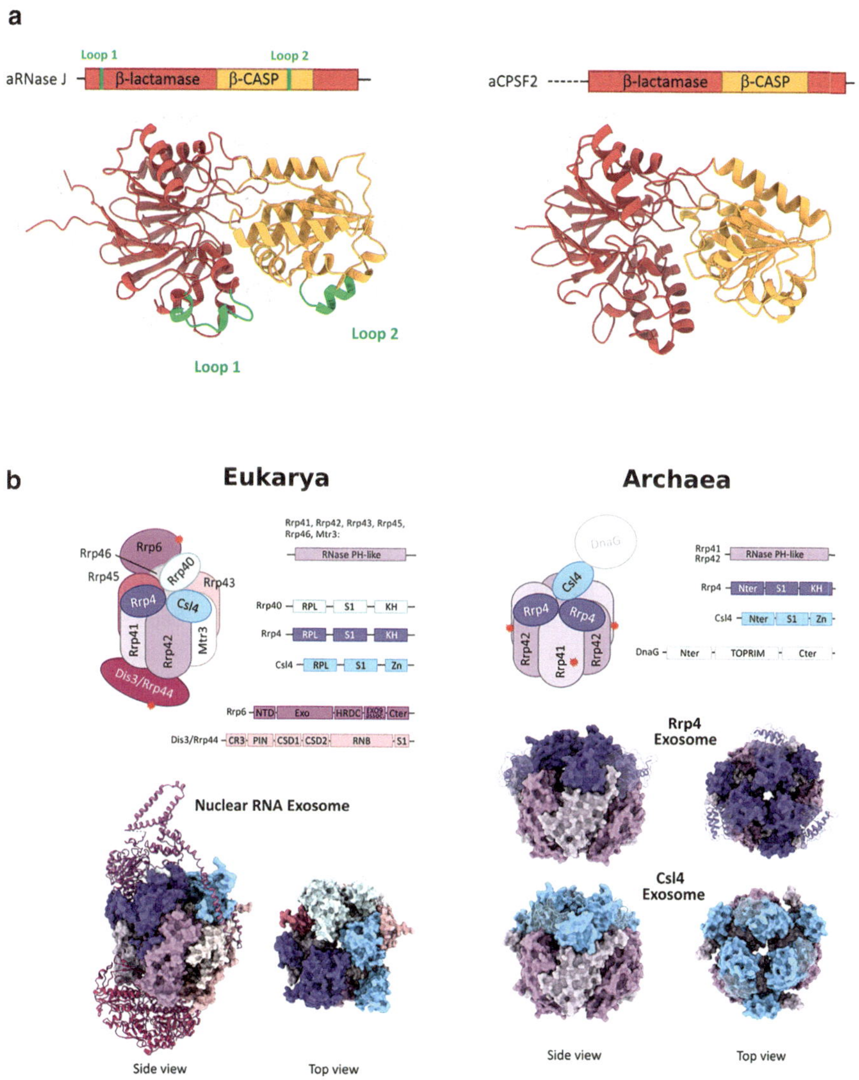

Fig. 6.3 Domain organizations and overall structures of the 5′-3′ exoRNases of the archaeal β-CASP ribonucleases and eukaryotic and archaeal RNA Exosomes. (**a**) AlphaFold3 modeling of *P. abyssi* aRNase J (Q9V076) and of *S. Solfataricus* aCPSF2 (Q980D0) restricted to the conserved core (amino acids 71 to 492): the N-terminal extension, represented in dotted lines, has been deleted due to low prediction confidence. Amino acids in green correspond to two archaea-specific insertions in aRNase J. (**b**) Left panel shows the structure and the schematic representation of the 12-subunit Exosome, EXO9-Rrp6-Dis3 complex, from *Saccharomyces cerevisiae* (PDB: 5C0W). EXO9 core component subunits are shown in surface format. The additional subunits Dis3/Rrp44 and Rrp6 (restricted to the nucleus) are shown in cartoon format. Catalytic subunits are indicated by red stars. Right panel shows the structure of an Archaeal Exosome core containing a homotrimeric Rrp4 (purple, PDB: 2BA0) or Csl4 (blue, PDB: 2BA1) cap from *Archaeoglobus fulgidus*, and is shown in surface format. The KH or Zn ribbon domains specific to Rrp4 or Csl4 are shown in cartoon format. The schematic representation illustrates the existence of exosomes carrying a heterotrimeric cap and the DnaG protein, whose structure has not been resolved. Catalytic subunits are indicated by red stars

known ribonucleases. Their vertical inheritance suggests that β-CASP enzymes have an ancient origin predating the separation of Bacteria, Eukarya and Archaea (Phung et al. 2013, 2017). In this context, β-CASP ribonucleases should also be critical for RNA decay and maturation in archaeal cells (Clouet-d'Orval et al. 2015). In Archaea, biochemical and phylogenomic approaches have clearly identified and characterised three separate groups of β-CASP RNases: aRNase J and aCPSF2, with a 5′–3′ exoribonucleolytic activity, and aCPSF1, with endoribonucleolytic activity (Phung et al. 2013; Dominski et al. 2013; Phung et al. 2017). The first listed enzyme is related to the bacterial RNase J, and the other two are related to the eukaryal transcription termination factor CPSF73.

The aCPSF1 factors from *Pyrococcus horikoshii*, *Methanosarcina mazei* and *Methanothermobacter thermautotrophicus* have a three-part architecture. KH domains are connected to the central core through a long linker (Nishida et al. 2010; Silva et al. 2011; Mir-Montazeri et al. 2011). In Thermococcales and Methanococcales, aCPSF1 has a dual activity: it is able to internally cleave single-stranded RNA at a CA site, and it possesses a 5′–3′ exoribonucleolytic activity that is restricted to 5′-monophosphorylated substrates (Levy et al. 2011; Phung et al. 2013). Its activity is comparable to that of CPSF73 in eukaryotes. The presence of aCPSF1 across archaeal genomes suggests that its activity is central to conserved metabolic pathways (Clouet-d'Orval et al. 2018). It has recently been shown that it is involved in transcription termination in *Thermococcus kodakarensis and Methanococcus maripaludis* (Phung et al. 2013; Sanders et al. 2020; Yue et al. 2020; Li et al. 2021; You et al. 2024).

The ribonuclease aCPSF2 is part of a less conserved cluster in archaea compared to the aCPSF1 cluster (Phung et al. 2013; Clouet-d'Orval et al. 2018). The members of this cluster, found in TACK/*Proteoarchaeota*, and sporadically in Euryarchaeota, are restricted to the MβL and β-CASP domains, except in *S. solfataricus*, where an N-terminal extension is present. aCPSF2 members are also found in Asgard genomes. Currently, no aCPSF2 structure has been experimentally determined. The structure presented in Fig. 6.3a (right panel) was predicted by AlphaFold with high prediction confidence. In *S. solfataricus*, the gene encoding aCPSF2 can be deleted without affecting cell doubling time (Levy et al. 2011; Märtens et al. 2013). The function of aCPSF2 has been investigated in *S. acidocaldarius* (Saci). Knocking out Saci-aCPSF2 gene altered the transcript abundance of 560 genes, suggesting a significant role for aCPSF2 in RNA metabolism (Märtens et al. 2013). Several transcripts were preferentially degraded at their 5′ end, indicating a role for aCPSF2 in the 5′–3′ directional decay of mRNA in crenarchaea. Notably, Saci-aCPSF2 has been demonstrated to degrade NAD-capped RNAs in *S. acidocaldarius*. NAD-capped RNA levels increased 4.5-fold in the Saci-aCPSF2 deletion mutant (Gomes-Filho et al. 2022). This highlights the important function of Saci-aCPSF2 in the turnover of NAD-capped RNAs and provides an additional layer of post-transcriptional regulation in archaeal cells.

aRNase J is restricted to *Euryarchaeota* and is highly conserved, except in the *Diafoarchaea* and Archeoglobi groups (Phung et al. 2013; Clouet-d'Orval et al.

2018). It is also restricted to the MβL and β-CASP domains (Fig. 6.3a, left panel). aRNase J differs from aCPSF2 via two specific insertions: a nine amino acid insertion in the β-CASP domain (Loop1) and a 12 amino acid insertion in the MβL domain (Loop2) (Clouet-d'Orval et al. 2010). *In vitro*, recombinant aRNase J proteins from Thermococcales and *Methanococci* exhibit highly processive 5′–3′ exoribonucleolytic activities, and no detectable endoribonucleolytic activity under the conditions tested. The two proteins also demonstrate a greatly reduced activity when degrading 5′-triphosphorylated RNA substrates (Clouet-d'Orval et al. 2010; Levy et al. 2011). The crystal structure of aRNase J from *Methanolobus psychrophilus* R15 shows that Loops 1 and 2 interact with the RNA molecule (Zheng et al. 2017). The loops are also important for the ribonucleolytic activity of aRNase J from *Pyrococcus abyssi* (Pab) (Clouet-d'Orval et al. 2010). Moreover, biochemical, proteomic and phylogenomic analyses suggest that *Pab*-aRNase J participates in mRNA degradation pathways associated with the ribosome, in coordination with the ASH-Ski2 helicase and/or the archaeal RNA exosome (Phung et al. 2020).

Both archaeal 5′–3′ exoribonucleases, aRNase J and aCPSF2, exhibit a preference for monophosphorylated RNA substrates (Clouet-d'Orval et al. 2010, 2018; Hasenöhrl et al. 2011; Phung et al. 2013). However, the molecular mechanism of 5′–3′ mRNA decay in Archaea remains unknown and an enzyme capable of converting the 5′-triphosphate ends of their mRNAs to monophosphate remains to be discovered (Fig. 6.2b). In eukaryotes, at least two mRNA decapping pyrophosphohydrolases of the NudiX superfamily, Dcp2 and Nudt16, are involved in removing the 5′-cap structure (Arribas-Layton et al. 2013). In Bacteria, the triphosphorylated 5′ termini are converted to monophosphorylated ends by the NudiX superfamily RppH (Celesnik et al. 2007). Although Nudix hydrolases are widely distributed among the three domains of life, archaeal Nudix hydrolases with 'decapping' activity have not yet been identified. Two Nudix candidate proteins from *S. solfataricus* were cloned and expressed in a heterologous system, but no pyrophosphohydrolase activity could be detected in *in vitro* assays (Hasenöhrl et al. 2011). Finally, one of four Nudix candidates from *S. acidocaldarius* was shown to produce 5′-end monophosphorylated RNAs, but from NAD/ADP-ribose RNA capped molecules (Gomes-Filho et al. 2023).

6.2.4 An Ubiquitous 3′–5′ Exoribonuclease in Archaea: The RNA exosome

In eukaryotes, there are two RNA exosome isoforms which differ in composition, cellular location and RNA substrates. The Exosome core, termed EXO9, consists of nine subunits: Rrp41, Rrp42, Rrp43, Rrp45, Rrp46 and Mtr3 constitute the barrel, whilst Rrp4, Rrp40 and Csl4 form the cap (Fig. 6.3b) (Liu et al. 2006; Zinder and Lima 2017). With the exception of the RNA exosome in plants, EXO9 is catalytically inactive in Eukarya, but remains essential as a molecular scaffold (Dziembowski et al. 2007; Lange and Gagliardi 2022). Within the nucleus, the Dis3 (also known as Rrp44) and Rrp6 subunits have been shown to be responsible for the hydrolytic catalytic activity required for 3′–5′ RNA degradation (Dziembowski et al. 2007; Wasmuth et al. 2017). It has been established that the nuclear RNA exosome is an integral component of the NEXT, PAXT and TRAMP complexes, which are instrumental in the degradation of non-coding, cryptic or defective RNAs (Chlebowski et al. 2013; Puno et al. 2019). Within the cytoplasm, EXO9 has been observed to interact exclusively with Dis3. This EXO9-Dis3 exosome is implicated in a number of cellular processes, including antiviral defence, RNA turnover, and the regulation of mRNA quality (Chlebowski et al. 2013; Puno et al. 2019). The quality control pathways denoted as NMD, NGD and NSD are responsible for degrading mRNAs containing premature stop codons, RNA internal ribosome arrest, or translation termination defects, respectively (Morris et al. 2021).

Most archaeal genomes contain genes encoding an evolutionarily-conserved phosphorolytic 3′–5′ RNA-degrading machinery, the RNA exosome (Koonin et al. 2001; Evguenieva-Hackenberg et al. 2017; Phung et al. 2017; Clouet-d'Orval et al. 2018) (Fig. 6.3b). The RNA exosome has been identified based on the detection of an operon that encodes homologues of the Eukaryal Rrp4, Rrp41 and Rrp42 proteins (Koonin et al. 2001). It is conserved in most archaeal genomes, except in species belonging to the Methanococci, Methanomicrobiales, Haloferacales and Halobacteriales (Evguenieva-Hackenberg et al. 2014; Clouet-d'Orval et al. 2018). It should be noted that the homolog of Csl4 is encoded in a locus, together with the gene of the ASH-Ski2 RNA helicase, in some archaeal genomes, such as *Pyrococcus abyssi*. The archaeal RNA exosome has been the subject of many studies. In particular, thorough characterisations of the RNA exosome of Sulfolobales and Pyrococcales species have determined their three-dimensional structures and their *in vitro* enzymatic activities (Hartung and Hopfner 2009; Chlebowski et al. 2013; Evguenieva-Hackenberg et al. 2014; Clouet-d'Orval et al. 2018).

In summary, the archaeal exosome typically comprises a nine-subunit core and manifests as a heterogeneous 250 kDa complex (Fig. 6.3b). Consistently, three Rrp41 and Rrp42 subunits belonging to the RNase PH family associate to form a hexameric barrel-like structure with a narrow RNA entrance pore. It has been established that a variable trimeric RNA-binding cap, consisting of Rrp4 and/or Csl4 subunits, binds tightly to the barrel. The precise stoichiometry of this cap remains to be elucidated. (Evguenieva-Hackenberg et al. 2003; Ramos et al. 2006; Witharana et al. 2012). However, it has been demonstrated that the two subunits Rrp4 and Csl4 could coexist within the same RNA exosome. Variation in the stoichiometry of these two proteins can result in the formation of heterogeneous caps, which may be linked to specific RNA decay pathways (Witharana et al. 2012). Indeed, the cap RNA binding subunits exhibit different combinations of RNA binding domains (S1 domain, KH motif, and Zn-ribbon) (Fig. 6.3b). Furthermore, the Rrp4 subunit, with its KH motif, has been demonstrated to confer to the RNA exosome a strong specificity to poly(A) molecules, while the Csl4 subunit, with its S1 domain, has been shown to allow the RNA exosome to bind the aDnaG protein (Witharana et al. 2012; Hou et al. 2013). Three active catalytic sites, carried by the Rrp41 subunits, are located in the lumen of the barrel structure and are responsible for the phosphorolytic degradation of single-stranded RNA from the 3′-end, releasing nucleoside 5′-diphosphates (NDPs) (Lorentzen et al. 2005, 2007; Hartung et al. 2010). It has been demonstrated that during the process of degradation, RNA moves from one catalytic site to another within the barrel. This mobility would stabilise its interaction with the RNA exosome, thereby facilitating faster and processive RNA degradation at the scale of the complex (Audin et al. 2016). It is noteworthy that the Rrp41 subunit of the eukaryal RNA exosome barrel has lost its phosphorolytic activity. Furthermore, the archaeal exosome is also able to perform the reverse reaction by using NDPs to synthetize heteropolymers at the 3′-end of RNAs, including adenine-rich tails (Portnoy et al. 2005). At present, this is the sole RNA-tailing enzyme that has been identified in archaeal cells. The 3′–5′ exoribonucleolytic and RNA-tailing activities can be modulated by changing the Mg^{2+} concentration (Evguenieva-Hackenberg et al. 2008).

The function of the RNA exosome in RNA surveillance and decay pathways in archaeal cells has been the focus of several studies in past years. Early indications of its role were observed in its interaction with the tRNA splicing enzyme EndA in *Methanothermobacter thermautotrophicus* (Farhoud et al. 2005), and with the translational elongation factor EF1α in *S. solfataricus* (Witharana et al. 2012). Additionally, Nop5, which is part of the C/D box RNA methylation complex, was shown to interact with Rrp4, while the conserved Sm-like protein, SmAP1 and SmAP2, were shown to interact with DnaG (Märtens et al. 2017; Gauernack et al. 2017). Furthermore, the heterogeneity of the complexes observed in *S. solfataricus* cells, together with their subcellular localization at the membrane, may reflect that their 3′-tailing or degradation activities are regulated in a multi-layered manner upon substrate recognition (Roppelt et al. 2010; Witharana et al. 2012).

A more recent study in Sulfolobales underscores the role of the RNA exosome in mRNA turnover, by determining binding sites to RNA on a global scale (Bathke et al. 2020). The data sets presented provide compelling evidence for the RNA exosome functioning as an exoribonuclease and RNA-tailing enzyme, with the capacity to interact with all RNA classes, including mRNAs, and circular 16S and 23S ribosomal RNA intermediates. However, the physiological relevance of the composite RNA-binding cap in the mechanism of substrate selection remains to be elucidated. It will be interesting to understand how and why DnaG is recruited to the RNA exosome. It is intriguing to note that certain species possess an incomplete set of RNA exosome subunits (Clouet-d'Orval et al. 2018). The analysis of RNA exosome samples from these species may contribute to the identification of the specific function of each cap subunit. Moreover, it is noteworthy that in species classified within the Methanomicrobiales, Haloferacales and Halobacteriales group, the absence of the exosome appears to be counterbalanced by the presence of additional 3′–5′ exoribonucleases, namely RNase R-like enzymes. However, in the case of the Methanococci group, no enzyme with an equivalent activity could be detected (Fig. 6.3b). This finding suggests that in these strains, 3′–5′ exoribonucleolytic activity may be nonexistent or may be compensated by other endo- or 5′–3′ exoribonucleolytic machineries (Clouet-d'Orval et al. 2015, 2018).

6.2.5 Emergence of a Connection Between the 5′–3′ Exo aRNase J and the RNA Exosome

In order to illuminate the role and composition of the major archaeal RNA-degrading machines operating in RNA surveillance and decay pathways, the protein-protein interaction networks of *P. abyssi* aRNase J and ASH-Ski2 were determined (Phung et al. 2020). These networks include key components of RNA metabolism which comprise the RNA exosome subunits, ribosomal proteins, and translation initiation factors. Interestingly, the aRNase J network comprises, in addition to ASH-Ski2, the Rrp41 and Rrp42 catalytic core subunits, and the Rrp4 cap subunit of the RNA exosome, which are among the most significant protein partners. In addition, co-purification assays showed that the exosome cap-subunit Csl4 is able to form a complex with aRNase J. For the ASH-Ski2 protein network aRNase J was ranked as the first candidate of the ASH-Ski2 partner list (Phung et al. 2020). In total, 21 protein partners were identified to be common to both aRNase J and ASH-Ski2 networks. Finally, an affinity pulldown assay using Rrp41 as a bait specifically detected the cap subunits Csl4 and DnaG, aRNase J, and ASH-Ski2. Overall, the tight cross-talk of aRNase J, ASH-Ski2 and the RNA exosome in *P. abyssi* cells hints at the existence of high-order β-CASP protein complexes in Thermococcales. These complexes with 5′–3′ and 3′–5′ exoribonucleolytic activities, and RNA helicase activity could play pivotal roles in archaeal RNA surveillance. Phylogenomic studies strengthen this hypothesis by showing that ASH-Ski2 is only encoded in those archaeal genomes that carry a gene for aRNase J, and that the amino acid resides of Csl4 that are involved in the *in vitro* interaction with aRNase J are only conserved in archaeal genomes where their encoding genes are both present (Phung et al. 2020).

6.3 Structural Diversity and Regulatory Function of Ski2 RNA Helicases, and How They Compare to Their Eukaryal Counterparts

6.3.1 The Family of Ski2-Like Helicases Is Defined By a Common Helicase Core Architecture

Helicases participate in all aspects of DNA and RNA metabolism (Singleton et al. 2007; Jankowsky and Fairman 2007; Pyle 2008; Linder and Jankowsky 2011). They perform unwinding reactions to separate double-stranded nucleic acids into a single-stranded form, and displacement reactions to dissociate nucleic acid-associated proteins. Helicases are classified into six helicase superfamilies (SF1–6) based on their amino acid sequence, oligomeric state (monomeric or hexameric), activity (single- or double-stranded DNA or RNA substrates; translocation activity) and 5′–3′ or 3′–5′ polarity (Singleton et al. 2007). SF1 and SF2 helicases share a catalytic core with high structural similarity and perform a wide spectrum of distinct functions on diverse substrates (Jankowsky and Fairman 2007; Lohman et al. 2008; Dillingham 2011). SF1 and SF2 members have been categorized in twelve families (Fairman-Williams et al. 2010). This classification highlights major structural and functional themes for these proteins (Gorbalenya and Koonin 1993; Jankowsky and Fairman 2007). Closer inspection of the sequences reveals at least 12 characteristic sequence motifs shared by both SF1–2. The helicase core consists of RecA1 and RecA2 domains with nine characteristic sequence motifs named Q, I, Ia, Ib, and II to VI, whose sequence can vary between SF1 and SF2. Among these motifs, the motifs I and II which bind NTPs are the most conserved (Gorbalenya and Koonin 1993, 1988; Koonin 1993). The helicase core provides the active site for ATP hydrolysis, binds nucleic acid and carries a basal unwinding activity. Although ATP-dependent unwinding of nucleic acid duplexes is their hallmark reaction, not all helicases catalyse unwinding *in vitro*, and disrupt duplexes *in vivo* (Fairman et al. 2004; Jankowsky and Bowers 2006). In the case of Archaea, a list of SF1 and SF2 members has been proposed based on a comprehensive *in silico* analysis (Chamieh et al. 2016). Most SF1 and SF2 members have been detected in archaea, with the exception of the bacterial RecG-like and the DEAH/RHA families, which are involved in DNA replication, recombination and repair, and RNA splicing and ribosome biogenesis, respectively (Chamieh et al. 2016). However, the relevant cellular processes of archaeal SF1 and SF2 members remain to be addressed.

SF2 is the largest and most diverse group of helicases, with more than ten families. SF2 members are non-hexameric helicases that share a conserved helicase core, and often contain N- and/or C-terminal accessory domains involved in regulation of their activities (Jankowsky and Fairman 2007; Fairman-Williams et al. 2010). The Ski2-like helicases represent a distinct family of SF2 helicases. Here, we highlight the common architectural themes and functional implications that have emerged from structural and biochemical studies of archaeal and eukaryal Ski2-like members. In addition, we discuss the role of Ski2-like helicases in larger complexes.

The Ski2-like helicase family was originally named after the RNA helicase Ski2 that works together with the eukaryal RNA exosome in mRNA turnover and quality control (Widner and Wickner 1993). The Ski2-like helicase family comprises members that possess the structure of the helicase core with motifs resembling those found in other families. A distinctive feature of this family is the large size of its members, which span from 100 kDa to over 200 kDa, These members function as both RNA and DNA helicases (Fairman-Williams et al. 2010; Johnson and Jackson 2013). The structures of several eukaryotic RNA helicases such as Mtr4, Brr2 and Ski2, as well of the archaeal DNA helicase Hel308 led to the identification of a hallmark domain for the Ski2-like family. Hence, the molecular core of all Ski2-like helicases is a ring-like four domain assembly of two adjacent RecA-like domains corresponding to the "DEAD" (RecA1) and "Helicase C" (RecA2) Pfam domains, a winged helix domain, and a ratchet domain (Weir et al. 2010; Johnson and Jackson 2013) (Fig. 6.4a, b).

Eukaryal species encode several representatives of Ski2-like helicases: the HelQ DNA helicase associated with RPA and RAD51 (Anand et al. 2022); the Ski2/Mtr4

RNA helicases associated with RNA exosome activities (Weick and Lima 2021); the Brr2 splicing factor (Absmeier et al. 2016); and the Ski2-like RNA helicase 1 (Slh1), also named rqt2, which is associated to ribosome quality control (RQC) (Matsuo et al. 2020; Best et al. 2023).

The cytoplasmic Ski2 and nuclear Mtr4 helicases are key components of RNA exosome cofactor networks. The two proteins are RNA-dependent ATPases that unwind RNA duplexes in a 3′–5′ direction after loading on a 3′ single-stranded overhang, and can remodel ribonucleoprotein complexes (Johnson and Jackson 2013). They have been proposed to act as hubs for RNA exosome-mediated RNA

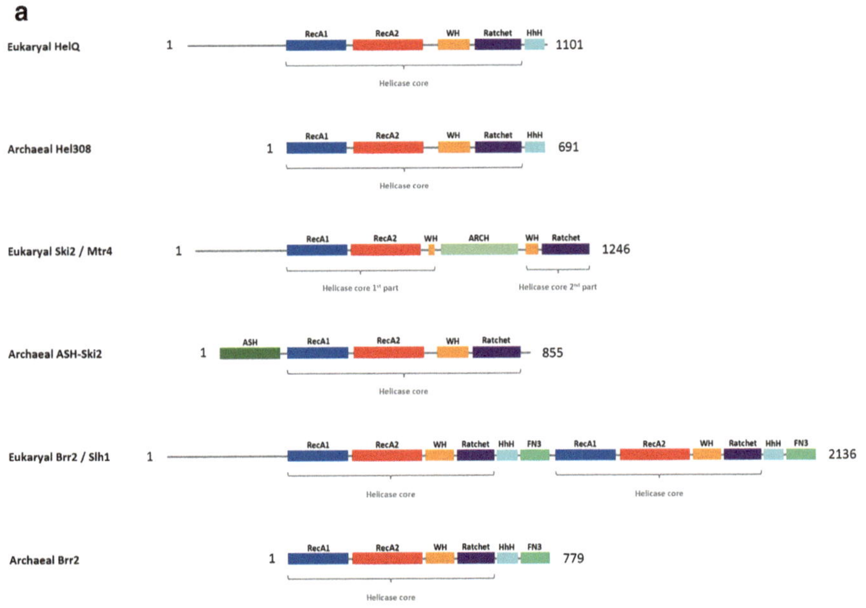

Fig. 6.4 Ski2-like helicase family member architectures. (**a**) Domain organization for the Ski2-like helicases. Amino acid residue numbers are from *Homo sapiens* (eukaryal HelQ, Ski2 and Brr2), *Archaeoglobus fulgidus* (Hel308), *Pyrococcus abyssi* (ASH-Ski2) and *Haloferax volcanii* (archaeal Brr2). The domains are to scale. Ski2-like helicases share a common helicase core consisting of 4 domains: RecA1 (blue), RecA2 (red), Winged Helix (WH, orange) and Ratchet (purple). The architectures also harbor specific accessory domains: Helix hairpin Helix (HhH, cyan), ARCH (green), Archaeal Specific Helicase (ASH, dark green), and Fibronectin type 3 (FN3, apple green). Eukaryal Brr2/Slh1 present 2 helicase core cassettes with accessory domains. (**b**) Superposition of eukaryal (left) and archaeal (right) Ski2-like helicase 3D structures. Structures resolved experimentally were taken from the PDB (*A. fulgidus* Hel308: 2P6R, human Ski2: 7QE0, human Brr2: 4F91) and the others were predicted using Alphafold 3 (human HelQ, *P. abyssi* ASH-Ski2 and *H. volcanii* Brr2). Domains have the same color in (a) and (b). For clarity, only one cassette of Brr2 is shown in (b) and the specific accessory domains have been removed in the top view. Experimentally uncharacterized N and C-terminal regions, or N and C-terminal regions predicted with poor confidence score are not indicated in (b) but are included in (a) as gray lines. Structures have been rendered using ChimeraX software

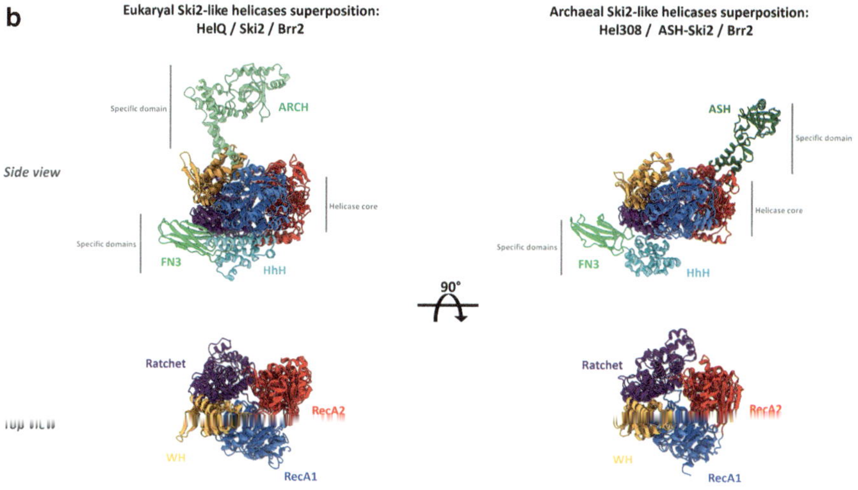

Fig. 6.4 (continued)

degradation by interacting with adaptor proteins to regulate the delivery of RNA to the RNA exosome exoribonuclease core (Weick and Lima 2021; Olsen and Johnson 2021). Mtr4 has been shown to interact with a variety of adapters, reflecting the diverse types of RNA exosome substrates in the nucleus, including ribosomal RNAs, transfer RNAs, precursor mRNAs, and various noncoding RNAs (Weick and Lima 2021). Mtr4 is a component of the NEXT, PAXT and TRAMP complexes: the nuclear exosome-targeting (NEXT) complex degrades short transcripts lacking mature (polyA) tails; the poly(A) exosome-targeting (PAXT) complex targets polyadenylated transcripts (Meola et al. 2016; Ogami et al. 2017; Wu et al. 2020); and the Trf4/Air2/Mtr4p polyadenylation (TRAM) complex interacts with the nuclear poly(A)-binding protein PABPN1 to assist in the surveillance of defective preribosomal RNAs, transfer RNAs, and other non-coding RNAs in the nucleus (Beaulieu et al. 2012). Although the NEXT and PAXT complexes share a large subset of substrates, they engage in mutually exclusive interactions with Mtr4 and exhibit differing substrate specificities (Meola et al. 2016; Ogami et al. 2017; Wu et al. 2020). In contrast to Mtr4, cytoplasmic Ski2 is embedded in a single complex, designated Ski2-Ski3-Ski8 (also named Ski238). This complex associates with the cytoplasmic RNA exosome and has a predominant substrate (i.e. mRNA) in this subcellular compartment (Weick and Lima 2021; Keidel et al. 2025). The Ski238 complex plays an essential role in the extraction of mRNA from stalled ribosome (Zinoviev et al. 2020). The process of extraction occurs in a 3′–5′ direction, nucleotide by nucleotide, and is strictly ATP-dependent. The process has been observed to occur on both pre- and post-translocation ribosomes. A minimum of 19 nucleotides downstream of the P site is required for efficient engagement. Structural analyses reveal that the Ski2 helicase adopts closed and open conformational states (Kögel et al. 2022). In the closed state, Ski2 traps the 3′ end of the RNA and obstructs the

RNA exit path. Subsequent to ATP binding, Ski2 adopts an open conformation that facilitates the extraction of ribosome-bound RNA. This open conformation also primes the helicase to channel the RNA towards the cytosolic RNA exosome for degradation.

Brr2 splicing factor plays a central role in spliceosome activation by catalysing the ATP-dependent unwinding of the U4/U6 snRNA duplex. Initial biochemical evidence demonstrated that Brr2 can mediate this process *in vitro*, establishing its role as the main motor responsible for initiating catalytic activation of the spliceosome (Raghunathan and Guthrie 1998). Subsequent mechanistic studies revealed that Brr2 only translocates a short distance along the U4 snRNA. Unwinding is completed through a conformational change in the partially displaced U6 snRNA, which disrupts the binding of all precatalytic binding partners but preserves other protein-binding sites on U4. This selective remodelling allows Brr2 to activate the spliceosome while maintaining the integrity of components necessary for U4/U6 reassembly after splicing (Theuser et al. 2016). In order to avoid premature activation, the activity of Brr2 is strictly controlled by its interaction with the spliceosomal scaffold protein, Prp8 (Mozaffari-Jovin et al. 2013). Collectively, these findings emphasise Brr2's dual functionality as both a pivotal catalytic factor and a regulatory entity, thereby ensuring temporal fidelity in pre-mRNA splicing.

Slh1 (Ski2-like helicase 1) is the helicase subunit of the ribosome quality control trigger (RQT) complex and plays a central role in resolving ribosomal collisions resulting from translation stalling on aberrant mRNAs. The RQT complex is recruited to polyubiquitinated complexes, which require an accessible 3′ mRNA region emerging from the lead ribosome, as well as the presence of a trailing ribosome (Matsuo et al. 2017). Cryo-electron microscopy revealed that RQT binds to the 40S subunit of the lead ribosome and adopts two distinct conformational states. This suggests that RQT interacts dynamically with the ribosome during the subunit splitting process. According to a model, the helicase Slh1 initiates mRNA-mediated mechanical remodelling of the stalled ribosome, which is a prerequisite for RQT-dependent subunit dissociation. This event provides a mechanistic framework for helicase-driven resolution of ribosomal collisions within the ribosome-associated quality control pathway (Best et al. 2023)

6.3.2 Archaeal Members of the Ski2-Like Helicase Family

The first archaeal homolog of a Ski2-like protein was reported in *Methanothermobacter thermautotrophicus* (Bolt and Guy 2003). Interestingly, *in silico* approaches proposed that this homolog is part of a predicted RNA exosome (Koonin et al. 2001). A more complete picture of the archaeal Ski2-helicase family has recently been proposed, building on the initial analysis performed 10 years ago (Chamieh et al. 2016; Batista et al. 2024). This new classification, which is based on an in-depth phylogenomic analysis of 294 archaeal genomes, proposes that there are three main groups of archaeal Ski2-like proteins: Hel308, ASH-Ski2 and BRR2-like (Fig. 6.4a).

Archaeal Hel308 is ubiquitously conserved among the archaeal clades and belongs to the same subgroup as eukaryal PolQ (Batista et al. 2024). This group contains proteins with DNA helicase activity. It was shown that Thermococcales (Euryarachaea) and Sulfolobales (TACK) Hel308 members are able to unwind various types of duplex DNA substrates in the 3′–5′ direction *in vitro* (Guy and Bolt 2005; Fujikane et al. 2006; Li et al. 2008; Zhai et al. 2018; Batista et al. 2024). Moreover, Hel308 activity appears to be essential for cell viability. Thus deleting Hel308 in *T. kodakaraensis* and *Sulfolobus islandicus* was detrimental to cell growth (Fujikane et al. 2010; Zhang et al. 2013). In *Sulfolobus tokodaii* and *S. islandicus*, Hel308 has been shown to interact with the Hjc endonuclease and the PINA ATPase, respectively. They are both involved in the resolution of Holliday junctions (Li et al. 2008; Zhai et al. 2018). Overall, archaeal Hel308 helicases are proposed to be major actors in DNA recombination and repair processes.

The group of archaeal ASH-Ski2 proteins is conserved in Euryarchaeota and was also identified in Asgard genomes (Phung et al. 2020; Batista et al. 2024). All ASH-Ski2 members are characterized by their specific N-terminal domain (ASH domain) with four strictly conserved cysteine residues forming a zinc finger domain (Fig. 6.4a). This group is monophyletic and is specific to the archaeal domain. This phylogenetic distribution suggests that ASH-Ski2 has a specific physiological role in Euryarchaeota cells. However, unlike the roles of eukaryal Ski2-like helicases in RNA homeostasis, the functions of the ASH-Ski2 members are poorly understood. The deletion of ASH-Ski2 has been reported not to affect the growth of *T. kodokarensis* and *T. barophilus* (Fujikane et al. 2010; Zhang et al. 2013; Phung et al. 2020; Batista et al. 2024). Moreover, a comparison of *T. barophilus* transcriptome compared with that of wild type cells could not identify a clear pathway involving ASH-Ski2 (Batista et al. 2024). *In vitro*, ASH-Ski2 from the archaeon *P. abyssi* was shown to support an unwinding activity with a 3′–5′ polarity and, surprisingly, an efficient annealing activity to form RNA duplexes (Batista et al. 2024). ASH-Ski2 from *T. kodakarensis* (also known as Eta) has been shown to exhibit translocase activity *in vitro* which is proposed to be involved in the rescue of transcriptional arrests (Walker et al. 2017; Maier and Marchfelder 2019; Marshall et al. 2022). The modelling of the interaction between ASH-Ski2 and RNA polymerase suggests that it involves residues from the ratchet domain, rather than the ASH domain. ASH-Ski2 has also been proposed to be part of an RNA-degrading complex involving the 5′–3′ exoribonuclease aRNase J (Phung et al. 2020). The ASH domain has been shown to be critical for the interaction with aRNase J, a key player in RNA metabolism in Euryarchaea (Phung et al. 2020). Interestingly, aRNase J and ASH-Ski2 are co-distributed in Archaea, except within the Asgard group where no homologues of aRNase J are conserved. This suggests that ASH-Ski2 and aRNase J have co-evolved and potentially have an ancient origin. Moreover, a close relationship between the ASH-Ski2/aRNase J complex, the RNA exosome and the ribosome was proposed in Thermococcales cells (Phung et al. 2020). Co-purification assays identified interaction between ASH-Ski2 and aRNase J, and between aRNase J and the exosomal subunit Csl4. Moreover, whole-cell extract fractionation assays highlighted a co-sedimentation of the ASH-Ski2/aRNase J complex and the RNA exosome in heavy fractions, together with the ribosome (Phung et al. 2020).

The archaeal Brr2 group, which is restricted to Halobacteria and Methanocellales, is thought to have been acquired through at least three independent horizontal gene transfer events (Batista et al. 2024). To our knowledge, no study reports cellular functions of any halobacteria nor methanocellales Brr2 Ski2-like members. In eukaryotes, a role of Brr2 RNA helicase in the dynamic regulation of the core splicing machinery has been proposed (Lauber et al. 1996; Zhang et al. 2015; Vester et al. 2020). However, no pre-mRNA splicing like events involving a spliceosome-like machinery have been described in Archaea. Therefore, it is difficult to predict if the archaeal Brr2-Ski2 members also act as RNA helicases in Halobacteria or Methanocellales.

6.3.3 Overview of Eukaryal Ski2-Like Protein Accessory Domains

Structural alignments of eukaryal and archaeal Ski2-like members highlight the emergence of accessory domains in these two lineages. These superpositions nicely highlight the conserved Ski2 specific core helicase structure formed by RecA1/RecA2 domains, and the Ratchet and WH motifs (Fig. 6.4b). The unique roles of Ski2-like members are derived from accessory domains that decorate the specific Ski2 helicase core (Johnson and Jackson 2013). A notable difference between eukaryal and archaeal Ski2 like-helicases is the presence of long unstructured N-terminal extensions for the eukaryal representatives (Fig. 6.4a). The extensions of Ski2 and Brr2 can interact with protein partners such as the Ski complex and the U4/U6·U5 tri-snRNPs, respectively (Kögel et al. 2024). They can also exert a regulatory effect, such as autoinhibition in the case of Brr2 (Absmeier et al. 2015).

Eukaryal HelQ as well as the archaeal Hel308 possess a specific $(HhH)_2$ domain flanking the C-terminus of the ratchet domain (Fig. 6.4a). The $(HhH)_2$ domain is a five-helical domain consisting of two consecutive Helix hairpin Helix (HhH) motifs linked by a connector helix (Shao and Grishin 2000) (Fig. 6.4b). In the crystal structure of archaeal Hel308 bound to unwound DNA, the HhH motifs were described as nucleic acid binders with no sequence specificity, interacting with the single strand nucleic acid (Büttner et al. 2007). It has been proposed that the HhH motif could act as a molecular brake controlling the helicase activity (Richard et al. 2008). Both eukaryal and archaeal Brr2 and eukaryal Slh1 possess a similar $(HhH)_2$ motif, but in these cases, it is followed by a fibronectin type III domain (FN3) (Fig. 6.4a). The FN3 domain is a β-sandwich fold with one four- stranded β-sheet and one three-stranded β-sheet (Fig. 6.4b). This domain has been shown to be involved in protein-protein interaction (Bloom and Calabro 2009). While eukaryotic Brr2 and Slh1 have a double helicase cassette architecture, the archaeal version of Brr2 has a single cassette. (Fig. 6.4a). The C-terminal cassette of Brr2 is catalytically inactive and has been proposed to have a regulatory effect on the active N-terminal cassette, whereas both cassettes seem to be active for Slh1 (Vester et al. 2020; Best et al. 2023). Both eukaryal Ski2 and Mtr4 possess a specific domain, named ARCH, which is inserted in the WH domain (Fig. 6.4a). This domain is formed of two anti-parallel coiled coils, extending over the Ski2 helicase core, forming an arch-like stalk, and terminating in a globular region called the fist (Fig. 6.4b). This domain has been shown to interact with protein partners and contribute to inhibition of the ATPase activity (Absmeier et al. 2020).

6.3.4 ASH-Ski2 in Euryarchaeota and Asgard Clades

An in-depth phylogenomic analysis identified archaeal ASH-Ski2 members in the Euryarchaea and Asgard clades, but not in the TACK clade (Batista et al. 2024). Aligning representative sequences from both clades highlights the differences between the Euryarchaea and Asgard members (Fig. 6.5a). The main differences are short insertions (generally less than 10 residues), which occur either at the N- or C-terminal ends of sequences, or in loops between secondary structure elements (Fig. 6.5a). Two specific features should be noted. Firstly, ASH-Ski2 proteins from the *Methanobrevibacter* genus have an insertion in their ratchet domain, ranging from 25 residues in *M. smithii* to 300 residues in *M. ruminantium*. Secondly, some variability is observed in the Stenosarchea ASH-Ski2 sequences. The Haloferales, Halobacteriales and Methanomicrobiales have a truncated ratchet domain, whereas the Methanocellales and Methanosarcinales have a complete one; ASH-Ski2 from *H. volcanii* lacks most of the ratchet domain, while ASH-Ski2 from *Methanocella paludicola* has a complete one (Fig. 6.5a). Remarkably, the Halobacteriales, Haloferales, Methanomicrobiales and Methanococci also lack genes encoding an RNA exosome complex (Clouet-d'Orval et al. 2018).

We also carried out structure predictions using the same representatives as above (Fig. 6.5b). In agreement with the sequence alignments, a highly conserved fold can be observed when comparing Asgard and Euryarchaeota structures, with only minimal differences. The Methanobrevibacter genus has a specific insertion in the ratchet domain. In addition, only the first two helices of the ratchet domain are present in *H. volcanii*. (Fig. 6.5b). Archaeal ASH-Ski2 present a N-terminal domain that we propose to call the ASH domain, for archaeal specific helicase domain (Fig. 6.5a). AlphaFold 3 predicts an ASH domain made up of alpha helices and beta strands, located on the same side of the helicase ring as the eukaryotic ARCH domain, but is inserted in the RecA2 domain (Fig. 6.5b).

The helicase core from Ski2-like helicase family members (eukaryal HelQ, Ski2, Mtr4, Brr2 and Slh1; archaeal Hel308, ASH-Ski2 and Brr2) based on experimental structures or AlphaFold3 models, were superimposed to highlight ASH-Ski2 specificities (Fig. 6.6). The first spot of diversity is located in the helicase core RecA2 domain, at the periphery of the helicase ring, close to the ASH domain insertion site. Compared to archaeal and eukaryal Brr2: ASH-Ski2 presents an extra beta strand;

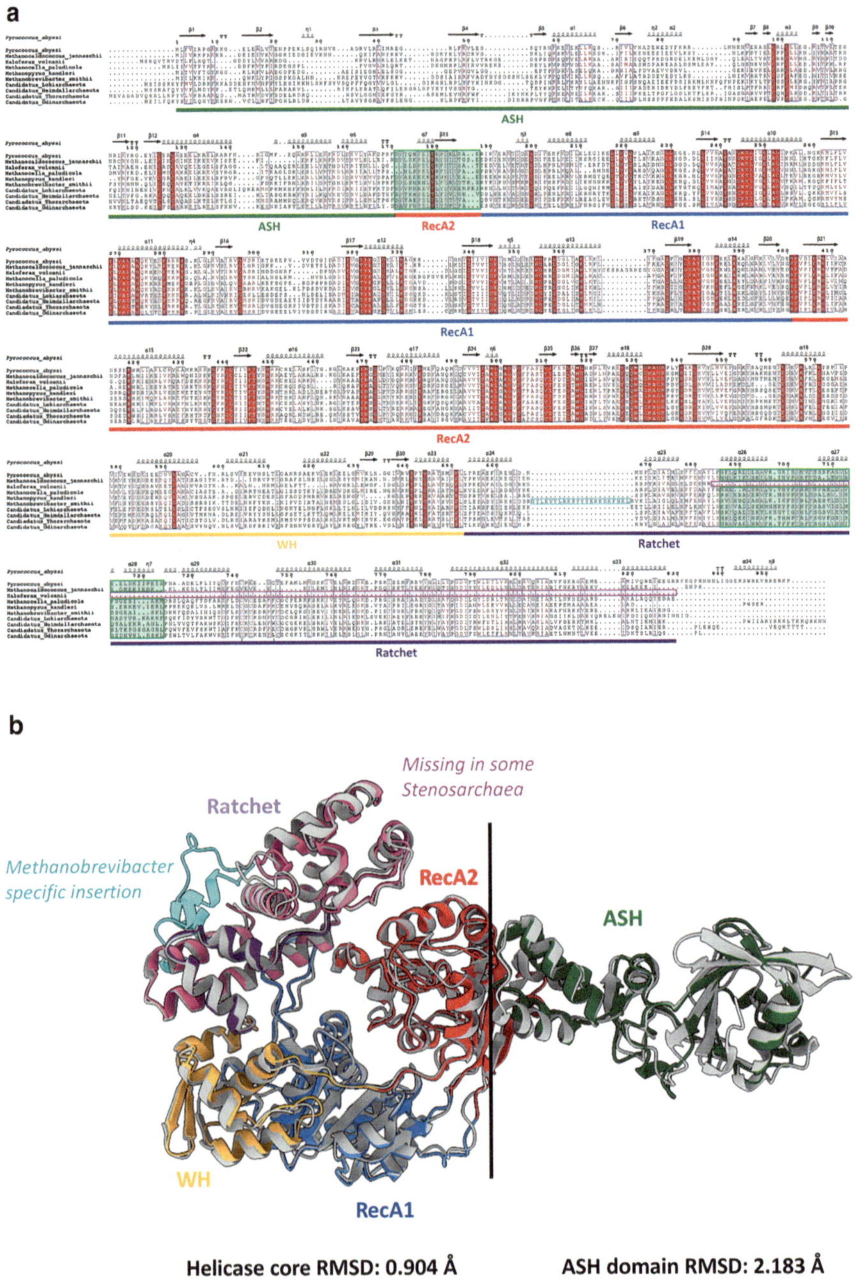

Fig. 6.5 Sequence conservation and specificities of ASH-Ski2 in Euryarchaeota and Asgard groups. (**a**) Alignment of sequences representative of different taxonomic clades in the Euryarchaeota and Asgard groups. Alignments have been done using Clustal Omega web server, manually corrected using Jalview software and rendered using ESPript 3 web server. For Asgards,

Hel308/HelQ presents two anti-parallel beta strands; and Ski2/Mtr4 presents a helix and a beta strand (Fig. 6.6). The second spot of diversity is located in the ratchet domain. Compared to both Brr2 and Hel308: ASH-Ski2 presents an insertion of 3 small helices oriented towards the ASH domain; Ski2/Mtr4 presents an insertion of a single long helix pointing towards the ARCH domain; and HelQ presents an insertion of one single small helix (Fig. 6.6). The third spot of diversity is located within the WH domain at the ARCH domain insertion site. All groups of the Ski2-like helicase exhibit insertions of different lengths, except ASH-Ski2, which has no insertion. Interestingly, Hel308 from Lokiarchaeota (Asgard clade) exhibits a very

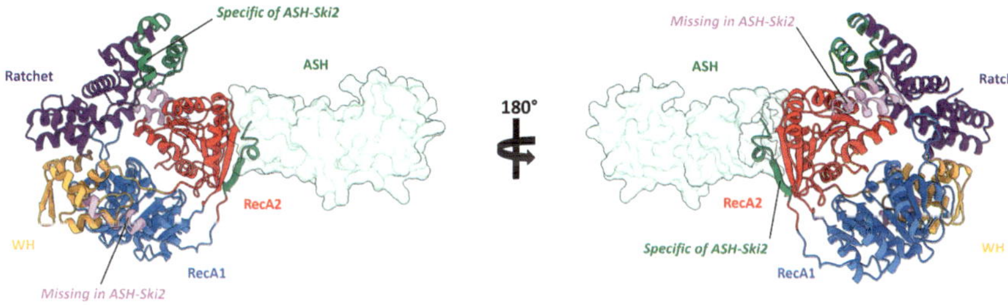

Fig. 6.6 Specificities of ASH-Ski2 helicase core. For clarity, both sides of the helicase core are shown, and the ASH domain is represented as a transparent surface. The different domains are indicated: ASH (dark green), RecA1 (blue), RecA2 (red), Winged Helix (WH, orange) and Ratchet (purple). ASH-Ski2 specific insertions in RecA2 and Ratchet domains are also highlighted (dark green) and correspond to those shown in Fig. 6.5a. Finally, structural elements present in other members of the Ski2-like helicase family but missing in ASH Ski2 are indicated (pink) and are taken from the *A. fulgidus* Hel308 structure (PDB 2P6R)

Fig. 6.5 (continued) one representative sequence of each of the groups were used (Lokiarchaeota, Thorarchaeota, Odinarchaeota and Heimdallarchaeota). For Euryarchaeota, either one representative sequence of the groups (*Pyrococcus abyssi* for Thermococci group and *Methanopyrus kandleri* for Methanopyri group) or two (*Methanocaldococcus jannaschii* and *Methanobrevibacter smithii* for Methanomada group, *Haloferax volcanii* and *Methanocella paludicola* for Stenosarchaea group) were used. Secondary structure elements from Alphafold 3 structure prediction of ASH-Ski2 from *P. abyssi* are displayed above the alignment. The different domains are highlighted below the alignment: ASH (dark green), RecA1 (blue), RecA2 (red), Winged Helix (WH, orange) and Ratchet (purple). Two specificities are also highlighted: a specific insertion of genus *Methanobrevibacter* (framed in cyan) and the absence of residues for the end of the Rachet domain on some Stenosarchaea, like *H. volcanii* (framed in pink). Two specific ASH-Ski2 specific insertions not present in other Ski2-like helicases are also indicated in the RecA2 and Ratchet domains (framed in green) and are also displayed in Fig. 6.6. (**b**) Superposition of Alphafold 3 predicted structures of ASH-Ski2 from a representative sequence of Euryarchaeota (*P. abyssi* structure with colored domains) and Asgard group (Odinarchaeota structure in grey). The RMSD values are given for the helicase core and the ASH domain separately since there is a hinge movement between them. RMSD values were calculated using Pymol software. The different domains are indicated: ASH (dark green), RecA1 (blue), RecA2 (red), Winged Helix (WH, orange) and Ratchet (purple and pink). Two Euryarcheaota specificities are also highlighted: the *Methanobrevibacter* genus specific insertion in the ratchet domain (cyan) and the missing part of the Ratchet domain in some Stenosarchaea (pink). Predicted N and C-terminal regions with poor confidence scores are not indicated

long unstructured loop which may reflect the events preceding the emergence of the ARCH domain in eukaryotes (Fig. 6.6). Finally, a distinguishing feature of ASH-Ski2 is the absence of a pair of helices from the RecA2 domain that is otherwise present in all other members of the Ski2-like helicase family (Fig. 6.6). These structural specificities are potential targets for further site-directed mutagenesis analyses to determine the contribution of the different domains to the activities and protein partners of Ski2-like helicases.

In Archaea, as previously mentioned, a study supports that aRNase J, a 5′–3′ exoribonuclease of the β CASP family, engages specifically with the ASH-Ski2 helicase and the RNA exosome to potentially exert control over RNA surveillance, in the vicinity of the ribosome (Phung et al. 2020). Similarly, it has been shown that the eukaryal Ski2 and Mtr4 helicases interact with the cytoplasmic and nuclear RNA exosome particles, respectively. These helicases are also essential for assembling RNA exosomes and ribosomes/preribosomes into supercomplexes (Keidel et al. 2025). The cryo-EM structure of the human ribosome/Ski2/RNA exosome supercomplex shows that Ski2 is located between the ribosome and the RNA exosome, acting as a physical bridge between the two, and guiding the mRNA from the ribosome to the RNA exosome (Fig. 6.7a) (Kögel et al. 2024). The RecA2 and ARCH domains of Ski2 interact with specific ribosome small subunit proteins (uS10, uS20, uS12 and uS3) and the 18S rRNA. Conversely, the RecA1, WH and Ratchet domains of Ski2 interact with RNA exosome cap proteins (Rrp4 and Rrp40) and Ski7.

To assess if ASH-Ski2 could also contact the RNA exosome and/or the ribosome, we superposed the structure of eukaryal Ski2 with the one of archaeal ASH-Ski2 structures (Fig. 6.7b). Interactions of ASH-SKi2 with the RNA exosome are conceivable. Indeed, the specific feature of the ASH-Ski2 helicase core specific to ASH-Ski2 are not found at the interface with the RNA exosome. Therefore, the RecA1, WH and Ratchet domains of ASH-Ski2 could also contact the RNA exosome. Interestingly, the few taxonomic groups that lack an RNA exosome encode ASH-Ski2 proteins that miss a complete ratchet domain. Finally, the Rrp4 cap protein that forms the largest interaction surface with eukaryal Ski2 is conserved in Archaea. As far as a potential interaction with the ribosome is concerned, a very different situation is presented. It is worth mentioning that the archaeal 70S ribosome appears to be a simplified version of the eukaryal 80S ribosome with a reduced protein number and shorter versions of rRNAs (Yutin et al. 2012; Schmitt et al. 2020; Londei and Ferreira-Cerca 2021). The ARCH domain, which interacts with the ribosome, is not present in ASH-SKi2, while the surface of the RecA2 domain involved in interactions with the ribosome in eukaryotes is not accessible due to the presence of the ASH domain. An interaction of ASH-Ski2 with the ribosome is still possible, but the modes of interaction will necessarily be different from those observed in eukaryotes. It could involve the ASH domain located on the same side of the helicase ring than the ARCH domain. Interestingly, affinity pulldown assays using ASH-Ski2 as the bait protein recovered several ribosomal proteins (Phung et al. 2020).

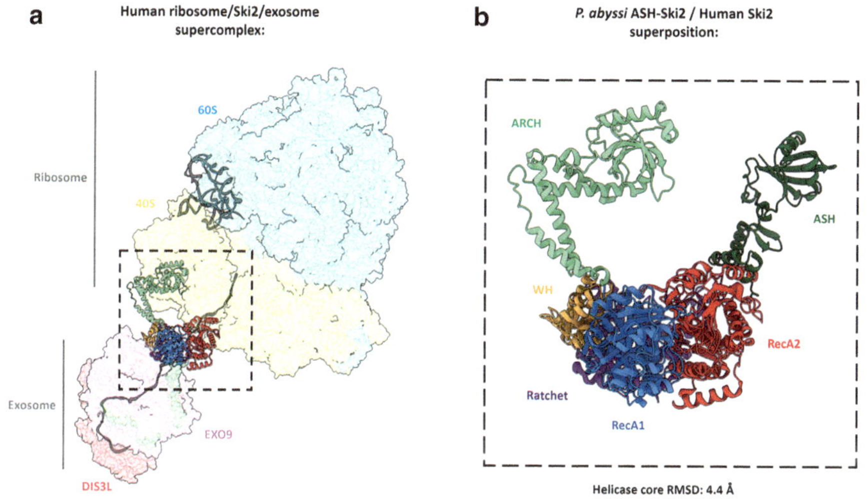

Fig. 6.7 Potential interactions of ASH-Ski2 with the exosome and ribosome. (**a**) Cryo-EM structure (code PDB 9G8M) of a human ribosome/Ski2/exosome supercomplex. For clarity, only the surfaces of the ribosome, the exosome and Ski7 are displayed, while Ski2 is displayed using a cartoon representation. The human cytoplasmic exosome consists of 9 subunits forming the catalytically inactive RNA binding cage (EXO9 in pink) and the processive 3′ to 5′ exoribonuclease DIS3L (light red). The messenger RNA (mRNA) and Ski7 are also indicated (in dark grey and light green respectively). The human 80S ribosome consists of a large ribosomal subunit (60S in light blue) and the small ribosomal subunit (40S in yellow). (**b**) Superposition of Alphafold 3 predicted structure of ASH-Ski2 from a representative sequence of Euryarchaeota (*P. abyssi*) superposed on human Ski2 from the supercomplex structure. For clarity, the exosome and ribosome are not shown in the superposition. The RMSD for the superposition of ASH-Ski2 and human Ski2 helicase core is 4.4 Å. The different domains of the helicases are indicated: ASH (dark green) or ARCH (green), RecA1 (blue), RecA2 (red), Winged Helix (WH, orange) and Ratchet (purple))

6.4 Conclusions & Perspectives

Coherent pictures of mRNA degradation and RNA quality control pathways that regulate gene expression at the post-transcriptional level are still missing in Archaea. However, archaeal families of ribonucleases and RNA helicases have been identified based on homologies with the bacterial and eukaryal domains (Clouet-d'Orval et al. 2018) (Fig. 6.8). The 3′–5′ exoribonucleolytic activity is performed by an RNA exosome that is homologous to the eukaryal ones, with the catalytic subunit belonging to the same RNase PH family as the bacterial PNPase (see also Chap. 5 by F. Briani in this volume) The archaeal RNA exosome is extremely well-conserved, and it is only missing from the Haloferales, Halobacteriales and Methanomicrobiales genomes. The 5′–3′ exoribonucleolytic activities are carried out by aRNase J and aCPSF2, two groups of β-CASP enzymes with bacterial (RNase J) and eukaryal (CPSF) homologies, respectively. While aCPSF2 can be found in all archaeal clades,

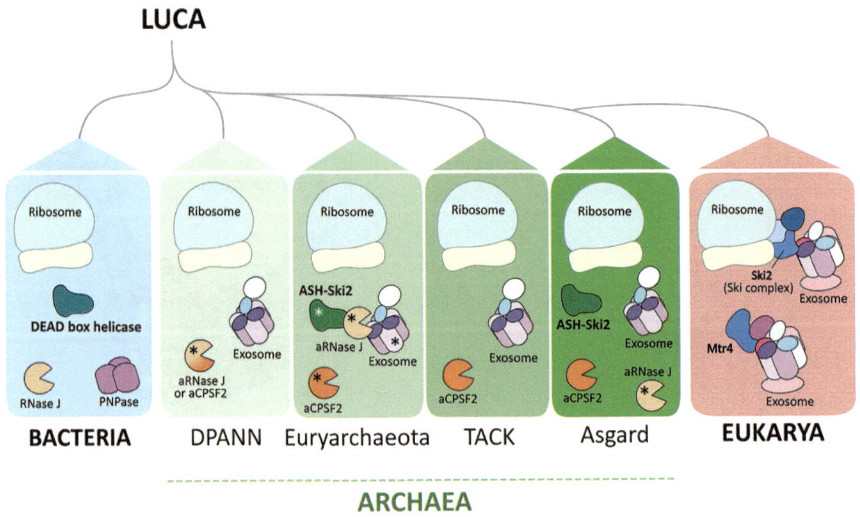

Fig. 6.8 Schematic view of the protein partners of Ski2 family helicases in different phyla. Direct interactions are represented by contacts between proteins. Proteins marked with an asterisk are only present in certain clades. In DPANN, aRNase J and aCPSF2 are mutually exclusive. In Asgard, a single aRNase J has been identified in the Candidatus Thorarchaeota archaeon (GenBank: MFW9880758.1). In Euryarchaeota, aCPSF2 is absent from Methanomicrobiales, Thermococci, Methanobacteria and Methanopyri; aRNase J is absent from Diaforarchaea and Archaeoglobi; ASH-Ski2 is absent from Thermoplasmatota and Archaeoglobi; the RNA Exosome is absent from Halophiles, Mathanococci and Methanomicrobiales

alone or together with aRNase J, aRNase J is mainly found in Euryarcheaota. Its distribution is the same than the one of ASH-Ski2, a 3′–5′ RNA helicase of the Ski2-like family (Batista et al. 2024).

Asgard archaeal genomes contain representatives of the RNA exosome, aCPSF2 and ASH-Ski2, while aRNase J is mostly missing. Nevertheless, it may be sporadically present, as at least one representative has been identified in the genome of the Candidatus Thorarchaeota archaeon (a meta-assembled genome, or MAG). More sequenced and annotated Asgard genomes are needed to confirm its presence and to determine its prevalence. Finally, the presence of the ASH-Ski2 RNA helicase in the Asgard clade—the Archaea clade most closely related to eukarya—suggests the hypothesis that a common ancestor of the Ski2-like RNA helicase existed prior the emergence of Eukarya. This early Ski2-like RNA helicase would have played a pivotal role in shaping the interaction between RNA-degrading enzymes and ribosomes within ancestral cells that predate the present day. Additionally, the highly conserved structure of the Ski2 core helicase suggests that a primitive RNA helicase of this type may have existed in a common ancestral organism, such as the Last Universal Common Ancestor (LUCA).

Most of the research on aRNase J and ASH-Ski2 has been conducted using Thermococcales species. Connections have been uncovered between aRNase J and ASH-Ski2, and between aRNase J and the RNA exosome (through the Csl4 cap-subunit) (Phung et al. 2020). Interestingly, the genes encoding ASH-Ski2 and Csl4 are often found adjacent to each other in these genomes. Therefore, the 5′ and 3′ RNA-degrading complexes may engage in an interplay in their handling of RNAs. Many questions remain open about their biological roles and their structures. Even more so given that these processes seem to occur in the vicinity of the ribosomes.

Acknowledgements We are indebted to P. Langendijk-Genevaux for her expertise in taxonomic identification of archaeal helicase members. This work is supported by the Centre National de la Recherche Scientifique (CNRS) and the University of Toulouse with additional funding from the Agence Nationale de la Recherche (ANR-22-CE12-0009-01 to BCO).

Bibliography

Absmeier E, Santos KF, Wahl MC (2016) Functions and regulation of the Brr2 RNA helicase during splicing. Cell Cycle 15:3362–3377. https://doi.org/10.1080/15384101.2016.1249549

Absmeier E, Santos KF, Wahl MC (2020) Molecular mechanism underlying inhibition of intrinsic ATPase activity in a Ski2-like RNA helicase. Structure 28:236–243.e3. https://doi.org/10.1016/j.str.2019.11.014

Absmeier E, Wollenhaupt J, Mozaffari-Jovin S et al (2015) The large N-terminal region of the Brr2 RNA helicase guides productive spliceosome activation. Genes Dev 29:2576–2587. https://doi.org/10.1101/gad.271528.115

Adam PS, Borrel G, Brochier-Armanet C, Gribaldo S (2017) The growing tree of archaea: new perspectives on their diversity, evolution and ecology. ISME J 11:2407–2425. https://doi.org/10.1038/ismej.2017.122

Anand R, Buechelmaier E, Belan O et al (2022) HELQ is a dual-function DSB repair enzyme modulated by RPA and RAD51. Nature 601:268–273. https://doi.org/10.1038/s41586-021-04261-0

Anantharaman V, Koonin EV, Aravind L (2002) Comparative genomics and evolution of proteins involved in RNA metabolism. Nucleic Acids Res 30:1427–1464. https://doi.org/10.1093/nar/30.7.1427

Andersson AF, Lundgren M, Eriksson S et al (2006) Global analysis of mRNA stability in the archaeon Sulfolobus. Genome Biol 7:R99. https://doi.org/10.1186/gb-2006-7-10-r99

Arkhipova V, Stolboushkina E, Kravchenko O et al (2015) Binding of the 5′-triphosphate end of mRNA to the γ-subunit of translation initiation factor 2 of the Crenarchaeon Sulfolobus solfataricus. J Mol Biol 427:3086–3095. https://doi.org/10.1016/j.jmb.2015.07.020

Arribas-Layton M, Wu D, Lykke-Andersen J, Song H (2013) Structural and functional control of the eukaryotic mRNA decapping machinery. Biochim Biophys Acta 1829:580–589. https://doi.org/10.1016/j.bbagrm.2012.12.006

Audin MJC, Wurm JP, Cvetkovic MA, Sprangers R (2016) The oligomeric architecture of the archaeal exosome is important for processive and efficient RNA degradation. Nucleic Acids Res 44:2962–2973. https://doi.org/10.1093/nar/gkw062

Babski J, Haas KA, Näther-Schindler D et al (2016) Genome-wide identification of transcriptional start sites in the haloarchaeon Haloferax volcanii based on differential RNA-Seq (dRNA-Seq). BMC Genomics 17:629. https://doi.org/10.1186/s12864-016-2920-y

Baker BJ, De Anda V, Seitz KW et al (2020) Diversity, ecology and evolution of Archaea. Nat Microbiol 5:887–900. https://doi.org/10.1038/s41564-020-0715-z

Bathke J, Gauernack AS, Rupp O et al (2020) iCLIP analysis of RNA substrates of the archaeal exosome. BMC Genomics 21:797. https://doi.org/10.1186/s12864-020-07200-x

Batista M, Langendijk-Genevaux P, Kwapisz M et al (2024) Evolutionary and functional insights into the Ski2-like helicase family in archaea: a comparison of Thermococcales ASH-Ski2 and Hel308 activities. NAR Genom Bioinform 6:lqae026. https://doi.org/10.1093/nargab/lqae026

Beaulieu YB, Kleinman CL, Landry-Voyer A-M et al (2012) Polyadenylation-dependent control of long noncoding RNA expression by the poly(A)-binding protein nuclear 1. PLoS Genet 8:e1003078. https://doi.org/10.1371/journal.pgen.1003078

Best K, Ikeuchi K, Kater L et al (2023) Structural basis for clearing of ribosome collisions by the RQT complex. Nat Commun 14:921. https://doi.org/10.1038/s41467-023-36230-8

Betts HC, Puttick MN, Clark JW et al (2018) Integrated genomic and fossil evidence illuminates life's early evolution and eukaryote origin. Nat Ecol Evol 2:1556–1562. https://doi.org/10.1038/s41559-018-0644-x

Bini E, Dikshit V, Dirksen K et al (2002) Stability of mRNA in the hyperthermophilic archaeon Sulfolobus solfataricus. RNA 8:1129–1136. https://doi.org/10.1017/s1355838202021052

Blombach F, Matelska D, Fouqueau T et al (2019) Key concepts and challenges in archaeal transcription. J Mol Biol 431:4184–4201. https://doi.org/10.1016/j.jmb.2019.06.020

Bloom L, Calabro V (2009) FN3: a new protein scaffold reaches the clinic. Drug Discov Today 14:949–955. https://doi.org/10.1016/j.drudis.2009.06.007

Bolt EL, Guy CP (2003) Homologous recombination in archaea: new holliday junction helicases. Biochem Soc Trans 31:703–705. https://doi.org/10.1042/bst0310703

Borrel G, Brugère J-F, Gribaldo S et al (2020) The host-associated archaeome. Nat Rev Microbiol 18:622–636. https://doi.org/10.1038/s41579-020-0407-y

Brochier-Armanet C, Forterre P, Gribaldo S (2011) Phylogeny and evolution of the archaea: one hundred genomes later. Curr Opin Microbiol 14:274–281. https://doi.org/10.1016/j.mib.2011.04.015

Büttner K, Nehring S, Hopfner K-P (2007) Structural basis for DNA duplex separation by a superfamily-2 helicase. Nat Struct Mol Biol 14:647–652. https://doi.org/10.1038/nsmb1246

Cao Y, Li J, Jiang N, Dong X (2014) Mechanism for stabilizing mRNAs involved in methanol-dependent methanogenesis of cold-adaptive Methanosarcina mazei zm-15. Appl Environ Microbiol 80:1291–1298. https://doi.org/10.1128/AEM.03495-13

Celesnik H, Deana A, Belasco JG (2007) Initiation of RNA decay in Escherichia coli by 5′ pyrophosphate removal. Mol Cell 27:79–90. https://doi.org/10.1016/j.molcel.2007.05.038

Chamieh H, Ibrahim H, Kozah J (2016) Genome-wide identification of SF1 and SF2 helicases from archaea. Gene 576:214–228. https://doi.org/10.1016/j.gene.2015.10.007

Chlebowski A, Lubas M, Jensen TH, Dziembowski A (2013) RNA decay machines: the exosome. Biochim Biophys Acta 1829:552–560. https://doi.org/10.1016/j.bbagrm.2013.01.006

Clouet-d'Orval B, Batista M, Bouvier M et al (2018) Insights into RNA-processing pathways and associated RNA-degrading enzymes in archaea. FEMS Microbiol Rev 42:579–613. https://doi.org/10.1093/femsre/fuy016

Clouet-d'Orval B, Phung DK, Langendijk-Genevaux PS, Quentin Y (2015) Universal RNA-degrading enzymes in archaea: prevalence, activities and functions of β-CASP ribonucleases. Biochimie 118:278–285. https://doi.org/10.1016/j.biochi.2015.05.021

Clouet-d'Orval B, Rinaldi D, Quentin Y, Carpousis AJ (2010) Euryarchaeal beta-CASP proteins with homology to bacterial RNase J Have 5′- to 3′-exoribonuclease activity. J Biol Chem 285:17574–17583. https://doi.org/10.1074/jbc.M109.095117

Dillingham MS (2011) Superfamily I helicases as modular components of DNA-processing machines. Biochem Soc Trans 39:413–423. https://doi.org/10.1042/BST0390413

Dombrowski N, Lee J-H, Williams TA et al (2019) Genomic diversity, lifestyles and evolutionary origins of DPANN archaea. FEMS Microbiol LettFEMS Microbiol Lett. https://doi.org/10.1093/femsle/fnz008

Dominski Z, Carpousis AJ, Clouet-d'Orval B (2013) Emergence of the β-CASP ribonucleases: highly conserved and ubiquitous metallo-enzymes involved in messenger RNA maturation and degradation. Biochim Biophys Acta 1829:532–551. https://doi.org/10.1016/j.bbagrm.2013.01.010

Durand S, Tomasini A, Braun F et al (2015) sRNA and mRNA turnover in gram-positive bacteria. FEMS Microbiol Rev 39:316–330. https://doi.org/10.1093/femsre/fuv007

Dziembowski A, Lorentzen E, Conti E, Séraphin B (2007) A single subunit, Dis3, is essentially responsible for yeast exosome core activity. Nat Struct Mol Biol 14:15–22. https://doi.org/10.1038/nsmb1184

Eme L, Spang A, Lombard J et al (2017) Archaea and the origin of eukaryotes. Nat Rev Microbiol 15:711–723. https://doi.org/10.1038/nrmicro.2017.133

Eme L, Tamarit D, Caceres EF et al (2023) Inference and reconstruction of the heimdallarchaeial ancestry of eukaryotes. Nature 618:992–999. https://doi.org/10.1038/s41586-023-06186-2

Evguenieva-Hackenberg E, Gauernack S, Klug G (2017) The archaeal exosome: degradation and tailing at the 3′-end of RNA. In: Clouet-d'Orval B (ed) RNA metabolism and gene expression in archaea. Springer International Publishing, Cham, pp 115–128

Evguenieva-Hackenberg E, Hou L, Glaeser S, Klug G (2014) Structure and function of the archaeal exosome. Wiley Interdiscip Rev RNA 5:623–635. https://doi.org/10.1002/wrna.1234

Evguenieva-Hackenberg E, Roppelt V, Finsterseifer P, Klug G (2008) Rrp4 and Csl4 are needed for efficient degradation but not for polyadenylation of synthetic and natural RNA by the archaeal exosome. Biochemistry 47:13158–13168. https://doi.org/10.1021/bi8012214

Evguenieva-Hackenberg E, Walter P, Hochleitner E et al (2003) An exosome-like complex in Sulfolobus solfataricus. EMBO Rep 4:889–893. https://doi.org/10.1038/sj.embor.embor929

Fairman ME, Maroney PA, Wang W et al (2004) Protein displacement by DExH/D "RNA helicases" without duplex unwinding. Science 304:730–734. https://doi.org/10.1126/science.1095596

Fairman-Williams ME, Guenther U-P, Jankowsky E (2010) SF1 and SF2 helicases: family matters. Curr Opin Struct Biol 20:313–324. https://doi.org/10.1016/j.sbi.2010.03.011

Farhoud MH, Wessels HJCT, Steenbakkers PJM et al (2005) Protein complexes in the archaeon Methanothermobacter thermautotrophicus analyzed by blue native/SDS-PAGE and mass spectrometry. Mol Cell Proteomics 4:1653–1663. https://doi.org/10.1074/mcp.M500171-MCP200

Fujikane R, Ishino S, Ishino Y, Forterre P (2010) Genetic analysis of DNA repair in the hyperthermophilic archaeon, Thermococcus kodakaraensis. Genes Genet Syst 85:243–257. https://doi.org/10.1266/ggs.85.243

Fujikane R, Shinagawa H, Ishino Y (2006) The archaeal Hjm helicase has recQ-like functions, and may be involved in repair of stalled replication fork. Genes Cells 11:99–110. https://doi.org/10.1111/j.1365-2443.2006.00925.x

Gauernack AS, Lassek C, Hou L et al (2017) Nop5 interacts with the archaeal RNA exosome. FEBS Lett 591:4039–4048. https://doi.org/10.1002/1873-3468.12915

Gomes-Filho JV, Breuer R, Morales-Filloy HG et al (2022) Identification of NAD-RNAs and ADPR-RNA decapping in the archaeal model organisms *Sulfolobus acidocaldarius* and *Haloferax volcanii*. BioRxiv. https://doi.org/10.1101/2022.11.02.514978

Gomes-Filho JV, Breuer R, Morales-Filloy HG et al (2023) Identification of NAD-RNA species and ADPR-RNA decapping in archaea. Nat Commun 14:7597. https://doi.org/10.1038/s41467-023-43377-x

Gorbalenya AE, Koonin EV (1988) One more conserved sequence motif in helicases. Nucleic Acids Res 16:7734. https://doi.org/10.1093/nar/16.15.7734

Gorbalenya AE, Koonin EV (1993) Helicases: amino acid sequence comparisons and structure-function relationships. Curr Opin Struct Biol 3:419–429. https://doi.org/10.1016/S0959-440X(05)80116-2

Guy CP, Bolt EL (2005) Archaeal Hel308 helicase targets replication forks in vivo and in vitro and unwinds lagging strands. Nucleic Acids Res 33:3678–3690. https://doi.org/10.1093/nar/gki685

Guy L, Ettema TJG (2011) The archaeal "TACK" superphylum and the origin of eukaryotes. Trends Microbiol 19:580–587. https://doi.org/10.1016/j.tim.2011.09.002

Hartung S, Hopfner K-P (2009) Lessons from structural and biochemical studies on the archaeal exosome. Biochem Soc Trans 37:83–87. https://doi.org/10.1042/BST0370083

Hartung S, Niederberger T, Hartung M et al (2010) Quantitative analysis of processive RNA degradation by the archaeal RNA exosome. Nucleic Acids Res 38:5166–5176. https://doi.org/10.1093/nar/gkq238

Hasenöhrl D, Konrat R, Bläsi U (2011) Identification of an RNase J ortholog in Sulfolobus solfataricus: implications for 5′-to-3′ directional decay and 5′-end protection of mRNA in Crenarchaeota. RNA 17:99–107. https://doi.org/10.1261/rna.2418211

Hasenöhrl D, Lombo T, Kaberdin V et al (2008) Translation initiation factor a/eIF2(-gamma) counteracts 5′ to 3′ mRNA decay in the archaeon Sulfolobus solfataricus. Proc Natl Acad Sci USA 105:2146–2150. https://doi.org/10.1073/pnas.0708894105

Heck AM, Wilusz J (2018) The interplay between the RNA decay and translation machinery in eukaryotes. Cold Spring Harb Perspect BiolCold Spring Harb Perspect Biol. https://doi.org/10.1101/cshperspect.a032839

Hennigan AN, Reeve JN (1994) mRNAs in the methanogenic archaeon Methanococcus vannielii: numbers, half-lives and processing. Mol Microbiol 11:655–670

Hou L, Klug G, Evguenieva-Hackenberg E (2013) The archaeal DnaG protein needs Csl4 for binding to the exosome and enhances its interaction with adenine-rich RNAs. RNA Biol 10:415–424. https://doi.org/10.4161/rna.23450

Inada T (2017) The ribosome as a platform for mRNA and nascent polypeptide quality control. Trends Biochem Sci 42:5–15. https://doi.org/10.1016/j.tibs.2016.09.005

Inada T (2013) Quality control systems for aberrant mRNAs induced by aberrant translation elongation and termination. Biochim Biophys Acta 1829:634–642. https://doi.org/10.1016/j.bbagrm.2013.02.004

Jäger A, Samorski R, Pfeifer F, Klug G (2002) Individual gvp transcript segments in Haloferax mediterranei exhibit varying half-lives, which are differentially affected by salt concentration and growth phase. Nucleic Acids Res 30:5436–5443

Jankowsky E, Bowers H (2006) Remodeling of ribonucleoprotein complexes with DExH/D RNA helicases. Nucleic Acids Res 34:4181–4188. https://doi.org/10.1093/nar/gkl410

Jankowsky E, Fairman ME (2007) RNA helicases--one fold for many functions. Curr Opin Struct Biol 17:316–324. https://doi.org/10.1016/j.sbi.2007.05.007

Johnson SJ, Jackson RN (2013) Ski2-like RNA helicase structures: common themes and complex assemblies. RNA Biol 10:33–43. https://doi.org/10.4161/rna.22101

Karousis ED, Mühlemann O (2019) Nonsense-mediated mRNA decay begins where translation ends. Cold Spring Harb Perspect BiolCold Spring Harb Perspect Biol 11. https://doi.org/10.1101/cshperspect.a032862

Keidel A, Long CL, Iwasa J, Conti E (2025) RNA-degrading exosome complexes: molecular mechanisms and structural insights. Annu Rev Cell Dev BiolAnnu Rev Cell Dev Biol. https://doi.org/10.1146/annurev-cellbio-111822-115115

Kögel A, Keidel A, Bonneau F et al (2022) The human SKI complex regulates channeling of ribosome-bound RNA to the exosome via an intrinsic gatekeeping mechanism. Mol Cell 82:756–769.e8. https://doi.org/10.1016/j.molcel.2022.01.009

Kögel A, Keidel A, Loukeri M-J et al (2024) Structural basis of mRNA decay by the human exosome-ribosome supercomplex. Nature 635:237–242. https://doi.org/10.1038/s41586-024-08015-6

Koonin EV (1993) A common set of conserved motifs in a vast variety of putative nucleic acid-dependent ATPases including MCM proteins involved in the initiation of eukaryotic DNA replication. Nucleic Acids Res 21:2541–2547. https://doi.org/10.1093/nar/21.11.2541

Koonin EV, Wolf YI, Aravind L (2001) Prediction of the archaeal exosome and its connections with the proteasome and the translation and transcription machineries by a comparative-genomic approach. Genome Res 11:240–252. https://doi.org/10.1101/gr.162001
Lange H, Gagliardi D (2022) Catalytic activities, molecular connections, and biological functions of plant RNA exosome complexes. Plant Cell 34:967–988. https://doi.org/10.1093/plcell/koab310
Lauber J, Fabrizio P, Teigelkamp S et al (1996) The HeLa 200 kDa U5 snRNP-specific protein and its homologue in Saccharomyces cerevisiae are members of the DEXH-box protein family of putative RNA helicases. EMBO J 15:4001–4015
Levy S, Portnoy V, Admon J, Schuster G (2011) Distinct activities of several RNase J proteins in methanogenic archaea. RNA Biol 8:1073–1083. https://doi.org/10.4161/rna.8.6.16604
Li J, Liang Y, Dong X (2025) Post-transcriptional regulation in archaea. Trends MicrobiolTrends Microbiol. https://doi.org/10.1016/j.tim.2025.02.004
Li J, Yue L, Li Z et al (2021) aCPSF1 cooperates with terminator U-tract to dictate archaeal transcription termination efficacy. Elife. https://doi.org/10.7554/eLife.70464
Li Z, Lu S, Hou G et al (2008) Hjm/Hel308A DNA helicase from Sulfolobus tokodaii promotes replication fork regression and interacts with Hjc endonuclease in vitro. J Bacteriol 190:3006–3017. https://doi.org/10.1128/JB.01662-07
Linder P, Jankowsky E (2011) From unwinding to clamping – the DEAD box RNA helicase family. Nat Rev Mol Cell Biol 12:505–516. https://doi.org/10.1038/nrm3154
Liu Q, Greimann JC, Lima CD (2006) Reconstitution, activities, and structure of the eukaryotic RNA exosome. Cell 127:1223–1237. https://doi.org/10.1016/j.cell.2006.10.037
Liu Y, Makarova KS, Huang W-C et al (2021) Expanded diversity of Asgard archaea and their relationships with eukaryotes. Nature 593:553–557. https://doi.org/10.1038/s41586-021-03494-3
Lohman TM, Tomko EJ, Wu CG (2008) Non-hexameric DNA helicases and translocases: mechanisms and regulation. Nat Rev Mol Cell Biol 9:391–401. https://doi.org/10.1038/nrm2394
Londei P (2005) Evolution of translational initiation: new insights from the archaea. FEMS Microbiol Rev 29:185–200. https://doi.org/10.1016/j.femsre.2004.10.002
Londei P, Ferreira-Cerca S (2021) Ribosome biogenesis in archaea. Front Microbiol 12:686977. https://doi.org/10.3389/fmicb.2021.686977
Lorentzen E, Dziembowski A, Lindner D et al (2007) RNA channelling by the archaeal exosome. EMBO Rep 8:470–476. https://doi.org/10.1038/sj.embor.7400945
Lorentzen E, Walter P, Fribourg S et al (2005) The archaeal exosome core is a hexameric ring structure with three catalytic subunits. Nat Struct Mol Biol 12:575–581. https://doi.org/10.1038/nsmb952
Maier L-K, Marchfelder A (2019) It's all about the T: transcription termination in archaea. Biochem Soc Trans 47:461–468. https://doi.org/10.1042/BST20180557
Marshall CJ, Qayyum MZ, Walker JE et al (2022) The structure and activities of the archaeal transcription termination factor Eta detail vulnerabilities of the transcription elongation complex. Proc Natl Acad Sci USA 119:e2207581119. https://doi.org/10.1073/pnas.2207581119
Märtens B, Amman F, Manoharadas S et al (2013) Alterations of the transcriptome of Sulfolobus acidocaldarius by exoribonuclease aCPSF2. PLoS One 8:e76569. https://doi.org/10.1371/journal.pone.0076569
Märtens B, Hou L, Amman F et al (2017) The SmAP1/2 proteins of the crenarchaeon Sulfolobus solfataricus interact with the exosome and stimulate A-rich tailing of transcripts. Nucleic Acids Res 45:7938–7949. https://doi.org/10.1093/nar/gkx437
Matsuo Y, Ikeuchi K, Saeki Y et al (2017) Ubiquitination of stalled ribosome triggers ribosome-associated quality control. Nat Commun 8:159. https://doi.org/10.1038/s41467-017-00188-1
Matsuo Y, Tesina P, Nakajima S et al (2020) RQT complex dissociates ribosomes collided on endogenous RQC substrate SDD1. Nat Struct Mol Biol 27:323–332. https://doi.org/10.1038/s41594-020-0393-9

Meola N, Domanski M, Karadoulama E et al (2016) Identification of a nuclear exosome decay pathway for processed transcripts. Mol Cell 64:520–533. https://doi.org/10.1016/j.molcel.2016.09.025

Mir-Montazeri B, Ammelburg M, Forouzan D et al (2011) Crystal structure of a dimeric archaeal cleavage and polyadenylation specificity factor. J Struct Biol 173:191–195. https://doi.org/10.1016/j.jsb.2010.09.013

Morris C, Cluet D, Ricci EP (2021) Ribosome dynamics and mRNA turnover, a complex relationship under constant cellular scrutiny. Wiley Interdiscip Rev RNA 12:e1658. https://doi.org/10.1002/wrna.1658

Mozaffari-Jovin S, Wandersleben T, Santos KF et al (2013) Inhibition of RNA helicase Brr2 by the C-terminal tail of the spliceosomal protein Prp8. Science 341:80–84. https://doi.org/10.1126/science.1237515

Nagarajan VK, Jones CI, Newbury SF, Green PJ (2013) XRN 5′→3′ exoribonucleases: structure, mechanisms and functions. Biochim Biophys Acta 1829:590–603. https://doi.org/10.1016/j.bbagrm.2013.03.005

Nishida Y, Ishikawa H, Baba S et al (2010) Crystal structure of an archaeal cleavage and polyadenylation specificity factor subunit from Pyrococcus horikoshii. Proteins 78:2395–2398. https://doi.org/10.1002/prot.22748

Offre P, Spang A, Schleper C (2013) Archaea in biogeochemical cycles. Ann Rev Microbiol 67:437–457. https://doi.org/10.1146/annurev-micro-092412-155614

Ogami K, Richard P, Chen Y et al (2017) An Mtr4/ZFC3H1 complex facilitates turnover of unstable nuclear RNAs to prevent their cytoplasmic transport and global translational repression. Genes Dev 31:1257–1271. https://doi.org/10.1101/gad.302604.117

Olsen KJ, Johnson SJ (2021) Mtr4 RNA helicase structures and interactions. Biol Chem 402:605–616. https://doi.org/10.1515/hsz-2020-0329

Peterson JR, Thor S, Kohler L et al (2016) Genome-wide gene expression and RNA half-life measurements allow predictions of regulation and metabolic behavior in Methanosarcina acetivorans. BMC Genomics 17:924. https://doi.org/10.1186/s12864-016-3219-8

Phung DK, Bouvier M, Clouet-d'Orval B (2017) An overview of ribonuclease repertoire and RNA processing pathways in archaea. In: Clouet-d'Orval B (ed) RNA metabolism and gene expression in archaea. Springer International Publishing, Cham, pp 89–114

Phung DK, Etienne C, Batista M et al (2020) RNA processing machineries in Archaea: the 5′-3′ exoribonuclease aRNase J of the β-CASP family is engaged specifically with the helicase ASH-Ski2 and the 3′-5′ exoribonucleolytic RNA exosome machinery. Nucleic Acids Res 48:3832–3847. https://doi.org/10.1093/nar/gkaa052

Phung DK, Rinaldi D, Langendijk-Genevaux PS et al (2013) Archaeal β-CASP ribonucleases of the aCPSF1 family are orthologs of the eukaryal CPSF-73 factor. Nucleic Acids Res 41:1091–1103. https://doi.org/10.1093/nar/gks1237

Portnoy V, Evguenieva-Hackenberg E, Klein F et al (2005) RNA polyadenylation in archaea: not observed in Haloferax while the exosome polynucleotidylates RNA in Sulfolobus. EMBO Rep 6:1188–1193. https://doi.org/10.1038/sj.embor.7400571

Puno MR, Weick E-M, Das M, Lima CD (2019) Snapshot: the RNA exosome. Cell 179:282–282.e1. https://doi.org/10.1016/j.cell.2019.09.005

Pyle AM (2008) Translocation and unwinding mechanisms of RNA and DNA helicases. Annu Rev Biophys 37:317–336. https://doi.org/10.1146/annurev.biophys.37.032807.125908

Qi L, Yue L, Feng D et al (2017) Genome-wide mRNA processing in methanogenic archaea reveals post-transcriptional regulation of ribosomal protein synthesis. Nucleic Acids Res 45:7285–7298. https://doi.org/10.1093/nar/gkx454

Raghunathan PL, Guthrie C (1998) RNA unwinding in U4/U6 snRNPs requires ATP hydrolysis and the DEIH-box splicing factor Brr2. Curr Biol 8:847–855. https://doi.org/10.1016/s0960-9822(07)00345-4

Ramos CRR, Oliveira CLP, Torriani IL, Oliveira CC (2006) The Pyrococcus exosome complex: structural and functional characterization. J Biol Chem 281:6751–6759. https://doi.org/10.1074/jbc.M512495200
Raymann K, Brochier-Armanet C, Gribaldo S (2015) The two-domain tree of life is linked to a new root for the archaea. Proc Natl Acad Sci USA 112:6670–6675. https://doi.org/10.1073/pnas.1420858112
Raymann K, Forterre P, Brochier-Armanet C, Gribaldo S (2014) Global phylogenomic analysis disentangles the complex evolutionary history of DNA replication in archaea. Genome Biol Evol 6:192–212. https://doi.org/10.1093/gbe/evu004
Richard DJ, Bolderson E, Cubeddu L et al (2008) Single-stranded DNA-binding protein hSSB1 is critical for genomic stability. Nature 453:677–681. https://doi.org/10.1038/nature06883
Rinke C, Chuvochina M, Mussig AJ et al (2021) A standardized archaeal taxonomy for the genome taxonomy database. Nat Microbiol 6:946–959. https://doi.org/10.1038/s41564-021-00918-8
Roppelt V, Klug G, Evguenieva-Hackenberg E (2010) The evolutionarily conserved subunits Rrp4 and Csl4 confer different substrate specificities to the archaeal exosome. FEBS Lett 584:2931–2936. https://doi.org/10.1016/j.febslet.2010.05.014
Sanders TJ, Wenck BR, Selan JN et al (2020) FttA is a CPSF73 homologue that terminates transcription in archaea. Nat Microbiol 5:545–553. https://doi.org/10.1038/s41564-020-0667-3
Schmidt C, Kowalinski E, Shanmuganathan V et al (2016) The cryo-EM structure of a ribosome-Ski2-Ski3-Ski8 helicase complex. Science 354:1431–1433. https://doi.org/10.1126/science.aaf7520
Schmitt E, Coureux P-D, Kazan R et al (2020) Recent advances in archaeal translation initiation. Front Microbiol 11:584152. https://doi.org/10.3389/fmicb.2020.584152
Seitz KW, Lazar CS, Hinrichs K-U et al (2016) Genomic reconstruction of a novel, deeply branched sediment archaeal phylum with pathways for acetogenesis and sulfur reduction. ISME J 10:1696–1705. https://doi.org/10.1038/ismej.2015.233
Shao X, Grishin NV (2000) Common fold in helix-hairpin-helix proteins. Nucleic Acids Res 28:2643–2650
Silva AP, Chechik M, Byrne RT et al (2011) Structure and activity of a novel archaeal beta-CASP protein with N-terminal KH domains. Structure 19:622–632
Singleton MR, Dillingham MS, Wigley DB (2007) Structure and mechanism of helicases and nucleic acid translocases. Annu Rev Biochem 76:23–50. https://doi.org/10.1146/annurev.biochem.76.052305.115300
Spang A, Eme L, Saw JH et al (2018) Asgard archaea are the closest prokaryotic relatives of eukaryotes. PLoS Genet 14:e1007080. https://doi.org/10.1371/journal.pgen.1007080
Spang A, Saw JH, Jørgensen SL et al (2015) Complex archaea that bridge the gap between prokaryotes and eukaryotes. Nature 521:173–179. https://doi.org/10.1038/nature14447
Stoecklin G, Mühlemann O (2013) RNA decay mechanisms: specificity through diversity. Biochim Biophys Acta 1829:487–490. https://doi.org/10.1016/j.bbagrm.2013.04.002
Tahon G, Geesink P, Ettema TJG (2021) Expanding archaeal diversity and phylogeny: past, present, and future. Ann Rev Microbiol 75:359–381. https://doi.org/10.1146/annurev-micro-040921-050212
Tesina P, Heckel E, Cheng J et al (2019) Structure of the 80S ribosome-Xrn1 nuclease complex. Nat Struct Mol Biol 26:275–280. https://doi.org/10.1038/s41594-019-0202-5
Theuser M, Höbartner C, Wahl MC, Santos KF (2016) Substrate-assisted mechanism of RNP disruption by the spliceosomal Brr2 RNA helicase. Proc Natl Acad Sci USA 113:7798–7803. https://doi.org/10.1073/pnas.1524616113
Tuck AC, Rankova A, Arpat AB et al (2020) Mammalian RNA decay pathways are highly specialized and widely linked to translation. Mol Cell 77:1222–1236.e13. https://doi.org/10.1016/j.molcel.2020.01.007
Vayssières M, Jüttner M, Haas K et al (2024) RNase W, a conserved ribonuclease family with a novel active site. Nucleic Acids Res 52:13386–13401. https://doi.org/10.1093/nar/gkae907

Vester K, Santos KF, Kuropka B et al (2020) The inactive C-terminal cassette of the dual-cassette RNA helicase BRR2 both stimulates and inhibits the activity of the N-terminal helicase unit. J Biol Chem 295:2097–2112. https://doi.org/10.1074/jbc.RA119.010964
Wagner EGH, Romby P (2015) Small RNAs in bacteria and archaea: who they are, what they do, and how they do it. Adv Genet 90:133–208. https://doi.org/10.1016/bs.adgen.2015.05.001
Walker JE, Luyties O, Santangelo TJ (2017) Factor-dependent archaeal transcription termination. Proc Natl Acad Sci USA 114:E6767–E6773. https://doi.org/10.1073/pnas.1704028114
Wang Y, Liu CL, Storey JD et al (2002) Precision and functional specificity in mRNA decay. Proc Natl Acad Sci USA 99:5860–5865. https://doi.org/10.1073/pnas.092538799
Wasmuth EV, Zinder JC, Zattas D et al (2017) Structure and reconstitution of yeast Mpp6-nuclear exosome complexes reveals that Mpp6 stimulates RNA decay and recruits the Mtr4 helicase. Elife 6:e29062. https://doi.org/10.7554/eLife.29062
Weick E-M, Lima CD (2021) RNA helicases are hubs that orchestrate exosome-dependent 3′-5′ decay. Curr Opin Struct Biol 67:86–94. https://doi.org/10.1016/j.sbi.2020.09.010
Weir JR, Bonneau F, Hentschel J, Conti E (2010) Structural analysis reveals the characteristic features of Mtr4, a DExH helicase involved in nuclear RNA processing and surveillance. Proc Natl Acad Sci USA 107:12139–12144. https://doi.org/10.1073/pnas.1004953107
Widner WR, Wickner RB (1993) Evidence that the SKI antiviral system of Saccharomyces cerevisiae acts by blocking expression of viral mRNA. Mol Cell Biol 13:4331–4341. https://doi.org/10.1128/mcb.13.7.4331
Williams TA, Cox CJ, Foster PG et al (2020) Phylogenomics provides robust support for a two-domains tree of life. Nat Ecol Evol 4:138–147. https://doi.org/10.1038/s41559-019-1040-x
Williams TA, Foster PG, Cox CJ, Embley TM (2013) An archaeal origin of eukaryotes supports only two primary domains of life. Nature 504:231–236. https://doi.org/10.1038/nature12779
Witharana C, Roppelt V, Lochnit G et al (2012) Heterogeneous complexes of the RNA exosome in Sulfolobus solfataricus. Biochimie 94:1578–1587. https://doi.org/10.1016/j.biochi.2012.03.026
Woese CR, Fox GE (1977) Phylogenetic structure of the prokaryotic domain: the primary kingdoms. Proc Natl Acad Sci USA 74:5088–5090. https://doi.org/10.1073/pnas.74.11.5088
Woese CR, Kandler O, Wheelis ML (1990) Towards a natural system of organisms: proposal for the domains Archaea, Bacteria, and Eucarya. Proc Natl Acad Sci USA 87:4576–4579. https://doi.org/10.1073/pnas.87.12.4576
Wolin SL, Maquat LE (2019) Cellular RNA surveillance in health and disease. Science 366:822–827. https://doi.org/10.1126/science.aax2957
Wu G, Schmid M, Rib L et al (2020) A two-layered targeting mechanism underlies nuclear RNA sorting by the human exosome. Cell Rep 30:2387–2401.e5. https://doi.org/10.1016/j.celrep.2020.01.068
You L, Wang C, Molodtsov V et al (2024) Structural basis of archaeal FttA-dependent transcription termination. Nature 635:229–236. https://doi.org/10.1038/s41586-024-07979-9
Yue L, Li J, Zhang B et al (2020) The conserved ribonuclease aCPSF1 triggers genome-wide transcription termination of archaea via a 3′-end cleavage mode. Nucleic Acids Res 48:9589–9605. https://doi.org/10.1093/nar/gkaa702
Yutin N, Puigbò P, Koonin EV, Wolf YI (2012) Phylogenomics of prokaryotic ribosomal proteins. PLoS One 7:e36972. https://doi.org/10.1371/journal.pone.0036972
Zaremba-Niedzwiedzka K, Caceres EF, Saw JH et al (2017) Asgard archaea illuminate the origin of eukaryotic cellular complexity. Nature 541:353–358. https://doi.org/10.1038/nature21031
Zhai B, DuPrez K, Han X et al (2018) The archaeal ATPase PINA interacts with the helicase Hjm via its carboxyl terminal KH domain remodeling and processing replication fork and holliday junction. Nucleic Acids Res 46:6627–6641. https://doi.org/10.1093/nar/gky451
Zhang C, Tian B, Li S et al (2013) Genetic manipulation in Sulfolobus islandicus and functional analysis of DNA repair genes. Biochem Soc Trans 41:405–410. https://doi.org/10.1042/BST20120285

Zhang J, Feng X, Li M et al (2025) Deep origin of eukaryotes outside Heimdallarchaeia within Asgardarchaeota. NatureNature 642. https://doi.org/10.1038/s41586-025-08955-7
Zhang L, Li X, Hill RC et al (2015) Brr2 plays a role in spliceosomal activation in addition to U4/U6 unwinding. Nucleic Acids Res 43:3286–3297. https://doi.org/10.1093/nar/gkv062
Zheng X, Feng N, Li D et al (2017) New molecular insights into an archaeal RNase J reveal a conserved processive exoribonucleolysis mechanism of the RNase J family. Mol Microbiol 106:351–366. https://doi.org/10.1111/mmi.13769
Zinder JC, Lima CD (2017) Targeting RNA for processing or destruction by the eukaryotic RNA exosome and its cofactors. Genes Dev 31:88–100. https://doi.org/10.1101/gad.294769.116
Zinoviev A, Ayupov RK, Abaeva IS et al (2020) Extraction of mRNA from stalled ribosomes by the Ski complex. Mol Cell 77:1340–1349.e6. https://doi.org/10.1016/j.molcel.2020.01.011

Part III
Evolution of Small RNA-Mediated Regulations

Chapter 7
RNase III Enzymes and Their Accessory Domains

Agnieszka Szczepanska, Klaudia Wojcik, and Anna Kurzynska-Kokorniak

Abstract Ribonucleases III (RNases III) are endoribonucleases that specifically cleave phosphodiester bonds present in double-stranded RNAs (dsRNAs). The RNase III family enzymes can be classified into three subfamilies (Classes 1–3) based on their structure and biological functions. Simple RNases belonging to Class 1 contain two domains: the RNase III domain, characteristic of all members of the RNase III family, and a dsRNA-binding domain. During the course of evolution, the RNase III enzymes have gained additional domains that allowed them to increase affinity for dsRNA substrates, improve the precision of substrate cleavage to generate products of defined length, and, overall, perform more and more complex functions. This way, Drosha (Class 2) and Dicer (Class 3) ribonucleases have appeared. Drosha is a key protein involved in the initiation of microRNA (miRNA) biogenesis in most animal cells by cleaving primary miRNA transcripts into ~70-nucleotide long pre-miRNAs. Dicer processes pre-miRNAs into functional ~21–23-nucleotide long miRNAs. Apart from their apparent function in miRNA biogenesis, ribonucleases Drosha and Dicer are also involved in other cellular pathways, which are dependent or independent of their cleavage activities. These non-canonical functions of Drosha and Dicer are often orchestrated by their accessory domains.

Keywords RNase III enzymes · Drosha · Dicer · RNA processing · RNase III domains · RNA-binding domains · Accessory domains · Conserved protein domains · Domain fusion

The authors Agnieszka Szczepanska and Klaudia Wojcik have contributed equally to this work.

A. Szczepanska · K. Wojcik · A. Kurzynska-Kokorniak (✉)
Department of Ribonucleoprotein Biochemistry, Institute of Bioorganic Chemistry, Polish Academy of Sciences, Poznan, Poland
e-mail: akurzyns@man.poznan.pl

A. Kanai, A. W. Nicholson (eds.), *Molecular Evolution of RNA Regulatory Enzymes and Their Systems*, Nucleic Acids and Molecular Biology 38,
https://doi.org/10.1007/978-3-032-03300-0_7

Abbreviations

Aa-RNase III	*Aquifex aeolicus* RNase III
ADAR-1	Adenosine deaminase acting on RNA 1
ADP	adenosine 5′-diphosphate
Ago	Argonaute protein
ATP	adenosine 5′-triphosphate
bp	base pair
BSB	Binding and stability box
C2H2	two cysteines, two histidines (motif characteristic of Zinc finger domains)
CAF	*CARPEL FACTORY* gene
CED	Central domain
CEB	Cleavage efficiency box
crRNA	CRISPR RNA
D	aspartic acid
DB	Distal box
DCL	Dicer-like protein
DDRNA/diRNA	damage-induced small RNA
DGCR8	DiGeorge syndrome critical region 8
DHX9	DExH-Box Helicase 9
DRB4	dsRNA-binding protein 4
DSBs	DNA double-strand breaks
dsDNA	double-stranded DNA
dsRBD	dsRNA-binding domain
dsRNA	double-stranded RNA
DUF283	Domain of unknown function (283)
E	glutamic acid
Ec-RNase III	*Escherichia coli* RNase III
ETS	External transcribed spacer
GYM motif	paired G, paired pyrimidine, and mismatched C or A
HEL	Helicase domain of Dicer-type proteins
HEN1	Hua enhancer 1
HERNA	Helicase with RNase motif
HST1	HASTY
HYL1	Hyponastic leaves 1
IBPB	Initial binding and positioning box
ITS	Internal transcribed spacer
KREN	Kinetoplastid RNA editing endonuclease
Loqs	Loquacious
miRNA	microRNA
ncRNA	non-coding RNA
NLS	nuclear localization signal
NRS	nascent RNA silencing

nt	nucleotide
NTD	N-terminal domain
OB-fold	Oligonucleotide/oligosaccharide binding fold
PACT	Protein activator of protein kinase R
PAZ	Piwi-Argonaute-Zwille domain
PB	Proximal box
P-rich	proline-rich
Pi	inorganic phosphate
PKR	Protein kinase RNA-activated
Pol II	RNA polymerase II
PPD	PAZ-Piwi Domain proteins
pre-miRNA	precursor miRNA
PUF	PUF protein family (Pumilio and FBF)
RIIID	RNase III domain
RNAi	RNA interference
rRNA	ribosomal RNA
RISC	RNA-induced silencing complex
RNase III	Ribonuclease III
RS-rich	arginine/serine-rich
SE	Serrate
SF2	Superfamily 2 helicases
siRNA	small interfering RNA
snRNA	small nuclear RNA
snoRNA	small nucleolar RNA
ssRNA	single-stranded RNA
tasiRNA	trans-acting siRNA
tRF	tRNA-derived fragment
tRNA	transfer RNA
TRBP	TAR RNA-binding protein
XPO5	Exportin 5
ZnF	Zinc finger domain
5′ UTR	5′ untranslated region

7.1 Introduction

Bi- or multidomain proteins have evolved by the fusion of two or more ancestral single-domain proteins or have emerged as a result of recombination events such as domain duplication or swapping (Marsh et al. 2013). New combinations of individual domains are important in protein evolution as they may ensure direct linkages between proteins that act in a coordinated manner; e.g., successive enzymes in the same pathway (Sapranauskas and Lubys 2005). Importantly, the consolidation of protein domains allows time and energy savings for the cell. Nevertheless, the repertoire of domain types is limited. The huge diversity of multidomain proteins found in nature seems to derive from occasional domain shuffling, followed by natural selection of the most effective multidomain variants. During evolution, unnecessary domains may be degenerated or lost (Sapranauskas and Lubys 2005).

In general, proteins that have evolved to exhibit specificity for double-stranded RNA (dsRNA) processing are members of the ribonuclease III (RNase III) family. dsRNA is structurally well-conserved and exists principally as an A-form double helix. The A-form double helix of RNA is characterized by a shallow, solvent-exposed minor groove and a narrow, deep major groove (Nicholson 1996, 2014). A fundamental principle underlying dsRNA substrate recognition by dsRNA-binding proteins is the difference in minor groove depth that distinguishes the A-form double helix of RNA from the B-form double helix of DNA, or from RNA/DNA hybrids (Kharrat et al. 1995; Ryter and Schultz 1998). The 2′ ribose hydroxyl groups that line the minor groove of the A-form dsRNA generate a polar, extensively hydrated surface and constitute 'specific signatures' recognized by dsRNA-binding proteins, including RNase III enzymes (Nicholson 2014). These 2′ hydroxyl groups are involved in interactions between dsRNA and dsRNA-binding proteins.

RNases III are endoribonucleases that specifically cleave phosphodiester bonds present in double-stranded regions of RNA. The substrates for these enzymes are dsRNAs and single-stranded (ss) RNAs adopting local dsRNA structures, such as hairpin-shaped RNAs. All RNases III share a unique fold (the RNase III domain) (Fig. 7.1) that dimerizes to form a catalytic valley that accommodates dsRNA (Blaszczyk et al. 2001; Gan et al. 2005, 2006). Within a catalytic valley, each subunit cleaves phosphodiester bonds on each strand of dsRNA, which results in a characteristic product with 5′-phosphate groups, 3′-hydroxyl ends, and two-base 3′-overhangs (Nicholson 2014; Gan et al. 2005, 2006). The cleavage activity of RNase III enzymes requires divalent metal ions, preferably Mg^{2+} (Blaszczyk et al. 2001; Gan et al. 2005, 2006) (Fig. 7.2). RNases III play a key role in many cellular pathways involving RNA-processing reactions, including, among others, generation of small regulatory RNAs such as microRNAs (miRNAs) and small interfering RNAs (siRNAs) (Bernstein et al. 2001; Zhang et al. 2004), processing of precursor molecules to produce ribosomal RNAs (rRNAs), small nucleolar RNAs (snoRNAs), small nuclear RNAs (snRNAs), and other non-coding RNAs (ncRNAs) (Young and Steitz 1978; King et al. 1984; Kufel et al. 1999; Allmang et al. 1999; Abou Elela and Ares Jr. 1998; Elela et al. 1996; Lejars et al. 2021).

RNase III enzymes are highly conserved from bacteria to higher eukaryotes and are only sporadically found in the Archaea (Nicholson 2014). It is believed that a simple RNase III, consisting of a single endonuclease domain, originated in early bacteria (Nicholson 2014), likely around the time when cellular RNA-based regulation mechanisms began to develop (2–3 billion years ago). Then, the RNase III ancestor entered the eukaryotic lineage, presumably via an endosymbiont (Nicholson 2014). The appearance of two RNase III domains within one protein was most likely caused by duplication of the RNase III coding gene (Conrad and Rauhut 2002). Then, the two RNase III domains diverged, giving rise to RNase IIIa and IIIb domains. The subsequent acquisition of additional domains led to the diversification of eukaryotic family members (Fig. 7.1). The accessory domains generally help the

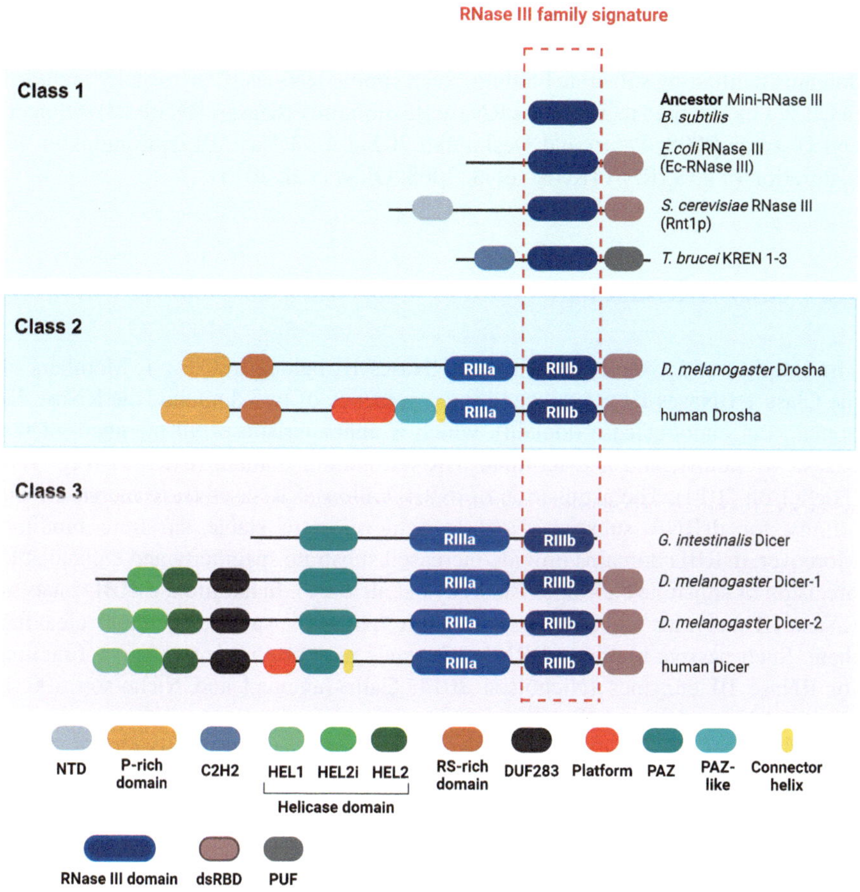

Fig. 7.1 Domain architecture of RNase III family members. The RNase III family members are classified into three classes based on domain organization. Class 1 is represented by *E. coli* RNase III and *S. cerevisiae* Rnt1p proteins. This family also includes the minimal RNase III, which is an ancestor of all RNases III. Class 2 is represented by *D. melanogaster* and human Drosha, while Class 3 is represented by *G. intestinalis, D. melanogaster*, and human Dicer. KREN RNase III enzymes (found in trypanosomatids) exhibit distinct structural features compared to bacterial and yeast RNases, Drosha, and Dicer, making their classification within the established RNase III classes challenging. Here, based on their strong similarity to bacterial and yeast RNases III, both characterized by a single RNase III domain, KREN nucleases are assigned to Class 1. Diagrams are not to scale. (The figure was generated in BioRender software)

RNase III enzymes in substrate recognition and binding, contribute to cleavage site selection, increase substrate range, serve as a platform for partner proteins, and provide regulation (Nicholson 1999, 2014).

The RNase III family contains a large group of enzymes that can be classified into three subfamilies (Class 1–3) based on their structure (the domain organization) and biological functions (Fig. 7.1). Presumably, the simplest RNase III is the ribonuclease found in *Bacillus subtilis*, also known as Mini-III or Bs-Mini III (Redko et al. 2008). This RNase has a single RNase III domain and no additional domains facilitating substrate binding. Such simple RNases III are usually included in Class 1 or assigned to a separate RNase III subfamily (Class 4 RNase III) (Olmedo and Guzman 2008; Paturi and Deshmukh 2021). Bs-RNase III contributes to the maturation of 23S rRNA (Redko et al. 2008; Glow et al. 2015).

7.2 Class 1 RNases III

The simplest and evolutionarily oldest RNases III belong to Class 1. Members of the Class 1 RNases III are defined by the presence of two domains: the RNase III domain (an endonuclease domain), which is characteristic of all members of the RNase III family, and a C-terminal dsRNA-binding domain (dsRBD) (Fig. 7.1) (Nicholson 2014). The acquisition of dsRBD allowed RNases III to increase their affinity for dsRNA substrates and to achieve more stable substrate binding. Moreover, dsRBD domains provide increased substrate specificity and enhance the precision of substrate cleavage (Blaszczyk et al. 2001). In addition, dsRBD enables RNase III enzymes to bind substrates in a "passive" way, i.e., without cleaving them. Such passive binding of RNA substrates may provide a regulatory function for RNase III enzymes (Nicholson 2014; Calin-Jageman and Nicholson 2003). Class 1 RNases III must dimerize to form a functional dsRNA-cleavage center (a catalytic valley) (Fig. 7.2) (Blaszczyk et al. 2001; Gan et al. 2005, 2006; Zhang et al. 2004).

7.2.1 *Bacterial RNases III*

The best-known examples of bacterial RNase III proteins are *Escherichia coli* RNase III (Ec-RNase III) and *Aquifex aeolicus* RNase III (Aa-RNase III). Ec-RNase III was the first dsRNA-specific endoribonuclease to be described, which took place in 1967 (Robertson et al. 1967, 1968). In 2001, the first structure of the RNase III domain was reported (the structure of Aa-RNase III with a deleted dsRBD; PDB entry: 1I4S) (Blaszczyk et al. 2001). Structural data show that the RNase III domain generally displays an all-α-helical fold (Fig. 7.3) (Blaszczyk et al. 2001). The RNase III domains contain a stretch of conserved negatively charged glutamic acid (E) and aspartic acid (D) residues, known as the RNase III signature motif. In the case of Ec-RNase III, these are: E38, E41, D45, E65, E100, D114, and E117 (Sun et al. 2004); side chains of E41, D114, and E117 coordinate divalent metal ions (Mg^{2+} or Mn^{2+}) (Sun et al. 2004). The binding of divalent metal ions is critical for the formation and stabilization of the catalytic valley (Blaszczyk et al. 2001, 2004); moreover, the two metal ions in each catalytic site of the RNase III domain are required for the cleavage of each phosphodiester bond (Fig. 7.2) (Gan et al. 2005, 2006).

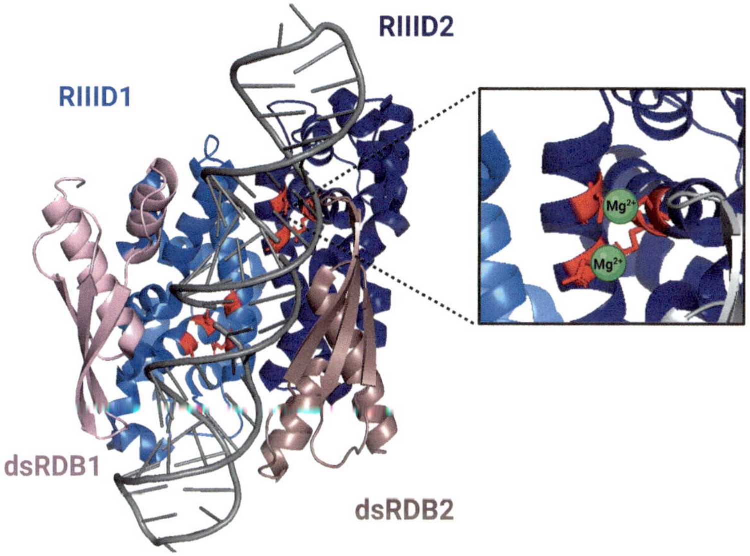

Fig. 7.2 Catalytic valley of the *E. coli* RNase III dimer (PDB: 7R97) (Dharavath et al. 2022). The RNase III domains (RIIIDs) are shown in blue, the dsRNA-binding domains (dsRBDs) in pink, and the RNA substrate in grey. Ion-binding residues (D41, D45, D114, E117) are marked in red, and magnesium ions in green. (The figure was generated in BioRender software)

In addition to the RNase III domain, Class 1 RNases III also contain a dsRBD domain (Fig. 7.1). The dsRBD domain is a small conserved protein domain found in eukaryotic, prokaryotic, and viral proteins. Presumably, dsRBDs were even present in the last common ancestor of metazoans (Kerner et al. 2011). The dsRBD domain adopts a conserved αβββα fold, in which two α-helices are positioned against three-stranded antiparallel β-sheets (Fig. 7.3) (Kharrat et al. 1995; Ryter and Schultz 1998; Fierro-Monti and Mathews 2000). The dsRBD's fold is one of the most abundant RNA-binding motifs involved in the recognition and binding of dsRNAs (Masliah et al. 2013). Proteins containing dsRBDs participate in many aspects of cellular life, such as RNA processing, RNA editing, RNA transport, RNA silencing, RNA metabolism, and even antiviral response (Gleghorn and Maquat 2014). The substrate recognition by dsRBD is structure- rather than sequence-dependent (Saunders and Barber 2003). Amino acid residues on the surface of dsRBD interact with A-form dsRNAs (St Johnston et al. 1992; Seeman et al. 1976), and ribose 2′-hydroxyl groups (the O2′ hydroxyls) of a dsRNA substrate are involved in these interactions (Saunders and Barber 2003). This explains why the dsRBD domain specifically interacts with dsRNA and not double-stranded DNA (dsDNA) or RNA/DNA hybrid structures (Ryter and Schultz 1998; Blaszczyk et al. 2001; Fierro-Monti and Mathews 2000; Seeman et al. 1976).

Numerous structural studies have shown that bacterial RNases III can adopt the catalytic and noncatalytic forms. In the catalytic state, each dsRBD of the dimer interacts with the same dsRNA molecule that is bound in the catalytic valley (Gan

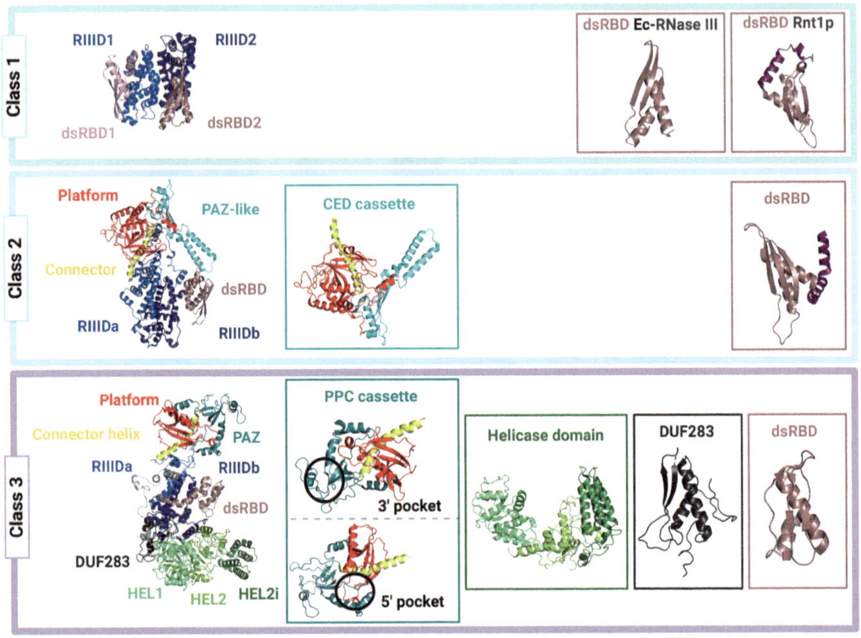

Fig. 7.3 The 3D structures of RNase III family members and their individual domains visualized by PyMOL. Class 1 is represented by *E. coli* RNase III (PDB: 7R97) (Dharavath et al. 2022). The RNase III domains (RIIIDs) are indicated in blue. The dsRBDs of *E. coli* RNase III and Rnt1p (PDB: 4OOG) (Liang et al. 2014) are indicated in pink. The dsRBD domain has an αβββα fold; additional helices of Rnt1p (α3 and α4) are marked in purple. Class 2 is represented by human Drosha (PDB: 9ASM) (Garg et al. 2024). The additional α motif of dsRBD is marked in purple. The Platform domain is indicated in red, PAZ-like in cyan, and the Connector helix in yellow. Platform, a PAZ-like domain, and Connector constitute the CED (central domain) cassette. Class 3 is represented by human Dicer (PDB: 5ZAL) (Liu et al. 2018b). The helicase domain (PDB: 5ZAL), with its three subdomains, is indicated in green. The Platform domain is indicated in red, PAZ in turquoise, and the Connector helix in yellow (PDB: 5ZAL). Platform, PAZ, and Connector constitute the PPC (Platform-PAZ-Connector helix) cassette. The 3′ pocket and 5′ pocket, which bind the pre-miRNA ends, are shown. The DUF283 domain (PBD: 7XW3) (Lee et al. 2023a) is indicated in grey. The dsRBD domain (PBD: 7XW3) (Lee et al. 2023a) is marked in pink. (The figure was generated in BioRender software)

et al. 2005; Blaszczyk et al. 2004). In the noncatalytic form, the two dsRBD domains bind different substrates outside the catalytic valley (Gan et al. 2005). These two different conformations can be adopted due to the presence of a 7-amino acid linker connecting the RNase III domain and dsRBD, which is flexible and allows dsRBD to rotate and modulate the binding of a substrate (Gan et al. 2005; Blaszczyk et al. 2004). In this way, the RNase III enzyme can act as an endonuclease, cleaving the substrate, or as a regulatory protein that passively binds dsRNA (without initiating its cleavage) (Gan et al. 2005; Calin-Jageman and Nicholson 2003).

Bacterial RNases III play an essential role in cleaving polycistronic rRNA transcripts, leading to the release of individual rRNA molecules, which are fundamental components of the bacterial ribosome (Young and Steitz 1978; King et al. 1984; Court et al. 2013). For example, Ec-RNase III processes rRNA transcripts derived from seven rRNA operons. Each of these operons produces a transcript that includes 16S rRNA, 23S rRNA, 5S rRNA, and a transfer RNA (tRNA), which is processed by RNase III during transcription (Young and Steitz 1978; Court et al. 2013; French and Miller Jr. 1989). The timing of transcription and RNase III activity is tightly coordinated to ensure the rapid assembly of rRNA and ribosomal subunit formation (French and Miller Jr. 1989; Bubunenko et al. 2013). Moreover, bacterial RNases III play essential roles in the processing and degradation of other RNA transcripts. Transcript cleavage may occur within the leader sequences (regulatory sequences) or the protein-coding sequences (King et al. 1984). One notable example is the autoregulation of its own transcript (the *rnc* operon transcript) by Ec-RNase III (Bardwell et al. 1989; Matsunaga et al. 1996). Ec-RNase III cleaves the *rnc* operon transcript within the double-stranded region of the *rnc* leader, which initiates its degradation by other RNases present in the cell (Matsunaga et al. 1996). In addition, bacterial RNases III are involved in CRISPR RNA (crRNA) maturation (Table 7.1) (Westra et al. 2012). In the CRISPR-Cas adaptive immune system of bacteria, RNase III processes precursor crRNAs into their mature forms, which are then utilized by the Cas proteins to target and degrade foreign genetic elements, such as plasmids and phages (Deltcheva et al. 2011).

7.2.2 Yeast RNase III

The most recognized yeast RNase III is *Saccharomyces cerevisiae* Rntp1, which was discovered in 1996 as an important factor required for pre-spliceosome formation (Wells et al. 1996). In addition to an RNase III domain and a dsRBD domain, Rnt1p also has an N-terminal domain (NTD) (Fig. 7.1). This NTD is unique, as it does not exhibit substantial similarity to other N-terminal domains found in the RNase III family proteins (Lamontagne et al. 2000). NTD is not essential for Rntp1 catalytic activity, but it enhances enzyme specificity (Lamontagne et al. 2000; Nagel and Ares Jr. 2000) and contributes to Rntp1 dimerization (Abou Elela and Ji 2019).

The Rnt1p dsRBD domain is more extensive compared to the bacterial RNase III dsRBDs. In addition to the conserved αβββα motif (typical of these domains), the dsRBD of Rntp1 also contains two additional helices at the C-terminus (Ryter and Schultz 1998). As a result, this domain adopts an αβββααα motif, which is characteristic of this particular RNase III (Fig. 7.3) (Liang et al. 2014). The addition of two helices to the C-terminal domain of the Rntp1 dsRBD results in specific recognition of the tetraloop region of the hairpin RNA substrate (Fig. 7.4) (Liang et al. 2014). Similarly to Ec-RNase III, Rntp1 has to dimerize to cleave a substrate. However, due to the distinct organization of the dsRBD domains, substrate binding by Rnt1p differs from that by bacterial RNases III. Specifically, at one end of Rntp1, the two NTDs dimerize and, together with one of the dsRBDs, interact with the tetraloop region of a substrate, while at the opposite end, the RNase III dimer, along with the

Table 7.1 Functions of RNase III enzymes

RNase III enzymes			
Class	Organism	Representative	Function
1	Bacteria	*B. substilis* Mini-RNase III	Plays an essential role in cleaving polycistronic rRNA transcripts, leading to the release of the individual rRNAs
		E. coli RNase III	Plays an essential role in cleaving polycistronic rRNA transcripts, leading to the release of the individual rRNAs Is involved in the regulation of transcript levels (processing and degradation of RNA transcripts; e.g., autoregulation of its own transcript, the *rnc* operon transcript) Is involved in the CRISPR RNA (crRNA) maturation
	Fungi (yeast)	Rnt1p	Plays an essential role in cleaving polycistronic rRNA transcripts, leading to the release of the individual rRNAs Is involved in the maturation of snRNAs and snoRNAs Is involved in the poly(A)-independent transcription termination Is involved in mRNA quality control
	Protista (Trypanosoma)	*T. brucei* KREN1–3	Are critical in editosome function (site-specific cleavage of mitochondrial pre-mRNAs - essential step in the generation of functional mitochondrial transcripts)
2	Animals	*D. melanogaster* Drosha	Recognizes and cuts pri-miRNAs from longer transcripts (in cooperation with Pasha, a DGCR8 homolog), generating pre-miRNAs Participates in the regulation of gene expression through the miRNA pathway
		Human Drosha	Recognizes and cuts pri-miRNAs from longer transcripts (in cooperation with DGCR8), generating pre-miRNAs Is involved in the termination of transcription of pri-miRNA transcripts Is involved in mRNA processing Contributes to transcriptional regulation and gene silencing by producing small ncRNAs that affect chromatin structure Is involved in producing diRNAs essential for repairing damaged DNA Is involved in specific cleavages during the large ribosome subunit rRNA maturation Is involved in the transcription of ribosomal protein genes Is involved in resolving R loops

(continued)

Table 7.1 (continued)

RNase III enzymes			
Class	Organism	Representative	Function
3	Protista (algae)	*C. reinhardtii* DCL1	Is involved in the post-transcriptional silencing of retrotransposons
		C. reinhardtii DCL2	Has not yet been well characterized
		C. reinhardtii DCL3	Generates miRNA-like small RNAs crucial for gene regulation, primarily through RNA silencing mechanisms (in *C. Reinhardtii*, these miRNAs may be derived from imperfect hairpin structures, similar to higher plants and animals)
	Fungi (yeast)	*S. pombe* Dcr1	Maintains centromeric heterochromatin Represses genome activity by a chromatin silencing mechanism referred to as cotranscriptional gene silencing Serves as a possible thermosensitive switch that regulates the nucleo–cytoplasmic distribution of Dcr1
	Plants	*A. thaliana* DCL1	Is involved in the production of miRNAs from pri-miRNA transcripts
		A. thaliana DCL2	Processes dsRNAs into siRNAs Is involved in antiviral response by producing 22-nt siRNAs from virus-derived dsRNAs (silencing of viral transcripts)
		A. thaliana DCL3	Processes dsRNA into siRNA Produces 24-nt siRNAs involved in heterochromatin formation and transcriptional gene silencing
		A. thaliana DCL4	Processes dsRNA into siRNA Is involved in antiviral defense Produces trans-acting siRNAs (tasiRNAs), a specialized class of endogenous small RNAs that regulate gene expression in a sequence-specific, post-transcriptional manner
	Animals (invertebrates)	*D. melanogaster* Dicer-1	Cleaves pre-miRNAs to produce miRNAs
		D. melanogaster Dicer-2	Processes dsRNA into siRNA Contributes to the antiviral response by cleaving viral dsRNAs
	Animal (vertebrates)	Human Dicer	Cleaves pre-miRNA to produce miRNA Is involved in the biogenesis of small RNAs derived from tRNAs, known as tRNA-derived fragments (tRFs) Is involved in producing diRNAs essential for repairing damaged DNA Participates in chromatin structure remodeling Restricts the deleterious accumulation of endogenous dsRNAs Is involved in resolving R loops Contributes to the antiviral response by binding proteins involved in the antiviral response

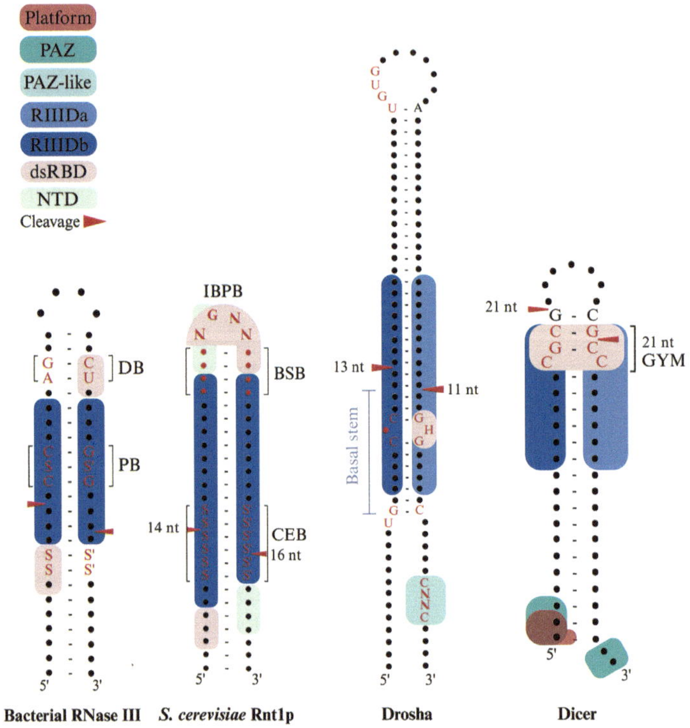

Fig. 7.4 RNase III enzymes as molecular rulers. The RNA substrates recognized by *E. coli* RNase III, yeast Rnt1p, Drosha, and Dicer are illustrated, highlighting the sequences and structural features (in red) that influence enzyme action. Bacterial RNase III recognizes the proximal box (PB) and the distal box (DB). Rnt1p recognizes the NGNN tetraloop within the initial binding and positioning box (IBPB), the binding and stability box (BSB), and the cleavage efficiency box (CEB). The ribonuclease Drosha, in concert with DGCR8, recognizes specific sequence motifs within pri-miRNAs, such as a GHG motif (recognized by the dsRBD domain of Drosha), an apical UGU/GUG (UGU) motif (recognized by the DGCR8 dimer), and a CNNC motif (recognized by the PAZ-like domain of Drosha). The basal UG motif is recognized by proteins partnering Drosha; e.g., splicing factors (Kim et al. 2018). Dicer recognizes the GYM motif (paired G, paired pyrimidine, and mismatched C or A) of pre-miRNA. In sequence notation: (N) or (•) represent any ribonucleotide; (S) corresponds to C or G, with (S′) being its complementary nucleotide; (H) represents any nucleotide except G. Cleavage sites are indicated by red triangles. Domains are color-coded: Platform (red), PAZ (cyan), PAZ-like (light cyan), RIIIDa (dark blue), RIIIDb (light blue), dsRBD (light pink), and NTD, eukaryotic N-terminal domain (light green). The figure was generated in BioRender software, based on (Li and Patel 2016; Nguyen et al. 2021; Le et al. 2020; Kwon et al. 2019; Lee et al. 2023a, b; Lamontagne and Elela 2004; Creugny et al. 2018; Yoshida et al. 2021; Podolska et al. 2013)

other dsRBD, binds the substrate ends (Abou Elela and Ji 2019; Liang et al. 2014). A flexible arrangement of Rntp1's domains is possible due to two linkers. The linker connecting the RNase III domain to the dsRBD is similar to that found in Ec-RNase III. The linker between NTD and the RNase III domain is 46-amino acids long and enables the domains located at opposite ends of the enzyme, the NTD and the dsRBD, to interact during substrate binding (Song et al. 2017; Lavoie and Abou Elela 2008). Rnt1p uses a "double-ruler" mechanism for substrate binding and cleavage (Liang et al. 2014). The first ruler is formed by the RNase III domain and the dsRBD belonging to the same dimer subunit. The second ruler consists of the two NTDs and the RNase III domain of the second subunit of the dimer. During the formation of the protein-RNA complex, these two rulers enable precise substrate cleavage by the RNase III domains at distances of 14 and 16 nucleotides (nt) from the tetraloop of a hairpin substrate (Fig. 7.4) (Liang et al. 2014). Beyond these differences, the interaction of the Rntp1 RNase III domain dimer with a substrate is identical to that of bacterial RNase III (Abou Elela and Ji 2019).

Similar to bacterial RNases III, yeast homologs are also involved in rRNA maturation. Unlike bacterial cells, eukaryotic cells have developed specialized compartments to efficiently perform their functions. The rRNA precursor synthesis and maturation occur in the nucleus (mostly in the nucleolus). Therefore, in yeast RNases III, a nuclear localization signal (NLS) must have appeared (Catala et al. 2004). Regarding Rnt1p, nuclear localization has been reported to be independently mediated via a canonical NLS located in the dsRBD domain and through a less-defined sequence in the NTD domain (Catala et al. 2004). In addition to processing pre-rRNAs, Rnt1 also functions in the maturation of both snRNAs (Abou Elela and Ares Jr. 1998; Chanfreau et al. 1997) and snoRNAs (Chanfreau et al. 1998). Cleavage by Rnt1p is also essential for poly(A)-independent transcription termination (Fernandez-Pevida et al. 2015; Ghazal et al. 2009). Additionally, yeast RNases III participate in mRNA quality control (Table 7.1) (Zer and Chanfreau 2005).

7.3 Class 2 RNases III

During the course of evolution, two RNase III domains were combined within one protein. Proteins containing two RNase III domains belong to Class 2 or Class 3 RNases III (Fig. 7.1). In contrast to the Class 1 RNase III enzymes, which must dimerize to acquire activity (Gan et al. 2006), Class 2 and Class 3 RNases III form an intramolecular dimer (Kwon et al. 2016; Liu et al. 2018a; Macrae et al. 2006).

7.3.1 Ribonuclease Drosha

Class 2 RNases III are represented by Drosha ribonucleases. Evolutionary studies suggest that Drosha and its homologs have appeared only in animals. However, Drosha is also found in *Ichthyosporea* (unicellular eukaryotes closely related to animals), which in turn indicates that Drosha evolved long before the last common ancestor of animals (Brate et al. 2018). A typical ribonuclease Drosha contains an N-terminal proline-rich (P-rich) domain, an arginine/serine-rich (RS-rich) domain, a central domain (CED, comprising Platform, a PAZ-like domain, and Connector), two RNase III domains (known as RIIIDa and RIIIDb or RNase IIIa and RNase

IIIb), and a dsRBD domain (Fig. 7.1) (Kwon et al. 2016; Son et al. 2023). The ribonuclease Drosha was first mentioned in 2000, in relation to pre-rRNA processing, at that time under the name "an RNase III enzyme" (Wu et al. 2000). It was demonstrated that inhibition of this RNase III enzyme led to the accumulation of 5.8S rRNA precursors (Wu et al. 2000; Liang and Crooke 2011), but its depletion did not significantly alter the amount of mature rRNAs (Chong et al. 2008). It was, therefore, assumed that the importance of this newly discovered RNase III in rRNA biosynthesis was negligible (Lee et al. 2006). The name Drosha first appeared in 2003 in the context of miRNA biogenesis (Lee et al. 2003). The RNase Drosha is a key protein involved in the initiation of miRNA biogenesis in most animal cells by cleaving primary miRNA transcripts (pri-miRNAs) into ~70-nt pre-miRNAs (Fig. 7.5) (Lee et al. 2003). Drosha functions in cooperation with an RNA-binding

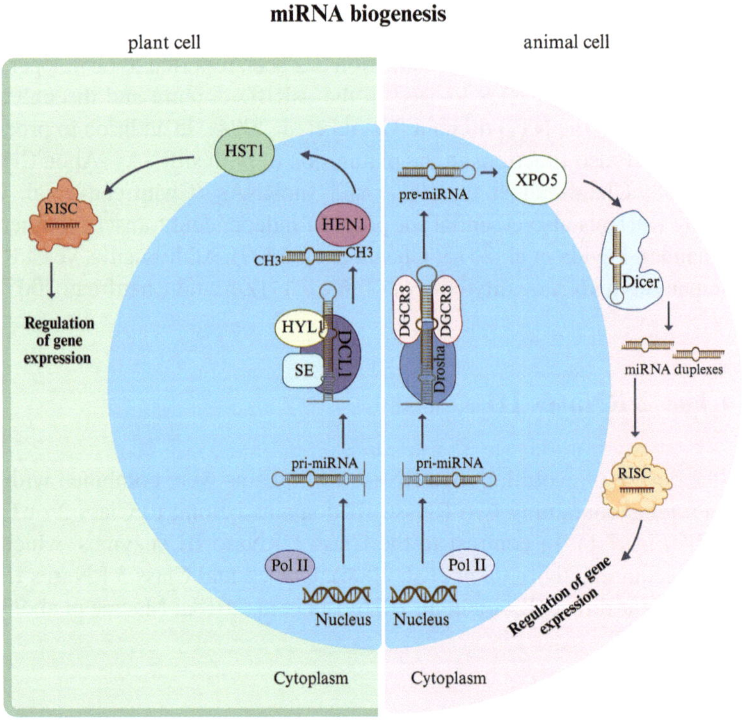

Fig. 7.5 The RNase III enzymes involved in miRNA biogenesis. The biosynthesis of miRNAs in animal and plant cells begins in the nucleus, where primary miRNAs (pri-miRNAs) are transcribed by DNA-dependent RNA polymerase II (Pol II). In animals, pri-miRNAs are processed into precursor miRNAs (pre-miRNAs) by the Microprocessor complex (Drosha/DGCR8), then exported to the cytoplasm by Exportin 5 (XPO5) and further processed by Dicer into miRNA duplexes. In plants, pri-miRNAs are processed by a complex comprising Dicer-like 1 (DCL1) protein, Hyponastic leaves 1 (HYL1), and Serrate (SE), generating pre-miRNAs and miRNA duplexes, which are methylated by miRNA methyltransferase Hua enhancer 1 (HEN1) and exported to the cytoplasm via the plant exportin 5 ortholog HASTY (HST1). In both systems, miRNA duplexes are loaded into the RNA-induced silencing complex (RISC), which regulates gene expression. (The figure was generated in BioRender software)

protein named "DiGeorge syndrome critical region 8" (DGCR8) (Denli et al. 2004; Gregory et al. 2004; Han et al. 2004). In the nucleus, one Drosha and two DGCR8 proteins form a Microprocessor complex (Denli et al. 2004; Gregory et al. 2004). Structural studies have revealed details of this heterotrimeric Microprocessor complex (Kwon et al. 2016) and its role in pri-miRNA substrate recognition, binding, and cleavage (Fig. 7.4) (Jin et al. 2020; Garg et al. 2024). Moreover, structural data for a pri-miR-16-1 substrate and the Drosha-DGCR8 core complex have revealed that the PAZ-like domain and the RNase IIIa domain recognize the single-stranded to double-stranded junction of pri-miRNA, while dsRBD makes extensive contacts with the pri-miRNA stem (Jin et al. 2020). Importantly, Microprocessor has the plasticity for accommodating a variety of pri-miRNA substrates (Garg et al. 2024). In general, pri-miRNA drives the repositioning of Microprocessor domains, orchestrating their organization on its scaffold to facilitate substrate cleavage (Garg et al. 2024); the main role of dsRBD is the recognition and stabilization of dsRNA before cleavage (Jin et al. 2020). The RNase IIIa and RNase IIIb domains form an intramolecular heterodimer with a catalytic valley (Kwon et al. 2016; Jin et al. 2020; Garg et al. 2024; Partin et al. 2020) where pri-miRNA cleavage takes place (Nguyen et al. 2020). The cleavage site is typically located ~11 base pairs (bp) from the basal junction and ~22 bp from the apical junction (Fig. 7.4) (Li and Patel 2016). The P-rich domain is known to preferentially promote the processing of "rapidly evolving" pri-miRNA hairpins located within introns (in contrast to pri-miRNAs located in exons) (Son et al. 2023). This domain displays high functional conservation across chordate Drosha homologs, despite their low sequence similarity (Nguyen et al. 2021). The RS-rich domain is mostly linked to the subcellular localization of Drosha (Lee et al. 2006; Tang et al. 2010, 2011; Link et al. 2016; Prabhakar et al. 2023), its stabilization (acetylation of lysine residues in the RS-rich domain stabilizes Drosha by preventing ubiquitination) (Tang et al. 2013), and response to stress (Link et al. 2016). The alternatively spliced Drosha transcripts, lacking a part of the RS-rich domain, are expressed in many human cells and can still bind DGCR8 and process pri-miRNAs (Link et al. 2016). Nevertheless, due to the high flexibility of the P-rich and RS-rich domains, which makes it difficult to obtain structural data, details of interactions between pri-miRNA and these domains are yet unknown.

The acquisition of accessory domains allowed RNase III enzymes to increase substrate specificity. During the course of evolution, hairpin-shaped RNA substrates have also evolved structural and sequence motifs that enable them to be more specifically recognized by the RNase III enzymes. The ribonuclease Drosha, in concert with DGCR8, recognizes specific sequence motifs within pri-miRNAs, such as a GHG motif (recognized by the dsRBD domain of Drosha) (Le et al. 2020; Kwon et al. 2019), an apical UGU/GUG (UGU) motif (recognized by the DGCR8 dimer) (Denzler and Stoffel 2015; Dang et al. 2020), and a CNNC motif (recognized by the PAZ-like domain of Drosha) (Garg et al. 2024; Auyeung et al. 2013; Kim et al. 2018; Zhang et al. 2021) (Fig. 7.4). Additionally, the internal loops in the lower stem of pri-miRNA (when present) promote a single cleavage by human Microprocessor. This single cleavage of pri-miRNA does not result in the pre-miRNA release (Nguyen et al. 2020).

7.3.2 Non-canonical Functions of Drosha

In addition to pri-miRNA cleavage, Drosha is also involved in mRNA processing (Chong et al. 2010). One of the earliest described examples of Drosha-mediated mRNA processing is the cleavage by the Drosha-DGCR8 complex of hairpin structures embedded within the 5′ untranslated region (5′ UTR) of DGCR8 mRNA (Han et al. 2009; Shenoy and Blelloch 2009; Triboulet et al. 2009). Substrates of Drosha are also other transcripts that may adopt stem-loop structures (Kim et al. 2017). These non-canonical substrates may be processed by Drosha and then by Dicer into miRNAs (Kim et al. 2017). Moreover, Drosha, along with other enzymes of the RNA interference (RNAi) pathway, contributes to transcriptional regulation and gene silencing by producing small ncRNAs that affect chromatin structure (Burger and Gullerova 2015). There are also reports showing that Drosha is recruited to DNA double-strand breaks (DSBs) via interactions with a DNA damage sensor MRN (a protein complex composed of Mre11, Rad50, and Nbs1) (Cabrini et al. 2021; Esposito et al. 2024; Sharma et al. 2021). The RNase III enzymes, Drosha and Dicer, process transcripts generated in response to DNA damage, producing damage-induced small RNAs (diRNAs or DDRNAs) essential for repairing damaged DNA (Francia et al. 2012; Wei et al. 2012; Francia et al. 2016). Despite the main nuclear localization (Lee et al. 2003), a small amount of Drosha can also be found in the cytoplasm (Link et al. 2016). Moreover, during viral infection, Drosha may translocate to the cytoplasm and recognize viral RNAs (Shapiro et al. 2012, 2014; Aguado et al. 2017; Noguchi et al. 2024). It is also known that a Microprocessor complex can be involved in the termination of transcription of pri-miRNA transcripts (Dhir et al. 2015). This way of transcription termination is an alternative to the canonical transcript cleavage followed by poly(A) tail addition (3′-end cleavage and polyadenylation).

Regarding pre-rRNA processing, mentioned earlier, Drosha has been shown to cleave structured RNA elements within the external transcribed spacers (ETS) and internal transcribed spacers (ITS) of pre-rRNA, which are critical for the proper maturation of the 18S, 5.8S, and 28S rRNA species. Depletion of Drosha does not alter the levels of 45S pre-rRNA (the precursor for the 18S, 5.8S, and 28S rRNA), indicating that it is not involved in the earliest stages of pre-rRNA transcription or initial cleavage events. However, knockdown of Drosha results in a significant accumulation of 32S and 12S pre-rRNA intermediates. These intermediates correspond to processing within the 5.8S–ITS2 region of pre-rRNA, suggesting that Drosha is required for specific cleavages during large subunit rRNA maturation (Wu et al. 2000). Although Drosha depletion leads to a significant accumulation of pre-5.8S rRNA, this processing defect appears to be tolerable, as the levels of mature rRNAs remain unchanged (Wu et al. 2000; Chong et al. 2008). Therefore, it is assumed that 5.8S rRNA production can also be maintained through alternative processing pathways that commonly operate during pre-rRNA maturation (Fatica and Tollervey 2002).

Additionally, the Microprocessor complex has been shown to potentiate the transcription of ribosomal protein genes by resolving RNA/DNA hybrid structures known as R loops (Jiang et al. 2021) (of note, R loops are three-stranded nucleic acid structures that consist of an RNA/DNA hybrid and a displaced ssDNA (Santos-Pereira and Aguilera 2015)). It has also been demonstrated that under conditions of nutrient deprivation, the nuclear export and subsequent degradation of Drosha via the E3 ubiquitin ligase lead to reduced ribosomal protein gene transcription and overall protein synthesis (Table 7.1) (Jiang et al. 2021).

Drosha also plays a significant role in transcription initiation by interacting with a nuclear cap-binding protein CBP80 and RNA polymerase II (Pol II). Drosha interacts with CBP80 through its N-terminal RS domain (Gromak et al. 2013). Moreover, Drosha plays a role in regulating the expression of its own gene by influencing the splicing process (Kim et al. 2017; Lee et al. 2017; Mechtler et al. 2017). By binding to the human *DROSHA* transcript, the Microprocessor complex, regardless of its catalytic activity, sterically hinders the splicing machinery from recognizing its cognate splice sites (Lee et al. 2017).

7.3.3 RNases Drosha and Dicer

The Drosha's structure is highly conserved across animal species, with differences occurring mainly in the length and sequence variation of the P-rich and RS-rich domains (Nguyen et al. 2015). However, CED, the RNase III domains, and dsRBD are highly conserved and show similarities in their amino acid sequences (Nguyen et al. 2021). Unlike animals, plants do not have Drosha or DGCR8 homologs (Chen 2005; Comella et al. 2008; Bhaskaran and Mohan 2014). In plants, pri-miRNA processing is accomplished by Dicer-like 1 (DCL1) proteins, which act as Drosha and Dicer, catalyzing both steps of pri-miRNA cleavage into pre-miRNA and then into mature miRNAs in the cell nucleus (Fig. 7.5) (Rogers and Chen 2013; Zhang et al. 2022; Rosatti et al. 2024). The ctenophore *Mnemiopsis leidyi* also lacks Drosha (Maxwell et al. 2012). Thus, it is assumed that *M. leidyi* may be the earliest separated evolutionary lineage among animals (Maxwell et al. 2012; Ryan et al. 2013). The absence of Drosha in *Mnemiopsis* (Maxwell et al. 2012) and DGCR8 in *Trichoplax* (Placozoa) (Grimson et al. 2008; Hertel et al. 2009) suggests that Drosha appeared later in the evolution of animal miRNAs, or that Drosha/DGCR8 were present in the ancestor of animals (of note, Drosha and DGCR8 homologs are present in *Ichthyosporea*, unicellular eukaryotes closely related to animals (Brate et al. 2018)) but were independently lost. Even though global and local sequence similarity shows that human Drosha (Kwon et al. 2016) and a simple Class 3 RNase III *Giardia* Dicer (Macrae et al. 2006) are not apparently homologous proteins, the overall tertiary structure of Drosha is similar to that of Dicer proteins (Fig. 7.3), which suggests the evolution of Drosha from a Dicer homolog (Kwon et al. 2016).

7.4 Class 3 RNases III

Class 3 RNases III comprise Dicer-type enzymes. Dicers are the most complex members of the RNase III family. Apart from the two RNase III domains and dsRBD, a typical Dicer protein is also equipped with accessory domains, such as a helicase domain, a domain of unknown function (DUF283), a Platform domain, a Piwi-Argonaute-Zwille (PAZ) domain, and a Connector helix (Fig. 7.1) (Zhang et al. 2004). The Platform-PAZ-Connector helix region is also known as "the PAZ cassette" (Tian et al. 2014) or "the PPC cassette" (Koralewska et al. 2021). However, there are examples of Dicers that are missing some of the accessory domains mentioned above; e.g., Dicer from *Giardia intestinalis* is devoid of the helicase, DUF283, and dsRBD domains (Macrae et al. 2006), *Arabidopsis thaliana* DCL3 protein is deficient in DUF283 (Johanson et al. 2013), and the green alga *Chlamydomonas reinhardtii* DCL3 lacks the PAZ domain (Valli et al. 2016). Similarly to Drosha, the two RNase III domains of Dicer must dimerize to form a catalytic valley where hydrolysis of the phosphodiester bonds of dsRNA occurs (Macrae et al. 2006); the access to the catalytic valley is restricted by the accessory domains of Dicer (Fig. 7.3).

In animals, the gene encoding Dicer ribonuclease was initially identified in 1994 during a screen of a *Xenopus* ovary cDNA expression library for dsRNA-binding proteins (Bass et al. 1994). In plants, the first documented identification of a Dicer-related gene dates back to 1999, with the characterization of the *CARPEL FACTORY (CAF)* gene in *A. thaliana* playing a role in regulating floral meristem determinacy (Jacobsen et al. 1999). The *CAF* gene is now known as *DCL1*, which encodes a well-characterized endonuclease responsible for the biogenesis of plant miRNAs (Meyers and Axtell 2019). In the same year, the human Dicer protein was first reported following its isolation via a yeast two-hybrid screen that used, as a bait, 5-lipoxygenase, an enzyme critical for leukotriene biosynthesis (Provost et al. 1999). Sequence analysis of the identified protein revealed strong homology to the hypothetical helicase K12H4.8 from *Caenorhabditis elegans*, as well as the presence of an RNase III signature motif and a dsRBD domain (Provost et al. 1999). In 2000, a comprehensive characterization of the *DICER1* gene, at that time referred to as "*HERNA*" (helicase with RNase motif), was published (Matsuda et al. 2000). The HERNA protein was shown to contain conserved RNA helicase features, including an ATP-binding motif, a DExD/H-box (Asp-Glu-x-Asp/His, where "x" can be any amino acid), and RNA-binding domains. Its expression was reported to be both ubiquitous and tissue-specific (Matsuda et al. 2000). The name "Dicer" was given in 2001, referring to the enzyme's ability to "dice" dsRNA into uniformly sized small RNAs, ~21–23 nt in length (Bernstein et al. 2001). As Dicer displayed specificity for dsRNAs, it was assigned to the family of RNase III endoribonucleases (Bernstein et al. 2001). Dicer ribonucleases are mainly known for their important role in the biogenesis of small regulatory RNAs, such as miRNAs (Fig. 7.5) and siRNAs.

7.4.1 Evolution of Dicer Proteins

Evolutionary studies show that Dicer and its homologs appeared only in eukaryotes (Mukherjee et al. 2013). The Dicer gene was duplicated during the emergence of the earliest multicellular organisms, and then Dicer genes differentiated independently in animals, plants, and fungi (Mukherjee et al. 2013). To meet the complex regulatory needs of gene expression and viral defense, animals evolved two Dicer proteins, Dicer-1 and Dicer-2. Dicer-1 cleaves pre-miRNA to produce miRNA, while Dicer-2 processes dsRNA into siRNA (Table 7.1) (Lee et al. 2004). One of the most well-studied Dicer-2 proteins is *Drosophila* Dicer-2, which contributes significantly to the antiviral response by cleaving viral dsRNAs (de Jong et al. 2009; Kurzynska-Kokorniak et al. 2015). Lineages that evolved complex immune systems, such as vertebrates, lost Dicer-2 (Mukherjee et al. 2013); however, *Xenopus laevis* expresses two Dicer forms, named L and S, which are encoded on long and short chromosomes, respectively (Koralewska et al. 2024). The Dicer gene has also been lost in some parasitic protozoa and some fungi; e.g., the model organism *Saccharomyces cerevisiae* and other closely related yeasts (Drinnenberg et al. 2011; Nicolas et al. 2013).

Regarding parasitic protozoa, it is important to mention that while canonical Dicer-like RNase III enzymes are largely absent or not involved in typical small RNA processing in trypanosomatids, RNase III-related activities are critical in editosome function (Carnes et al. 2005; Trotter et al. 2005). The editosome is a large multi-protein complex found in kinetoplastid parasites (trypanosomatids) that catalyzes the insertion and deletion of uridine residues in mitochondrial mRNAs, a process known as RNA editing. This post-transcriptional modification is essential for the generation of functional mitochondrial transcripts. The editosome harbors several RNase III-like endonucleases, notably KREN1, KREN2, and KREN3 (after: kinetoplastid RNA editing endonucleases). Although these proteins diverge from classical RNase III enzymes, they retain RNase III-like domains (Fig. 7.1) and are functionally adapted for site-specific cleavage of mitochondrial pre-mRNAs. The selection of editing sites is directed by trans-acting, non-coding guide RNAs (gRNAs). A region of each gRNA is complementary to specific clusters of editing sites, enabling the formation of an intermolecular RNA duplex that guides endonucleolytic cleavage of the pre-mRNA at the single-stranded/double-stranded junction. This cleavage is a critical step in the RNA editing mechanism (Carnes et al. 2005; Trotter et al. 2005). In contrast to canonical RNase III enzymes, the KREN1–3 proteins contain a PUF domain instead of the typical dsRBD (Nicholson 2014). The PUF (Pumilio/FBF) domain is a conserved RNA-binding domain found in Pumilio proteins (Edwards et al. 2001; Zamore et al. 1999), which are involved in post-transcriptional regulation of gene expression. Given that PUF domains in other proteins are known to recognize RNA in a sequence-specific manner, it is likely that this domain contributes to the selection of cleavage sites (of note, a dsRBD recognizes RNA based on its structure rather than sequence). In addition to the RNase III and PUF domains, KREN1–3 proteins also comprise a C2H2 zinc finger (ZnF) domain (Fig. 7.1) (Nicholson 2014). All known RNase III endonucleases operate as dimers, either as intermolecular homodimers like the classic Ec-RNase III or as intramolecular heterodimers like Drosha or Dicer. Interestingly, editing endonucleases KREN1–3 form intermolecular heterodimers with the catalytically inactive editosome's proteins, KREPB4 and KREPB5, that are essential for editosome integrity (Carnes et al. 2012). Considering the distinct structural features of KREN nucleases compared to bacterial RNases, Drosha, and Dicer, their classification within the established RNase III classes is challenging. However, due to their greatest similarity to bacterial and yeast nucleases, which possess a single RNase III domain, we have here assigned KREN nucleases to Class 1 (Fig. 7.1).

The plant Dicer protein family likely diversified early in the evolution of land plants when the Dicer gene underwent a four-way duplication, giving rise to four distinct Dicer groups (DCL1 to DCL4) (Mukherjee et al. 2013). DCL1 is found in the nucleus and is involved in the production of miRNAs from pri-miRNA transcripts (Fig. 7.5), whereas DCL2, DCL3, and DCL4 generate siRNAs (Xie et al. 2004). Specifically, DCL2, similarly to its insect homolog Dicer-2, is involved in antiviral response by producing 22-nt siRNAs from virus-derived dsRNAs. DCL2-generated siRNAs are crucial in recognizing and silencing viral transcripts (Bouché et al. 2006; Mlotshwa et al. 2008). DCL3 produces 24-nt siRNAs involved in heterochromatin formation and transcriptional gene silencing (Chan et al. 2004). DCL4, similarly to DCL2, is involved in antiviral defense (Bouché et al. 2006). In addition, DCL4 produces trans-acting siRNAs (tasiRNAs) (Table 7.1) (Allen et al. 2005). Nevertheless, certain plants have more than four DCL proteins. For example, in *Medicago truncatula*, six DCL genes are found: *DCL1*, *DCL2*, *DCL3*, *DCL4*, and two *DCL2* homologs, including the variant encoding a truncated DCL2 protein (Tworak et al. 2016). All these DCL genes have been found to be expressed at the mRNA level in *M. truncatula*, with expression levels varying across different tissues, including cotyledons, mature leaves, stems, shoot tips, roots, and seeds (Tworak et al. 2016).

The DCL family was already present in early land plants, such as mosses (*Physcomitrella patens*), but it was relatively simpler with fewer members (Liu et al. 2009). Moreover, Dicer proteins were already present in algae (Liu et al. 2009). For example, *C. reinhardtii*, a single-cell green alga, developed RNA silencing machinery involving three DCL proteins (DCL1–3) (Table 7.1) (Merchant et al. 2007; Casas-Mollano et al. 2008). However, the miRNA-based silencing system of algae is distinct from that of land plants and shares some similarities with the miRNA silencing mechanism observed in animals. *C. reinhardtii* DCL3 is known to generate the majority of miRNAs, despite the absence of the PAZ domain (Valli et al. 2016). A distinct characteristic of DCL3 is a P-rich domain located on the N-terminal side of the RNase III domains. A similar domain is found in a related protein, a ribonuclease Drosha, that also lacks a typical PAZ domain (Valli et al. 2016; Ha and Kim 2014). It might be possible that *C. reinhardtii* DCL3 serves the roles of both Drosha and Dicer, participating in the subsequent stages of miRNA biogenesis (Valli et al. 2016). The coupled functions of Drosha and Dicer are also characteristic of plant DCL1 (Fig. 7.5).

The plant and animal miRNA pathways are very different (Fig. 7.5), so it is possible that these two pathways evolved separately from the ancestral RNA silencing pathway (Ghildiyal and Zamore 2009; Axtell et al. 2011). The miRNA pathway in algae is also distinct from that of higher plants, and two evolutionary hypotheses could explain these differences. The first hypothesis assumes that animal, higher plant, and algal miRNA pathways evolved independently of each other. The second hypothesis assumes that the animal-like miRNA pathway evolved early and persisted in lower plant lineages, including green algae (photosynthetic protists), although it was not retained in higher plants. The second hypothesis is more likely because the *C. reinhardtii* and animal miRNA pathways share the presence of a Drosha-like structure, i.e., the presence of the P-rich domain and the absence of the PAZ domain in the miRNA-generating enzyme (Valli et al. 2016).

While evolutionary studies indicate that Dicer proteins are absent from bacteria and archaea, the helicase domain of Dicer displays high sequence similarity to its prokaryotic orthologs, specifically the archaeal Hef proteins that are part of Superfamily 2 (SF2) helicases (Aravind et al. 1999; Lestini et al. 2015). Additionally, the RNase III domains of Dicer are highly similar to the RNase III domains of bacteria (Shabalina and Koonin 2008). The integration of the SF2 helicase and the RNase III domains within a single protein is a eukaryotic signature and is believed to be one of the key early events that contributed to the consolidation of the eukaryotic RNAi system (Shabalina and Koonin 2008).

7.4.2 *Accessory Domains of Dicer Proteins*

7.4.2.1 The Helicase Domain

With a high degree of sequence similarity to the helicases of the superfamily SF2 (Aravind et al. 1999; Lestini et al. 2015; Shabalina and Koonin 2008), the helicase domain is regarded as one of the most conserved regions across all Dicer proteins (Kidwell et al. 2014). Based on sequence homology, the SF2 superfamily is categorized into several families, including, among others, the DExD/H-box RNA and DNA helicases and RIG-I-like helicases (Byrd and Raney 2012). The DExD/H-box motif (also known as the Walker B motif) is crucial for adenosine 5′-triphosphate (ATP) hydrolysis and is present in the helicases that unwind double-stranded regions of nucleic acids (Singleton et al. 2007). The RIG-I motif is characteristic of ATP-dependent dsRNA translocases (Byrd and Raney 2012). The helicase domains of Dicer-type proteins contain three subdomains known as HEL1, HEL2i (insertion domain), and HEL2 (Kidwell et al. 2014). These three subdomains are structurally organized into three lobes (Fig. 7.3) (Liu et al. 2018a). HEL1 includes the DExD/H-box motif, HEL2i contains the RIG-I motif, whereas HEL2 (also called "helicase C") is conserved in all helicases from the DExD/H and the RIG-I-like families (Zou et al. 2009). However, not all Dicer proteins have all three helicase subdomains, and some may lack completely the helicase domain; e.g., *D. melanogaster* miRNA generating Dicer-1 has a degenerated HEL1 subdomain (Lee et al. 2004), HEL2i is missing in the Dicer-type protein from the fungus *Magnaporthe oryzae* (Kadotani et al. 2004), while *Giardia* Dicer lacks the entire helicase domain (Macrae et al. 2006).

A comparative analysis of the helicase domains of *A. thaliana* DCL1 (Wei et al. 2021), *D. melanogaster* Dicer-2 (Su et al. 2022a), and mammalian Dicers (Liu et al. 2018a; Lee et al. 2023a; Zapletal et al. 2022) indicates that both *A. thaliana* DCL1 and *D. melanogaster* Dicer-2 have a wide cleft within their helicase domains that can accommodate dsRNA regions of substrates. The helicase domains of *D. melanogaster* Dicer-2 or plant DCL proteins can also bind long dsRNAs, such as viral dsRNAs (van Rij et al. 2006; Ahmad and Hur 2015). This ability of the helicase domains is characteristic of Dicers, which contribute to viral defense mechanisms. These Dicers also display ATP-dependent translocation activity (Tworak et al. 2016; Liu et al. 2012; Fukudome and Fukuhara 2017). In contrast, the helicase domain of mammalian Dicers has a much narrower cleft that cannot accommodate dsRNAs (Liu et al. 2018a; Lee et al. 2023a; Zapletal et al. 2022). Deleting the helicase domain in mammalian Dicer increases the efficiency of pre-siRNA substrate processing (Kennedy et al. 2015). A naturally occurring Dicer isoform that lacks the HEL1 subdomain is expressed in mouse oocytes. This truncated isoform efficiently produces small RNAs from long dsRNA substrates (Zapletal et al. 2022; Buccheri et al. 2024; Flemr et al. 2013). Additionally, some mammalian cell lines produce a truncated Dicer variant devoid of the HEL2i subdomain; this Dicer variant protects tissue stem cells from RNA viruses by cleaving viral dsRNA (Poirier et al. 2021).

ATPase Activity of the Helicase Domains

In general, the unwinding and translocase activities of DExD/H-box helicases and RIG-I-like helicases, respectively, are fueled by the hydrolysis of ATP (Jankowsky and Fairman 2007). However, the presence of the DExD/H-box and RIG-I motifs does not guarantee that the protein will unwind dsRNA or translocate. Examples of such proteins are vertebrate Dicers, whose helicase domains possess DExD/H-box and RIG-I motifs but lack RNA unwinding and translocase activities. The ATPase activity of vertebrate Dicers is also a matter of debate. The helicase domains of vertebrate Dicers have well-conserved ATP-binding (Walker A) and ATP hydrolysis (Walker B) motifs; however, it is suggested that ancestral Dicers in the human Dicer lineage had decreased ATP-binding affinity and thus lost ATPase activity, in contrast to plant or arthropod Dicers (Aderounmu et al. 2023). Our studies show, however, that the helicase domain of human Dicer does hydrolyze ATP (Ciechanowska et al. 2024). *In vitro*, the ATPase activity of human Dicer can only be observed under a low substrate-to-enzyme ratio (Ciechanowska et al. 2024). We hypothesize that under the *in vitro* conditions, product dissociation (ADP or Pi, or both) may be a reaction-limiting step. In living cells, a product release is presumably stimulated by a yet unknown factor or factors missing in standard *in vitro* ATP hydrolysis assays (Ciechanowska et al. 2024).

In contrast to vertebrate Dicers, the ATP hydrolysis activity has been well-documented for *D. melanogaster* Dicer-2 (Cenik et al. 2011; Su et al. 2022b; Welker et al. 2011), *C. elegans* Dicer (Welker et al. 2011), plant DCL proteins (Tworak et al. 2016; Wei et al. 2021; Liu et al. 2012; Fukudome and Fukuhara 2017), and *Schizosaccharomyces pombe* (fission yeast) Dicer (Colmenares et al. 2007). These Dicer-type proteins have functional RIG-I type subdomains and act as viral RNA sensors that initiate appropriate effector RNAi pathways that trigger viral RNA degradation (van Rij et al. 2006). It is suggested that the helicase domains of vertebrate Dicers lost their ability to recognize dsRNA because vertebrates evolved other RNA virus sensors, such as RIG-I helicases, which act as receptors for the innate immune system (Hansen et al. 2019). Therefore, the helicase domains of vertebrate and invertebrate Dicer proteins might develop different activities (Hansen et al. 2019).

ATP hydrolysis by the helicase domain is critical to the successful processing of a long dsRNA into siRNA duplexes (Cenik et al. 2011; Su et al. 2022b; Welker et al. 2011). Nevertheless, in plant DCL1 proteins (miRNA-generating enzymes), the ATPase function of the helicase domain has been retained (Wei et al. 2021). Structural data collected for *Arabidopsis* DCL1 complexed with a pri-miRNA and a pre-miRNA suggest that the helicase domain serves as an engine that transfers the substrate between two sequential cleavage events (Wei et al. 2021).

In organisms with a single Dicer, such as nematodes and mammals, the helicase domain distinguishes between pre-miRNA and pre-siRNA substrates (Ma et al. 2008; Chakravarthy et al. 2010). However, the discriminating mechanism appears to differ for nematode and mammalian Dicers. The human Dicer's helicase domain, by interacting with the apical loop of a pre-miRNA, can distinguish between pre-miRNA and pre-siRNA (Ma et al. 2012; Taylor et al. 2013). Nevertheless, most invertebrates segregate the processing of RNA substrates between two Dicer proteins, Dicer-1 (miRNA-generating enzyme) and Dicer-2 (siRNA-generating enzyme) (Mukherjee et al. 2013; Lee et al. 2004). Additionally, invertebrate Dicers cleave pre-siRNA differently, depending on the substrate termini: blunt-ended dsRNAs are progressively cleaved in an ATP-hydrolysis dependent manner, while substrates with 3′-overhanging termini undergo distributive cleavage that does not require ATP (Kidwell et al. 2014; Welker et al. 2011). A cryo-EM structure of *Drosophila* Dicer-2 in complex with a blunt-ended dsRNA revealed that this substrate binds to the helicase domain (Sinha et al. 2018), while substrates with 2-nt 3′-overhanging termini are bound by the PAZ and Platform domains (Tian et al. 2014).

The Helicase Domain as a Protein Binding Platform

The helicase domain provides a binding platform for proteins cooperating with Dicer ribonucleases. For example, the helicase domain of human Dicer may bind the TAR RNA-binding protein (TRBP) or the protein activator of protein kinase R (PACT) (Kedersha and Rome 1986). Both TRBP and PACT contain dsRBD domains and are important regulatory factors that contribute to substrate specificity during miRNA and siRNA production by influencing substrate discrimination (Lee et al. 2013). TRBP and PACT may also impact the selection of the cleavage site by Dicer, thereby triggering the generation of different-sized miRNAs (iso-miRs) (Fukunaga et al. 2012; Lee and Doudna 2012; Wilson et al. 2015). Alternative pre-miRNA processing can significantly alter the repertoire of targeted mRNAs and, consequently, the silenced genes. In other species, the Dicer helicase domain interacts with various partner proteins that perform similar functions to human TRBP or PACT. In *Drosophila*, the Loquacious (Loqs) and R2D2 proteins bind to Dicer-1 and Dicer-2 proteins (Hansen et al. 2019; Liu et al. 2003, 2006, 2007). Loqs and R2D2 differently impact small regulatory RNA biogenesis (Fukunaga et al. 2012; Hartig et al. 2009; Zhou et al. 2009). Additionally, Loqs isoforms may tune the length of mature miRNAs (Fukunaga et al. 2012).

During viral infection, the helicase domain of human Dicer may specifically interact with a number of dsRNA-binding proteins and RNA helicases (Montavon et al. 2021). Specifically, the helicase domain may interact with DExH-Box Helicase 9 (DHX9), adenosine deaminase acting on RNA 1 (ADAR-1), and protein kinase RNA-activated (PKR) (Montavon et al. 2021). In this way, the helicase domain may control the recruitment of different factors that diversify the functions of Dicer proteins.

7.4.2.2 The DUF283 Domain

The name "DUF" stands for ***d**omain of **u**nknown **f**unction* and is given to protein domains with uncharacterized functions, followed by the number assigned to a particular domain. The DUF283 domain of Dicer-type proteins adopts the αβββα fold (Fig. 7.3) (Liu et al. 2018a; Dlakic 2006; Qin et al. 2010), similar to that of dsRBD that can be found in prokaryotic, eukaryotic, and viral proteins (Gleghorn and Maquat 2014). The role of DUF283 in the small RNA biogenesis pathway is still elusive. Although the DUF283 domain adopts a structural motif characteristic of the canonical dsRBD motifs involved in dsRNA-binding (Fig. 7.3), experimental data have shown only trace dsRNA-binding activity of the *A. thaliana* DCL4 DUF283 domain (Qin et al. 2010), and there are no reports on dsRNA-binding activity for the DUF283 domains of other Dicer proteins (Kurzynska-Kokorniak et al. 2016). Results of our studies indicate, however, that the human Dicer DUF283 domain can bind ssRNAs and ssDNAs (Kurzynska-Kokorniak et al. 2016). Moreover, DUF283 of human Dicer, by supporting the base pairing of complementary sequences found in nucleic acids, may perform a function characteristic of nucleic acid annealers (chaperone-like proteins) (Kurzynska-Kokorniak et al. 2016; Pokornowska et al. 2020). Nevertheless, the biological relevance of this activity has not yet been found.

Due to its involvement in binding Dicer partner proteins, the DUF283 domain is also known as the "dimerization domain". For example, DUF283 from *A. thaliana* DCL1 and DCL4 can bind HYL1 (hyponastic leaves 1) and DRB4 (dsRNA-binding protein 4), respectively (Qin et al. 2010), while the DUF283 domain of human Dicer interacts with ADAR-1 (Ota et al. 2013). Interactions between DUF283 and ADAR-1 support both the efficiency of pre-miRNA cleavage by Dicer and miRNA transfer to Argonaute (Ago) proteins (Ota et al. 2013).

7.4.2.3 The PAZ Domain

The PAZ and two RNase III domains constitute a functional core of canonical Dicer-type proteins (Fig. 7.3). These domains also compose a "minimal Dicer" such as *Giardia* Dicer (Fig. 7.1) (Macrae et al. 2006). The Dicer's PAZ domain shows high structural similarity to the PAZ domains of other proteins involved in small RNA processing and the effector stages of RNAi (Tahbaz et al. 2004). The PAZ domain can be found in two protein families: the PAZ-Piwi Domain Proteins family (PPD) and the Dicer family (Kolb et al. 2005). Well-known PPD proteins are Ago proteins. Structural studies indicate that PAZ consists of an oligonucleotide/oligosaccharide binding fold (OB-fold)-like structure that may bind ssRNAs or ssDNAs (Lingel et al. 2003; Song et al. 2003). The PAZ domain of human Ago1 anchors the 2-nt 3′-overhang of a siRNA duplex within a highly conserved binding pocket (Ma et al. 2004), which is similar to the PAZ domains of Dicer proteins. The initial recognition and binding of a pre-miRNA substrate by Dicer take place within the Platform–PAZ–Connector helix cassette (Tian et al. 2014). This region contains two adjacent pockets, a 2-nt 3′-overhang-binding pocket (3′ pocket) located within the PAZ domain and a phosphate-binding pocket (5′ pocket) located within the Platform and PAZ domains (Fig. 7.3) (Tian et al. 2014). These two pockets in the tertiary structure are placed in close proximity to ensure simultaneous accommodation of the substrate ends (Park et al. 2011). In contrast to pre-siRNAs, the pre-miRNA hairpins usually have only partially paired ends, which are relatively easy to split up (Tian et al. 2014). Consequently, the 5′-end of pre-miRNA can be easily anchored within the 5′ pocket. The 5′ pocket is highly conserved across most miRNA-producing Dicer homologs, and is missing in Dicers found in lower eukaryotes, lacking the miRNA pathway, such as fungi and *Giardia* (Park et al. 2011). The miRNA-generating plant DCL1 enzymes are only partially conserved in the region spanning the 5′ pocket (Park et al. 2011). Additionally, siRNA-generating *Drosophila* Dicer-2 has a unique 5′ pocket, which is important for the fidelity of 21-nt long siRNA production (Kandasamy and Fukunaga 2016). In contrast to the PAZ domain of Dicer proteins, the PAZ-like domain of Drosha does not bind substrate (pri-miRNA) ends (Jin et al. 2020; Garg et al. 2024).

Some Dicer enzymes lack the PAZ domain. A very simple Dicer-like protein found in the early diverging protozoan *Trypanosoma brucei* comprises only two RNase III domains (Shi et al. 2006). Dicer of another protozoan parasite, *Entamoeba histolytica*, has a single RNase III domain, but it lacks a typical dsRBD domain (Yu et al. 2017). The budding yeast *S. cerevisiae* does not have the ribonuclease Dicer, but other yeasts such as *Saccharomyces castellii*, *Candida albicans* (Drinnenberg et al. 2009), and *Kluyveromyces polysporus* (Weinberg et al. 2011), comprise simple

Dicer proteins composed of only one RNase III domain and dsRBD. Dicers that are devoid of the PAZ domain cannot bind the substrate ends. Instead, such "non-canonical" Dicers initiate processing in the interior and work outward (Weinberg et al. 2011). The PAZ-deficient Dicers, as dimers, bind cooperatively along the dsRNA substrate, and the distance between the successive active sites determines the length of the generated products (~23-nt RNAs) (Weinberg et al. 2011). Regarding canonical Dicers, miRNA or siRNA length is determined by the distance from the 3′ and 5′ pockets to the catalytic center of the enzyme, i.e. RNase IIIa and IIIb domains, respectively (Fig. 7.4) (Liu et al. 2018a; MacRae and Doudna 2007; Lau et al. 2012).

Recognition and cleavage of a substrate by plant DCL proteins differ from the processes observed for animal Dicers. Plant pri-miRNAs are much more variable in size and shape than animal miRNA precursors (Rogers and Chen 2013; Axtell et al. 2011; Bologna and Voinnet 2014). In addition, plant miRNA precursors are processed by only one enzyme, the DCL1 protein (Fig. 7.5). Structural data for *Arabidopsis* DCL1 in complex with pri-miRNA and pre-miRNA have revealed that a huge plasticity of DCL1 is provided by the PAZ domain, which is critical for the recognition of both pri-miRNA and pre-miRNA substrates (Wei et al. 2021). Plant miRNA precursors can be processed in two different modes: a "base-to-loop" mode and a "loop-to-base" mode (Moro et al. 2018). In a "base-to-loop" processing mode, pri-miRNA is bound ~15–17 nt below the lower stem, which determines the position of the first cut by DCL1; the second cut takes place 21 nt away from the first one, releasing the miRNA duplex (Moro et al. 2018; Mateos et al. 2010; Zhu et al. 2013; Bologna et al. 2013). This mechanism is similar to the processing of animal pre-miRNAs (Chorostecki et al. 2017). In a "loop-to-base" processing mode, the miRNA precursor processing starts from the cleavage below the loop region of the pri-miRNA, and further processing continues towards the base to release the mature miRNA (Moro et al. 2018; Bologna et al. 2013; Addo-Quaye et al. 2009).

7.4.2.4 The dsRBD Domain

A simple RNase III enzyme, such as Ec-RNase III, is composed of only a single RNase III domain and a single dsRBD, and it forms a homodimer to generate ~11-bp products from dsRNAs or hairpin substrates (Jin et al. 2019). The proposed catalytic pathway for simple RNase III enzymes assumes that dsRBD plays an important function by binding the dsRNA substrate and handing it over to the RNase III domains. After the cleavage, dsRBD takes the product over and releases it, making the enzyme ready for another round of cleavage (Nicholson 2014; Hansen et al. 2019). In the case of more complex RNase III enzymes, like Dicers, it has been demonstrated that the C-terminal fragment of human Dicer, comprising two RNase III domains and dsRBD, cleaves dsRNA into ~15-bp products (Ma et al. 2012). The reconstituted human Dicer, containing both the N-terminal fragment (comprising DUF283, Platform, PAZ) and the C-terminal fragment, restores the production of the characteristic ~21–23-bp products (Ma et al. 2012). Recent structural data have revealed that the dsRBD domain of human Dicer recognizes the "GYM motif" (paired G, paired pyrimidine, and mismatched C or A) located in pre-miRNA near the cleavage site (Fig. 7.4) (Lee et al. 2023a). The GYM motif promotes the processing of a pre-miRNA substrate at a specific position (Lee et al. 2023a).

Mammalian Dicers are mainly found in the cytoplasm, where they produce miRNAs (Fig. 7.5); however, these proteins can also be found in the nucleus (Sinkkonen et al. 2010; Yadav et al. 2020). The shuttling properties of human Dicer are attributed to the NLS comprising a number of positively charged amino acid residues located on the surface of dsRBD (Doyle et al. 2013; Banerjee and Barraud 2014). The dsRBD-dependent transport of human Dicer into the nucleus is mediated by importins (Importin-β, Importin-7, and Importin-8) (Doyle et al. 2013). It is suggested that the NLS present in the dsRBD domain can be masked by other Dicer's domains; e.g., the helicase domain (Doyle et al. 2013) or RNA molecules associated with dsRBD (Pong and Gullerova 2018). The dsRBD domain of *S. pombe* Dicer has also been reported to comprise NLS. In contrast to human Dicer, *S. pombe* Dicer localizes predominantly in the nucleus, near the nuclear pore complexes (Banerjee and Barraud 2014). The dsRBD domain of *S. pombe* Dicer has a C-terminal extension composed of an α-helical turn followed by a zinc-coordination motif (Banerjee and Barraud 2014). This extension is thought to provide a protein-protein interaction surface for an as-yet unidentified nuclear protein, potentially contributing to the nuclear retention of *S. pombe* Dicer. Additionally, the extended dsRBD of the *S. pombe* Dicer is a temperature-sensitive domain. When the temperature rises above 34 °C, the nuclear retention properties of the dsRBD domain change, and the *S. pombe* Dicer migrates to the cytoplasm to control the expression of stress-dependent genes (Woolcock et al. 2012).

The dsRBD domain of human Dicer has also been shown to bind directly to 5-lipoxygenase (Dincbas-Renqvist et al. 2009), and this interaction influences Dicer's processing of miRNA precursors (Uebbing et al. 2021) (of note, human Dicer cDNA clones were isolated from a yeast two-hybrid screen using 5-lipoxygenase as a bait (Provost et al. 1999)).

7.4.3 *Dicers as Multitasking Proteins*

Dicer ribonucleases are mostly known for their essential functions in the biogenesis of miRNAs (Fig. 7.5) and siRNAs (Kurzynska-Kokorniak et al. 2015; Ciechanowska et al. 2021). This canonical role of Dicer is associated with its cytoplasmic localization (Bernstein et al. 2001). In addition to miRNA and siRNA production, Dicer proteins have been implicated in the biogenesis of small RNAs derived from tRNAs, known as tRNA-derived fragments (tRFs) (Cole et al. 2009). Some tRNA molecules can fold into structures resembling canonical Dicer substrates. Therefore, it is possible that Dicer-dependent tRF processing has evolved in some organisms as a mechanism of biological relevance (Pong and Gullerova 2018). Indeed, there are reports showing that Dicer-generated tRFs may be involved in the regulation of genome stability through the targeting of endogenous transposable element transcripts produced from transcriptionally active sites (Martinez et al. 2017). Moreover, Dicer-dependent tRFs have been shown to promote gene silencing through a mechanism distinct from post-transcriptional gene silencing and transcriptional gene silencing (Di Fazio et al. 2022). These tRFs can guide Ago2 proteins to nascent protein-coding and long non-coding transcripts in the nucleus, promoting so-called nascent RNA silencing (NRS). Aberrant expression of NRS-regulated genes has been linked to multiple diseases, underscoring the physiological relevance of this pathway (Di Fazio et al. 2022).

Dicer can also function in the nucleus. It is already known that nuclear Dicer can participate in chromatin structure remodeling (Fukagawa et al. 2004; Gullerova and Proudfoot 2012; Verdel et al. 2004; Volpe et al. 2002; Noma et al. 2004), restrict the deleterious accumulation of endogenous dsRNAs (White et al. 2014), or contribute to DNA damage response activation by generating, along with Drosha, damage-induced small RNAs, diRNAs (Kurzynska-Kokorniak et al. 2016; Pokornowska et al. 2020; Burger et al. 2017). Nuclear Dicer is also implicated in the transcription-independent, global-genomic nucleotide excision repair pathway, in which it mediates the recruitment of a histone methyltransferase complex to the DNA damage sites (Chitale and Richly 2018). In this way, Dicer presumably facilitates chromatin decondensation, allowing repair factors to access the damaged sites (Chitale and Richly 2017). Moreover, recent data show that human Dicer may be involved in cleaving RNA within R loops (Table 7.1) (Camino et al. 2023). R loops form naturally in cells, mainly as intermediates during replication or as by-products during transcription. They are involved in the regulation of many essential cellular processes, such as replication, transcription, centromere stabilization, protection of chromosome ends, and control of the telomere length. Unscheduled R loops are linked to many diseases, including cancer and neurodegenerative or inflammatory disorders (Wojcik et al. 2025). Dicer proteins may play an important role in the regulation of R loops, preventing their accumulation in cells (Camino et al. 2023).

7.5 Summary

The gain and loss of protein domains are important events that have occurred frequently during the evolution of species. The RNase III enzymes gained domains to improve the precision of substrate cleavage and generate products of defined lengths. In the case of Dicer-type proteins, the combination of PAZ and RNase III domains within one protein resulted in the formation of a molecular ruler responsible for generating small regulatory RNAs of ~21–23 nt in size. If small regulatory RNAs were longer, instead of playing a regulatory role, they would activate intracellular RNA virus sensors, such as RIG-I helicases, which would consequently induce the interferon pathway and cause inflammation in vertebrates or activate an antiviral response in lower organisms (Ciechanowska et al. 2021). A minimal Dicer of *G. intestinalis* has only PAZ and two RNase III domains, sufficient to specifically bind dsRNA and generate small regulatory RNAs of a specific length (Macrae et al. 2006).

Acquiring a helicase domain allowed invertebrate and plant Dicer-type proteins to function as cytosolic sensors of foreign RNA, which effectively translocate and cut dsRNAs in an ATP-hydrolysis-dependent manner. However, the helicase domain of vertebrate Dicers does not seem to be involved in recognizing long dsRNAs, such as viral RNAs. Instead, it helps Dicer proteins to discriminate among pre-miRNA substrates and between pre-miRNAs and pre-siRNAs. The helicase domain is also involved in the passive binding of some cellular transcripts (Ciechanowska et al. 2024; Rybak-Wolf et al. 2014). Such passive substrate binding by Dicer may sequester its cleavage activity, but the biological role of this activity has yet to be elucidated. Additionally, recent findings of the ATPase activity of the helicase domain of human Dicer (Ciechanowska et al. 2024) lead us to speculate on the potential biological significance of this activity. We hypothesize that activation of the ATP-hydrolyzing function may trigger a conformational rearrangement of human Dicer domains. Once stimulated, Dicer may transmit signals downstream to activate or repress as yet undefined cellular pathways. This hypothesis postulates that human Dicer may be directly involved in signal transduction, which may also apply to other vertebrate Dicers.

Gaining the helicase and DUF283 domains (Gleghorn and Maquat 2014) has allowed Dicer proteins to develop a binding platform for protein-protein interactions. By choosing the protein partner, Dicer can control the specificity of substrate and cleavage site selection (Fukunaga et al. 2012; Lee and Doudna 2012; Wilson et al. 2015). Additionally, the helicase domain of human Dicer has been shown to sequester proteins involved in the antiviral response, such as DHX9, ADAR-1, and PKR kinase (Montavon et al. 2021). By binding PKR kinase, human Dicer can be engaged in PKR-dependent regulation of immune signaling pathways, especially by preventing unwanted activation of antiviral and inflammatory pathways (Baldaccini et al. 2024). In addition, interactions between Dicer and 5-lipoxygenase, involving the dsRBD of Dicer, may also influence the processing of miRNA precursors (Uebbing et al. 2021).

The acquisition of domains equipped with the NLS, such as NTD (yeast RNases III) (Catala et al. 2004), RS-rich domains (Drosha) (Son et al. 2023), and dsRBD (Dicers) (Doyle et al. 2013; Banerjee and Barraud 2014), has enabled RNase III enzymes to shuttle between the cytoplasm and the nucleus upon specific signals.

Regarding domain loss or degeneration, it can be noticed that the helicase domain of the *Drosophila* miRNA-producing Dicer-1 has degenerated and cannot hydrolyze ATP (Lee et al. 2004). However, this scenario seems to be energy-efficient and beneficial for the cell, as ATPase activity appears to be dispensable for pre-miRNA processing.

Some domains can also be excluded from Dicer enzymes by alternative splicing, transcription from alternative promoters, or proteolytic cleavage processes. For example, a truncated Dicer variant lacking a fragment of the helicase domain is produced via alternative splicing in mammalian stem cells (Poirier et al. 2021). Similarly, a Dicer isoform that lacks the HEL1 subdomain is expressed in mouse oocytes from an alternative, intronic retrotransposon promoter (Zapletal et al. 2022; Buccheri et al. 2024; Flemr et al. 2013). These processes remove the steric hindrance in Dicer, ensuring efficient dsRNA cleavage (the helicase domain of vertebrate Dicers restricts the access of long dsRNAs to the catalytic valley). Moreover, the N-terminal fragment of *C. elegans* Dicer can be removed by death-promoting caspases (Nakagawa et al. 2010). The remaining Dicer fragment, containing one active RNase III domain and the dsRBD domain, no longer functions as an RNase III enzyme. Interestingly, despite the presence of the dsRBD domain (which specifically recognizes dsRNA), this truncated *C. elegans* Dicer acts as a DNase that nicks chromosomal DNA during apoptosis (Nakagawa et al. 2010).

Certainly, we do not know yet all the benefits that the accessory domains of RNase III enzymes may provide. However, it is undoubtedly interesting to speculate in which direction the RNase III enzymes will develop further.

Acknowledgments This work was supported by the National Science Centre, Poland [OPUS 2021/41/B/NZ2/03781 to A.K-K., and MINIATURA 2024/08/X/NZ1/00899 to A.Sz.]

References

Abou Elela S, Ares M Jr (1998) Depletion of yeast RNase III blocks correct U2 3′ end formation and results in polyadenylated but functional U2 snRNA. EMBO J 17:3738–3746

Abou Elela S, Ji X (2019) Structure and function of Rnt1p: an alternative to RNAi for targeted RNA degradation. Wiley Interdiscip Rev RNA 10:e1521

Addo-Quaye C, Snyder JA, Park YB, Li YF, Sunkar R, Axtell MJ (2009) Sliced microRNA targets and precise loop-first processing of MIR319 hairpins revealed by analysis of the Physcomitrella patens degradome. RNA 15:2112–2121

Aderounmu AM, Aruscavage PJ, Kolaczkowski B, Bass BL (2023) Ancestral protein reconstruction reveals evolutionary events governing variation in dicer helicase function. elife 12

Aguado LC, Schmid S, May J, Sabin LR, Panis M, Blanco-Melo D, Shim JV, Sachs D, Cherry S, Simon AE, Levraud JP, tenOever BR (2017) RNase III nucleases from diverse kingdoms serve as antiviral effectors. Nature 547:114–117

Ahmad S, Hur S (2015) Helicases in antiviral immunity: dual properties as sensors and effectors. Trends Biochem Sci 40:576–585

Allen E, Xie Z, Gustafson AM, Carrington JC (2005) microRNA-directed phasing during trans-acting siRNA biogenesis in plants. Cell 121:207–221

Allmang C, Kufel J, Chanfreau G, Mitchell P, Petfalski E, Tollervey D (1999) Functions of the exosome in rRNA, snoRNA and snRNA synthesis. EMBO J 18:5399–5410

Aravind L, Walker DR, Koonin EV (1999) Conserved domains in DNA repair proteins and evolution of repair systems. Nucleic Acids Res 27:1223–1242

Auyeung VC, Ulitsky I, McGeary SE, Bartel DP (2013) Beyond secondary structure: primary-sequence determinants license pri-miRNA hairpins for processing. Cell 152:844–858

Axtell MJ, Westholm JO, Lai EC (2011) Vive la différence: biogenesis and evolution of microRNAs in plants and animals. Genome Biol 12:221

Baldaccini M, Gaucherand L, Chane-Woon-Ming B, Messmer M, Gucciardi F, Pfeffer S (2024) The helicase domain of human dicer prevents RNAi-independent activation of antiviral and inflammatory pathways. EMBO J 43:806–835

Banerjee S, Barraud P (2014) Functions of double-stranded RNA-binding domains in nucleocytoplasmic transport. RNA Biol 11:1226–1232

Bardwell JC, Regnier P, Chen SM, Nakamura Y, Grunberg-Manago M, Court DL (1989) Autoregulation of RNase III operon by mRNA processing. EMBO J 8:3401–3407

Bass BL, Hurst SR, Singer JD (1994) Binding properties of newly identified Xenopus proteins containing dsRNA-binding motifs. Curr Biol 4:301–314

Bernstein E, Caudy AA, Hammond SM, Hannon GJ (2001) Role for a bidentate ribonuclease in the initiation step of RNA interference. Nature 409:363–366

Bhaskaran M, Mohan M (2014) MicroRNAs: history, biogenesis, and their evolving role in animal development and disease. Vet Pathol 51:759–774

Blaszczyk J, Tropea JE, Bubunenko M, Routzahn KM, Waugh DS, Court DL, Ji X (2001) Crystallographic and modeling studies of RNase III suggest a mechanism for double-stranded RNA cleavage. Structure 9:1225–1236

Blaszczyk J, Gan J, Tropea JE, Court DL, Waugh DS, Ji X (2004) Noncatalytic assembly of ribonuclease III with double-stranded RNA. Structure 12:457–466

Bologna NG, Voinnet O (2014) The diversity, biogenesis, and activities of endogenous silencing small RNAs in Arabidopsis. Annu Rev Plant Biol 65:473–503

Bologna NG, Schapire AL, Zhai J, Chorostecki U, Boisbouvier J, Meyers BC, Palatnik JF (2013) Multiple RNA recognition patterns during microRNA biogenesis in plants. Genome Res 23:1675–1689

Bouché N, Lauressergues D, Gasciolli V, Vaucheret H (2006) An antagonistic function for Arabidopsis DCL2 in development and a new function for DCL4 in generating viral siRNAs. EMBO J 25:3347–3356

Brate J, Neumann RS, Fromm B, Haraldsen AAB, Tarver JE, Suga H, Donoghue PCJ, Peterson KJ, Ruiz-Trillo I, Grini PE, Shalchian-Tabrizi K (2018) Unicellular origin of the animal MicroRNA machinery. Curr Biol 28(3288–3295):e5

Bubunenko M, Court DL, Al Refaii A, Saxena S, Korepanov A, Friedman DI, Gottesman ME, Alix JH (2013) Nus transcription elongation factors and RNase III modulate small ribosome subunit biogenesis in Escherichia coli. Mol Microbiol 87:382–393

Buccheri V, Pasulka J, Malik R, Loubalova Z, Taborska E, Horvat F, Roos Kulmann MI, Jenickova I, Prochazka J, Sedlacek R, Svoboda P (2024) Functional canonical RNAi in mice expressing a truncated dicer isoform and long dsRNA. EMBO Rep 25:2896–2913

Burger K, Gullerova M (2015) Swiss army knives: non-canonical functions of nuclear Drosha and dicer. Nat Rev Mol Cell Biol 16:417–430

Burger K, Schlackow M, Potts M, Hester S, Mohammed S, Gullerova M (2017) Nuclear phosphorylated dicer processes double-stranded RNA in response to DNA damage. J Cell Biol 216:2373–2389

Byrd AK, Raney KD (2012) Superfamily 2 helicases. Front Biosci (Landmark Ed) 17:2070–2088

Cabrini M, Roncador M, Galbiati A, Cipolla L, Maffia A, Iannelli F, Sabbioneda S, d'Adda di Fagagna F, Francia S (2021) DROSHA is recruited to DNA damage sites by the MRN complex to promote non-homologous end joining. J Cell Sci 134

Calin-Jageman I, Nicholson AW (2003) RNA structure-dependent uncoupling of substrate recognition and cleavage by Escherichia coli ribonuclease III. Nucleic Acids Res 31:2381–2392

Camino LP, Dutta A, Barroso S, Pérez-Calero C, Katz JN, García-Rubio M, Sung P, Gómez-González B, Aguilera A (2023) DICER ribonuclease removes harmful R-loops. Mol Cell 83:3707–3719.e5

Carnes J, Trotter JR, Ernst NL, Steinberg A, Stuart K (2005) An essential RNase III insertion editing endonuclease in Trypanosoma brucei. Proc Natl Acad Sci USA 102:16614–16619

Carnes J, Schnaufer A, McDermott SM, Domingo G, Proff R, Steinberg AG, Kurtz I, Stuart K (2012) Mutational analysis of Trypanosoma brucei editosome proteins KREPB4 and KREPB5 reveals domains critical for function. RNA 18:1897–1909

Casas-Mollano JA, Rohr J, Kim EJ, Balassa E, van Dijk K, Cerutti H (2008) Diversification of the core RNA interference machinery in Chlamydomonas reinhardtii and the role of DCL1 in transposon silencing. Genetics 179:69–81

Catala M, Lamontagne B, Larose S, Ghazal G, Elela SA (2004) Cell cycle-dependent nuclear localization of yeast RNase III is required for efficient cell division. Mol Biol Cell 15:3015–3030

Cenik ES, Fukunaga R, Lu G, Dutcher R, Wang Y, Tanaka Hall TM, Zamore PD (2011) Phosphate and R2D2 restrict the substrate specificity of Dicer-2, an ATP-driven ribonuclease. Mol Cell 42:172–184

Chakravarthy S, Sternberg SH, Kellenberger CA, Doudna JA (2010) Substrate-specific kinetics of dicer-catalyzed RNA processing. J Mol Biol 404:392–402

Chan SW, Zilberman D, Xie Z, Johansen LK, Carrington JC, Jacobsen SE (2004) RNA silencing genes control de novo DNA methylation. Science 303:1336

Chanfreau G, Elela SA, Ares M, Guthrie C (1997) Alternative 3′-end processing of U5 snRNA by RNase III. Genes Dev 11:2741–2751

Chanfreau G, Rotondo G, Legrain P, Jacquier A (1998) Processing of a dicistronic small nucleolar RNA precursor by the RNA endonuclease Rnt1. EMBO J 17:3726–3737

Chen X (2005) MicroRNA biogenesis and function in plants. FEBS Lett 579:5923–5931

Chitale S, Richly H (2017) DICER and ZRF1 contribute to chromatin decondensation during nucleotide excision repair. Nucleic Acids Res 45:5901–5912

Chitale S, Richly H (2018) DICER- and MMSET-catalyzed H4K20me2 recruits the nucleotide excision repair factor XPA to DNA damage sites. J Cell Biol 217:527–540

Chong MM, Rasmussen JP, Rudensky AY, Littman DR (2008) The RNAseIII enzyme Drosha is critical in T cells for preventing lethal inflammatory disease. J Exp Med 205:2005–2017

Chong MM, Zhang G, Cheloufi S, Neubert TA, Hannon GJ, Littman DR (2010) Canonical and alternate functions of the microRNA biogenesis machinery. Genes Dev 24:1951–1960

Chorostecki U, Moro B, Rojas AML, Debernardi JM, Schapire AL, Notredame C, Palatnik JF (2017) Evolutionary footprints reveal insights into plant MicroRNA biogenesis. Plant Cell 29:1248–1261

Ciechanowska K, Pokornowska M, Kurzynska-Kokorniak A (2021) Genetic insight into the domain structure and functions of dicer-type ribonucleases. Int J Mol Sci 22

Ciechanowska K, Szczepanska A, Szpotkowski K, Wojcik K, Urbanowicz A, Kurzynska-Kokorniak A (2024) The human dicer helicase domain is capable of ATP hydrolysis and single-stranded nucleic acid binding. BMC Biol 22:287

Cole C, Sobala A, Lu C, Thatcher SR, Bowman A, Brown JW, Green PJ, Barton GJ, Hutvagner G (2009) Filtering of deep sequencing data reveals the existence of abundant dicer-dependent small RNAs derived from tRNAs. RNA 15:2147–2160

Colmenares SU, Buker SM, Buhler M, Dlakic M, Moazed D (2007) Coupling of double-stranded RNA synthesis and siRNA generation in fission yeast RNAi. Mol Cell 27:449–461

Comella P, Pontvianne F, Lahmy S, Vignols F, Barbezier N, Debures A, Jobet E, Brugidou E, Echeverria M, Sáez-Vásquez J (2008) Characterization of a ribonuclease III-like protein required for cleavage of the pre-rRNA in the 3'ETS in Arabidopsis. Nucleic Acids Res 36:1163–1175

Conrad C, Rauhut R (2002) Ribonuclease III: new sense from nuisance. Int J Biochem Cell Biol 34:116–129

Court DL, Gan J, Liang YH, Shaw GX, Tropea JE, Costantino N, Waugh DS, Ji X (2013) RNase III: genetics and function; structure and mechanism. Annu Rev Genet 47:405–431

Creugny A, Fender A, Pfeffer S (2018) Regulation of primary microRNA processing. FEBS Lett 592:1980–1996

Dang TL, Le CT, Le MN, Nguyen TD, Nguyen TL, Bao S, Li S, Nguyen TA (2020) Select amino acids in DGCR8 are essential for the UGU-pri-miRNA interaction and processing. Commun Biol 3:344

de Jong D, Eitel M, Jakob W, Osigus HJ, Hadrys H, Desalle R, Schierwater B (2009) Multiple dicer genes in the early-diverging metazoa. Mol Biol Evol 26:1333–1340

Deltcheva E, Chylinski K, Sharma CM, Gonzales K, Chao Y, Pirzada ZA, Eckert MR, Vogel J, Charpentier E (2011) CRISPR RNA maturation by trans-encoded small RNA and host factor RNase III. Nature 471:602–607

Denli AM, Tops BB, Plasterk RH, Ketting RF, Hannon GJ (2004) Processing of primary microRNAs by the microprocessor complex. Nature 432:231–235

Denzler R, Stoffel M (2015) The Long, the short, and the unstructured: a unifying model of miRNA biogenesis. Mol Cell 60:4–6

Dharavath S, Shaw GX, Ji X (2022) Structural basis for dicer-like function of an engineered RNase III variant and insights into the reaction trajectory of two-Mg(2+)-ion catalysis. RNA Biol 19:908–915

Dhir A, Dhir S, Proudfoot NJ, Jopling CL (2015) Microprocessor mediates transcriptional termination of long noncoding RNA transcripts hosting microRNAs. Nat Struct Mol Biol 22:319–327

Di Fazio A, Schlackow M, Pong SK, Alagia A, Gullerova M (2022) Dicer dependent tRNA derived small RNAs promote nascent RNA silencing. Nucleic Acids Res 50:1734–1752

Dincbas-Renqvist V, Pepin G, Rakonjac M, Plante I, Ouellet DL, Hermansson A, Goulet I, Doucet J, Samuelsson B, Radmark O, Provost P (2009) Human dicer C-terminus functions as a 5-lipoxygenase binding domain. Biochim Biophys Acta 1789:99–108

Dlakic M (2006) DUF283 domain of dicer proteins has a double-stranded RNA-binding fold. Bioinformatics 22:2711–2714

Doyle M, Badertscher L, Jaskiewicz L, Guttinger S, Jurado S, Hugenschmidt T, Kutay U, Filipowicz W (2013) The double-stranded RNA binding domain of human dicer functions as a nuclear localization signal. RNA 19:1238–1252

Drinnenberg IA, Weinberg DE, Xie KT, Mower JP, Wolfe KH, Fink GR, Bartel DP (2009) RNAi in budding yeast. Science 326:544–550

Drinnenberg IA, Fink GR, Bartel DP (2011) Compatibility with killer explains the rise of RNAi-deficient fungi. Science 333:1592

Edwards TA, Pyle SE, Wharton RP, Aggarwal AK (2001) Structure of Pumilio reveals similarity between RNA and peptide binding motifs. Cell 105:281–289

Elela SA, Igel H, Ares M Jr (1996) RNase III cleaves eukaryotic preribosomal RNA at a U3 snoRNP-dependent site. Cell 85:115–124

Esposito F, Capozzo I, Modafferi S, Gioia U, Manfredi L, Iannelli F, Colantoni A, Riccardi A, di Lillo A, Tavella S, Cabrini M et al (2024) DROSHA, DICER and damage-induced long ncRNA control BMI1-dependent transcriptional repression at DNA double-strand break, *bioRxiv*, 2024.08.07.606960

Fatica A, Tollervey D (2002) Making ribosomes. Curr Opin Cell Biol 14:313–318

Fernandez-Pevida A, Kressler D, de la Cruz J (2015) Processing of preribosomal RNA in Saccharomyces cerevisiae. Wiley Interdiscip Rev RNA 6:191–209

Fierro-Monti I, Mathews MB (2000) Proteins binding to duplexed RNA: one motif, multiple functions. Trends Biochem Sci 25:241–246

Flemr M, Malik R, Franke V, Nejepinska J, Sedlacek R, Vlahovicek K, Svoboda P (2013) A retrotransposon-driven dicer isoform directs endogenous small interfering RNA production in mouse oocytes. Cell 155:807–816

Francia S, Michelini F, Saxena A, Tang D, de Hoon M, Anelli V, Mione M, Carninci P, d'Adda di Fagagna F (2012) Site-specific DICER and DROSHA RNA products control the DNA-damage response. Nature 488:231–235

Francia S, Cabrini M, Matti V, Oldani A, d'Adda di Fagagna F (2016) DICER, DROSHA and DNA damage-response RNAs are necessary for the secondary recruitment of DNA damage response factors. J Cell Sci

French SL, Miller OL Jr (1989) Transcription mapping of the Escherichia coli chromosome by electron microscopy. J Bacteriol 171:4207–4216

Fukagawa T, Nogami M, Yoshikawa M, Ikeno M, Okazaki T, Takami Y, Nakayama T, Oshimura M (2004) Dicer is essential for formation of the heterochromatin structure in vertebrate cells. Nat Cell Biol 6:784–791

Fukudome A, Fukuhara T (2017) Plant dicer-like proteins: double-stranded RNA-cleaving enzymes for small RNA biogenesis. J Plant Res 130:33–44

Fukunaga R, Han BW, Hung JH, Xu J, Weng Z, Zamore PD (2012) Dicer partner proteins tune the length of mature miRNAs in flies and mammals. Cell 151:912

Gan J, Tropea JE, Austin BP, Court DL, Waugh DS, Ji X (2005) Intermediate states of ribonuclease III in complex with double-stranded RNA. Structure 13:1435–1442

Gan J, Tropea JE, Austin BP, Court DL, Waugh DS, Ji X (2006) Structural insight into the mechanism of double-stranded RNA processing by ribonuclease III. Cell 124:355–366

Garg A, Shang R, Cvetanovic T, Lai EC, Joshua-Tor L (2024) The structural landscape of microprocessor-mediated processing of pri-let-7 miRNAs. Mol Cell 84:4175–4190.e6

Ghazal G, Gagnon J, Jacques PE, Landry JR, Robert F, Elela SA (2009) Yeast RNase III triggers polyadenylation-independent transcription termination. Mol Cell 36:99–109

Ghildiyal M, Zamore PD (2009) Small silencing RNAs: an expanding universe. Nat Rev Genet 10:94–108

Gleghorn ML, Maquat LE (2014) Black sheep' that don't leave the double-stranded RNA-binding domain fold. Trends Biochem Sci 39:328–340

Glow D, Pianka D, Sulej AA, Kozlowski LP, Czarnecka J, Chojnowski G, Skowronek KJ, Bujnicki JM (2015) Sequence-specific cleavage of dsRNA by mini-III RNase. Nucleic Acids Res 43:2864–2873

Gregory RI, Yan KP, Amuthan G, Chendrimada T, Doratotaj B, Cooch N, Shiekhattar R (2004) The microprocessor complex mediates the genesis of microRNAs. Nature 432:235–240

Grimson A, Srivastava M, Fahey B, Woodcroft BJ, Chiang HR, King N, Degnan BM, Rokhsar DS, Bartel DP (2008) Early origins and evolution of microRNAs and Piwi-interacting RNAs in animals. Nature 455:1193–1197

Gromak N, Dienstbier M, Macias S, Plass M, Eyras E, Cáceres JF, Proudfoot NJ (2013) Drosha regulates gene expression independently of RNA cleavage function. Cell Rep 5:1499–1510

Gullerova M, Proudfoot NJ (2012) Convergent transcription induces transcriptional gene silencing in fission yeast and mammalian cells. Nat Struct Mol Biol 19:1193–1201

Ha M, Kim VN (2014) Regulation of microRNA biogenesis. Nat Rev Mol Cell Biol 15:509–524

Han J, Lee Y, Yeom KH, Kim YK, Jin H, Kim VN (2004) The Drosha-DGCR8 complex in primary microRNA processing. Genes Dev 18:3016–3027

Han J, Pedersen JS, Kwon SC, Belair CD, Kim YK, Yeom KH, Yang WY, Haussler D, Blelloch R, Kim VN (2009) Posttranscriptional crossregulation between Drosha and DGCR8. Cell 136:75–84

Hansen SR, Aderounmu AM, Donelick HM, Bass BL (2019) Dicer's helicase domain: a meeting place for regulatory proteins. Cold Spring Harb Symp Quant Biol 84:185–193

Hartig JV, Esslinger S, Bottcher R, Saito K, Forstemann K (2009) Endo-siRNAs depend on a new isoform of loquacious and target artificially introduced, high-copy sequences. EMBO J 28:2932–2944

Hertel J, de Jong D, Marz M, Rose D, Tafer H, Tanzer A, Schierwater B, Stadler PF (2009) Non-coding RNA annotation of the genome of Trichoplax adhaerens. Nucleic Acids Res 37:1602–1615

Jacobsen SE, Running MP, Meyerowitz EM (1999) Disruption of an RNA helicase/RNAse III gene in Arabidopsis causes unregulated cell division in floral meristems. Development 126:5231–5243

Jankowsky E, Fairman ME (2007) RNA helicases – one fold for many functions. Curr Opin Struct Biol 17:316–324

Jiang X, Prabhakar A, Van der Voorn SM, Ghatpande P, Celona B, Venkataramanan S, Calviello L, Lin C, Wang W, Black BL, Floor SN, Lagna G, Hata A (2021) Control of ribosomal protein synthesis by the microprocessor complex. Sci Signal 14

Jin L, Song H, Tropea JE, Needle D, Waugh DS, Gu S, Ji X (2019) The molecular mechanism of dsRNA processing by a bacterial dicer. Nucleic Acids Res 47:4707–4720

Jin W, Wang J, Liu CP, Wang HW, Xu RM (2020) Structural basis for pri-miRNA recognition by Drosha. Mol Cell 78:423–433.e5

Johanson TM, Lew AM, Chong MM (2013) MicroRNA-independent roles of the RNase III enzymes Drosha and dicer. Open Biol 3:130144

Kadotani N, Nakayashiki H, Tosa Y, Mayama S (2004) One of the two dicer-like proteins in the filamentous fungi Magnaporthe oryzae genome is responsible for hairpin RNA-triggered RNA silencing and related small interfering RNA accumulation. J Biol Chem 279:44467–44474
Kandasamy SK, Fukunaga R (2016) Phosphate-binding pocket in Dicer-2 PAZ domain for high-fidelity siRNA production. Proc Natl Acad Sci USA 113:14031–14036
Kedersha NL, Rome LH (1986) Isolation and characterization of a novel ribonucleoprotein particle: large structures contain a single species of small RNA. J Cell Biol 103:699–709
Kennedy EM, Whisnant AW, Kornepati AV, Marshall JB, Bogerd HP, Cullen BR (2015) Production of functional small interfering RNAs by an amino-terminal deletion mutant of human dicer. Proc Natl Acad Sci USA 112:E6945–E6954
Kerner P, Degnan SM, Marchand L, Degnan BM, Vervoort M (2011) Evolution of RNA-binding proteins in animals: insights from genome-wide analysis in the sponge Amphimedon queenslandica. Mol Biol Evol 28:2289–2303
Kharrat A, Macias MJ, Gibson TJ, Nilges M, Pastore A (1995) Structure of the dsRNA binding domain of E. Coli RNase III. EMBO J 14:3572–3584
Kidwell MA, Chan JM, Doudna JA (2014) Evolutionarily conserved roles of the dicer helicase domain in regulating RNA interference processing. J Biol Chem 289:28352–28362
Kim B, Jeong K, Kim VN (2017) Genome-wide mapping of DROSHA cleavage sites on primary MicroRNAs and noncanonical substrates. Mol Cell 66:258–269.e5
Kim K, Nguyen TD, Li S, Nguyen TA (2018) SRSF3 recruits DROSHA to the basal junction of primary microRNAs. RNA 24:892–898
King TC, Sirdeshmukh R, Schlessinger D (1984) RNase III cleavage is obligate for maturation but not for function of Escherichia coli pre-23S rRNA. Proc Natl Acad Sci USA 81:185–188
Kolb FA, Zhang H, Jaronczyk K, Tahbaz N, Hobman TC, Filipowicz W (2005) Human dicer: purification, properties, and interaction with PAZ PIWI domain proteins. Methods Enzymol 392:316–336
Koralewska N, Szczepanska A, Ciechanowska K, Wojnicka M, Pokornowska M, Milewski MC, Gudanis D, Baranowski D, Nithin C, Bujnicki JM, Gdaniec Z, Figlerowicz M, Kurzynska-Kokorniak A (2021) RNA and DNA G-quadruplexes bind to human dicer and inhibit its activity. Cell Mol Life Sci 78:3709–3724
Koralewska N, Corradi E, Milewski MC, Masante L, Szczepanska A, Kierzek R, Figlerowicz M, Baudet ML, Kurzynska-Kokorniak A (2024) Short 2'-O-methyl/LNA oligomers as highly-selective inhibitors of miRNA production in vitro and in vivo. Nucleic Acids Res
Kufel J, Dichtl B, Tollervey D (1999) Yeast Rnt1p is required for cleavage of the pre-ribosomal RNA in the 3' ETS but not the 5' ETS. RNA 5:909–917
Kurzynska-Kokorniak A, Koralewska N, Pokornowska M, Urbanowicz A, Tworak A, Mickiewicz A, Figlerowicz M (2015) The many faces of dicer: the complexity of the mechanisms regulating dicer gene expression and enzyme activities. Nucleic Acids Res 43:4365–4380
Kurzynska-Kokorniak A, Pokornowska M, Koralewska N, Hoffmann W, Bienkowska-Szewczyk K, Figlerowicz M (2016) Revealing a new activity of the human dicer DUF283 domain in vitro. Sci Rep 6:23989
Kwon SC, Nguyen TA, Choi YG, Jo MH, Hohng S, Kim VN, Woo JS (2016) Structure of human DROSHA. Cell 164:81–90
Kwon SC, Baek SC, Choi YG, Yang J, Lee YS, Woo JS, Kim VN (2019) Molecular basis for the single-nucleotide precision of primary microRNA processing. Mol Cell 73:505–518.e5
Lamontagne B, Elela SA (2004) Evaluation of the RNA determinants for bacterial and yeast RNase III binding and cleavage. J Biol Chem 279:2231–2241
Lamontagne B, Tremblay A, Abou Elela S (2000) The N-terminal domain that distinguishes yeast from bacterial RNase III contains a dimerization signal required for efficient double-stranded RNA cleavage. Mol Cell Biol 20:1104–1115
Lau PW, Guiley KZ, De N, Potter CS, Carragher B, MacRae IJ (2012) The molecular architecture of human dicer. Nat Struct Mol Biol 19:436–440
Lavoie M, Abou Elela S (2008) Yeast ribonuclease III uses a network of multiple hydrogen bonds for RNA binding and cleavage. Biochemistry 47:8514–8526

Le CT, Nguyen TL, Nguyen TD, Nguyen TA (2020) Human disease-associated single nucleotide polymorphism changes the orientation of DROSHA on pri-mir-146a. RNA 26:1777–1786
Lee HY, Doudna JA (2012) TRBP alters human precursor microRNA processing in vitro. RNA 18:2012–2019
Lee Y, Ahn C, Han J, Choi H, Kim J, Yim J, Lee J, Provost P, Radmark O, Kim S, Kim VN (2003) The nuclear RNase III Drosha initiates microRNA processing. Nature 425:415–419
Lee YS, Nakahara K, Pham JW, Kim K, He Z, Sontheimer EJ, Carthew RW (2004) Distinct roles for drosophila Dicer-1 and Dicer-2 in the siRNA/miRNA silencing pathways. Cell 117:69–81
Lee Y, Han J, Yeom KH, Jin H, Kim VN (2006) Drosha in primary microRNA processing. Cold Spring Harb Symp Quant Biol 71:51–57
Lee HY, Zhou K, Smith AM, Noland CL, Doudna JA (2013) Differential roles of human dicer-binding proteins TRBP and PACT in small RNA processing. Nucleic Acids Res 41:6568–6576
Lee D, Nam JW, Shin C (2017) DROSHA targets its own transcript to modulate alternative splicing. RNA 23:1035–1047
Lee YY, Lee H, Kim H, Kim VN, Roh SH (2023a) Structure of the human DICER-pre-miRNA complex in a dicing state. Nature 615:331–338
Lee YY, Kim H, Kim VN (2023b) Sequence determinant of small RNA production by DICER. Nature 615:323–330
Lejars M, Kobayashi A, Hajnsdorf E (2021) RNase III, ribosome biogenesis and beyond. Microorganisms:9
Lestini R, Delpech F, Myllykallio H (2015) DNA replication restart and cellular dynamics of Hef helicase/nuclease protein in Haloferax volcanii. Biochimie 118:254–263
Li S, Patel DJ (2016) Drosha and dicer: slicers cut from the same cloth. Cell Res 26:511–512
Liang XH, Crooke ST (2011) Depletion of key protein components of the RISC pathway impairs pre-ribosomal RNA processing. Nucleic Acids Res 39:4875–4889
Liang YH, Lavoie M, Comeau MA, Abou Elela S, Ji X (2014) Structure of a eukaryotic RNase III postcleavage complex reveals a double-ruler mechanism for substrate selection. Mol Cell 54:431–444
Lingel A, Simon B, Izaurralde E, Sattler M (2003) Structure and nucleic-acid binding of the drosophila Argonaute 2 PAZ domain. Nature 426:465–469
Link S, Grund SE, Diederichs S (2016) Alternative splicing affects the subcellular localization of Drosha. Nucleic Acids Res 44:5330–5343
Liu Q, Rand TA, Kalidas S, Du F, Kim HE, Smith DP, Wang X (2003) R2D2, a bridge between the initiation and effector steps of the drosophila RNAi pathway. Science 301:1921–1925
Liu X, Jiang F, Kalidas S, Smith D, Liu Q (2006) Dicer-2 and R2D2 coordinately bind siRNA to promote assembly of the siRISC complexes. RNA 12:1514–1520
Liu X, Park JK, Jiang F, Liu Y, McKearin D, Liu Q (2007) Dicer-1, but not loquacious, is critical for assembly of miRNA-induced silencing complexes. RNA 13:2324–2329
Liu Q, Feng Y, Zhu Z (2009) Dicer-like (DCL) proteins in plants. Funct Integr Genomics 9:277–286
Liu C, Axtell MJ, Fedoroff NV (2012) The helicase and RNaseIIIa domains of Arabidopsis dicer-Like1 modulate catalytic parameters during microRNA biogenesis. Plant Physiol 159:748–758
Liu Z, Wang J, Cheng H, Ke X, Sun L, Zhang QC, Wang HW (2018a) Cryo-EM structure of human dicer and its complexes with a pre-miRNA substrate. Cell 173:1549–1550
Liu Z, Wang J, Cheng H, Ke X, Sun L, Zhang QC, Wang H-W (2018b) Cryo-EM structure of human dicer and its complexes with a pre-miRNA substrate. Cell 173:1191–1203.e12
Ma JB, Ye K, Patel DJ (2004) Structural basis for overhang-specific small interfering RNA recognition by the PAZ domain. Nature 429:318–322
Ma E, MacRae IJ, Kirsch JF, Doudna JA (2008) Autoinhibition of human dicer by its internal helicase domain. J Mol Biol 380:237–243
Ma E, Zhou K, Kidwell MA, Doudna JA (2012) Coordinated activities of human dicer domains in regulatory RNA processing. J Mol Biol 422:466–476
MacRae IJ, Doudna JA (2007) Ribonuclease revisited: structural insights into ribonuclease III family enzymes. Curr Opin Struct Biol 17:138–145

Macrae IJ, Zhou K, Li F, Repic A, Brooks AN, Cande WZ, Adams PD, Doudna JA (2006) Structural basis for double-stranded RNA processing by dicer. Science 311:195–198

Marsh JA, Hernandez H, Hall Z, Ahnert SE, Perica T, Robinson CV, Teichmann SA (2013) Protein complexes are under evolutionary selection to assemble via ordered pathways. Cell 153:461–470

Martinez G, Choudury SG, Slotkin RK (2017) tRNA-derived small RNAs target transposable element transcripts. Nucleic Acids Res 45:5142–5152

Masliah G, Barraud P, Allain FH (2013) RNA recognition by double-stranded RNA binding domains: a matter of shape and sequence. Cell Mol Life Sci 70:1875–1895

Mateos JL, Bologna NG, Chorostecki U, Palatnik JF (2010) Identification of microRNA processing determinants by random mutagenesis of Arabidopsis MIR172a precursor. Curr Biol 20:49–54

Matsuda S, Ichigotani Y, Okuda T, Irimura T, Nakatsugawa S, Hamaguchi M (2000) Molecular cloning and characterization of a novel human gene (HERNA) which encodes a putative RNA-helicase. Biochim Biophys Acta 1490:163–169

Matsunaga J, Simons EL, Simons RW (1996) RNase III autoregulation: structure and function of rncO, the posttranscriptional "operator". RNA 2:1228–1240

Maxwell EK, Ryan JF, Schnitzler CE, Browne WE, Baxevanis AD (2012) MicroRNAs and essential components of the microRNA processing machinery are not encoded in the genome of the ctenophore Mnemiopsis leidyi. BMC Genomics 13:714

Mechtler P, Johnson S, Slabodkin H, Cohanim AB, Brodsky L, Kandel ES (2017) The evidence for a microRNA product of human DROSHA gene. RNA Biol 14:1508–1513

Merchant SS, Prochnik SE, Vallon O, Harris EH, Karpowicz SJ, Witman GB, Terry A, Salamov A, Fritz-Laylin LK, Maréchal-Drouard L, Marshall WFQ, Nelson LH, Sanderfoot DR, Spalding AA, Kapitonov MH, Ren VV, Ferris Q, Lindquist P, Shapiro E, Lucas H, Grimwood SM, Schmutz J, Cardol J, Cerutti P, Chanfreau H, Chen G, Cognat CL, Croft V, Dent MT, Dutcher R, Fernández S, Fukuzawa E, González-Ballester H, González-Halphen D, Hallmann D, Hanikenne A, Hippler M, Inwood M, Jabbari W, Kalanon K, Kuras M, Lefebvre R, Lemaire PA, Lobanov SD, Lohr AV, Manuell M, Meier A, Mets I, Mittag L, Mittelmeier M, Moroney T, Moseley JV, Napoli J, Nedelcu C, Niyogi AM, Novoselov K, Paulsen SV, Pazour IT, Purton G, Ral S, Riaño-Pachón JP, Riekhof DM, Rymarquis W, Schroda L, Stern M, Umen D, Willows J, Wilson R, Zimmer N, Allmer SL, Balk J, Bisova J, Chen K, Elias CJ, Gendler M, Hauser K, Lamb C, Ledford MR, Long H, Minagawa JC, Page J, Pan MD, Pootakham J, Roje W, Rose S, Stahlberg A, Terauchi E, Yang AM, Ball P, Bowler S, Dieckmann C, Gladyshev CL, Green VN, Jorgensen P, Mayfield R, Mueller-Roeber S, Rajamani B, Sayre S, Brokstein RT et al (2007) The Chlamydomonas genome reveals the evolution of key animal and plant functions. Science 318:245–250

Meyers BC, Axtell MJ (2019) MicroRNAs in plants: key findings from the early years. Plant Cell 31:1206–1207

Mlotshwa S, Pruss GJ, Peragine A, Endres MW, Li J, Chen X, Poethig RS, Bowman LH, Vance V (2008) DICER-LIKE2 plays a primary role in transitive silencing of transgenes in Arabidopsis. PLoS One 3:e1755

Montavon TC, Baldaccini M, Lefevre M, Girardi E, Chane-Woon-Ming B, Messmer M, Hammann P, Chicher J, Pfeffer S (2021) Human DICER helicase domain recruits PKR and modulates its antiviral activity. PLoS Pathog 17:e1009549

Moro B, Chorostecki U, Arikit S, Suarez IP, Hobartner C, Rasia RM, Meyers BC, Palatnik JF (2018) Efficiency and precision of microRNA biogenesis modes in plants. Nucleic Acids Res 46:10709–10723

Mukherjee K, Campos H, Kolaczkowski B (2013) Evolution of animal and plant dicers: early parallel duplications and recurrent adaptation of antiviral RNA binding in plants. Mol Biol Evol 30:627–641

Nagel R, Ares M Jr (2000) Substrate recognition by a eukaryotic RNase III: the double-stranded RNA-binding domain of Rnt1p selectively binds RNA containing a 5'-AGNN-3′ tetraloop. RNA 6:1142–1156

Nakagawa A, Shi Y, Kage-Nakadai E, Mitani S, Xue D (2010) Caspase-dependent conversion of dicer ribonuclease into a death-promoting deoxyribonuclease. Science 328:327–334

Nguyen TA, Jo MH, Choi YG, Park J, Kwon SC, Hohng S, Kim VN, Woo JS (2015) Functional anatomy of the human microprocessor. Cell 161:1374–1387
Nguyen TL, Nguyen TD, Bao S, Li S, Nguyen TA (2020) The internal loops in the lower stem of primary microRNA transcripts facilitate single cleavage of human microprocessor. Nucleic Acids Res 48:2579–2593
Nguyen TL, Nguyen TD, Nguyen TA (2021) The conserved single-cleavage mechanism of animal DROSHA enzymes. Commun Biol 4:1332
Nicholson AW (1996) Structure, reactivity, and biology of double-stranded RNA. Prog Nucleic Acid Res Mol Biol 52:1–65
Nicholson AW (1999) Function, mechanism and regulation of bacterial ribonucleases. FEMS Microbiol Rev 23:371–390
Nicholson AW (2014) Ribonuclease III mechanisms of double-stranded RNA cleavage. Wiley Interdiscip Rev RNA 5:31–48
Nicolas FE, Torres-Martinez S, Ruiz-Vazquez RM (2013) Loss and retention of RNA interference in fungi and parasites. PLoS Pathog 9:e1003089
Noguchi S, Ohkura S, Negishi Y, Tozawa S, Takizawa T, Morita R, Takahashi H, Ohkuchi A, Takizawa T (2024) Cytoplasmic and nuclear DROSHA in human villous trophoblasts. J Reprod Immunol 162:104189
Noma K, Sugiyama T, Cam H, Verdel A, Zofall M, Jia S, Moazed D, Grewal SI (2004) RITS acts in cis to promote RNA interference-mediated transcriptional and post-transcriptional silencing. Nat Genet 36:1174–1180
Olmedo G, Guzman P (2008) Mini-III, a fourth class of RNase III catalyses maturation of the Bacillus subtilis 23S ribosomal RNA. Mol Microbiol 68:1073–1076
Ota H, Sakurai M, Gupta R, Valente L, Wulff BE, Ariyoshi K, Iizasa H, Davuluri RV, Nishikura K (2013) ADAR1 forms a complex with dicer to promote MicroRNA processing and RNA-induced gene silencing. Cell 153:575–589
Park JE, Heo I, Tian Y, Simanshu DK, Chang H, Jee D, Patel DJ, Kim VN (2011) Dicer recognizes the 5′ end of RNA for efficient and accurate processing. Nature 475:201–205
Partin AC, Zhang K, Jeong BC, Herrell E, Li S, Chiu W, Nam Y (2020) Cryo-EM structures of human Drosha and DGCR8 in complex with primary MicroRNA. Mol Cell 78:411–422.e4
Paturi S, Deshmukh MV (2021) A glimpse of "dicer biology" through the structural and functional perspective. Front Mol Biosci 8:643657
Podolska K, Sedlak D, Bartunek P, Svoboda P (2013) Fluorescence-based high-throughput screening of dicer cleavage activity. J Biomol Screen 19
Poirier EZ, Buck MD, Chakravarty P, Carvalho J, Frederico B, Cardoso A, Healy L, Ulferts R, Beale R, Reis e Sousa, C. (2021) An isoform of dicer protects mammalian stem cells against multiple RNA viruses. Science 373:231–236
Pokornowska M, Milewski MC, Ciechanowska K, Szczepanska A, Wojnicka M, Radogostowicz Z, Figlerowicz M, Kurzynska-Kokorniak A (2020) The RNA-RNA base pairing potential of human dicer and Ago2 proteins. Cell Mol Life Sci 77:3231–3244
Pong SK, Gullerova M (2018) Noncanonical functions of microRNA pathway enzymes – Drosha, DGCR8, Dicer and Ago proteins. *FEBS Lett* 592:2973–2986
Prabhakar A, Hu S, Tang J, Ghatpande P, Lagna G, Jiang X, Hata A (2023) Essential role of the amino-terminal region of Drosha for the Microprocessor function. *iScience* 26:107971
Provost P, Samuelsson B, Radmark O (1999) Interaction of 5-lipoxygenase with cellular proteins. Proc Natl Acad Sci USA 96:1881–1885
Qin H, Chen F, Huan X, Machida S, Song J, Yuan YA (2010) Structure of the Arabidopsis thaliana DCL4 DUF283 domain reveals a noncanonical double-stranded RNA-binding fold for protein-protein interaction. RNA 16:474–481
Redko Y, Bechhofer DH, Condon C (2008) Mini-III, an unusual member of the RNase III family of enzymes, catalyses 23S ribosomal RNA maturation in B. subtilis. Mol Microbiol 68:1096–1106
Robertson HD, Webster RE, Zinder ND (1967) A nuclease specific for double-stranded RNA. Virology 32:718–719

Robertson HD, Webster RE, Zinder ND (1968) Purification and properties of ribonuclease III from Escherichia coli. J Biol Chem 243:82–91

Rogers K, Chen X (2013) Biogenesis, turnover, and mode of action of plant microRNAs. Plant Cell 25:2383–2399

Rosatti S, Rojas AML, Moro B, Suarez IP, Bologna NG, Chorostecki U, Palatnik JF (2024) Principles of miRNA/miRNA* function in plant MIRNA processing. Nucleic Acids Res 52:8356–8369

Ryan JF, Pang K, Schnitzler CE, Nguyen AD, Moreland RT, Simmons DK, Koch BJ, Francis WR, Havlak P, Smith SA, Putnam NH, Haddock SH, Dunn CW, Wolfsberg TG, Mullikin JC, Martindale MQ, Baxevanis AD (2013) The genome of the ctenophore Mnemiopsis leidyi and its implications for cell type evolution. Science 342:1242592

Rybak-Wolf A, Jens M, Murakawa Y, Herzog M, Landthaler M, Rajewsky N (2014) A variety of dicer substrates in human and C-elegans. Cell 159:1153–1167

Ryter JM, Schultz SC (1998) Molecular basis of double-stranded RNA-protein interactions: structure of a dsRNA-binding domain complexed with dsRNA. EMBO J 17:7505–7513

Santos-Pereira JM, Aguilera A (2015) R loops: new modulators of genome dynamics and function. Nat Rev Genet 16:583–597

Sapranauskas R, Lubys A (2005) Random gene dissection: a tool for the investigation of protein structural organization. BioTechniques 39:395–402

Saunders LR, Barber GN (2003) The dsRNA binding protein family: critical roles, diverse cellular functions. FASEB J 17:961–983

Seeman NC, Rosenberg JM, Rich A (1976) Sequence-specific recognition of double helical nucleic acids by proteins. Proc Natl Acad Sci USA 73:804–808

Shabalina SA, Koonin EV (2008) Origins and evolution of eukaryotic RNA interference. Trends Ecol Evol 23:578–587

Shapiro JS, Langlois RA, Pham AM, Tenoever BR (2012) Evidence for a cytoplasmic microprocessor of pri-miRNAs. RNA 18:1338–1346

Shapiro JS, Schmid S, Aguado LC, Sabin LR, Yasunaga A, Shim JV, Sachs D, Cherry S, tenOever BR (2014) Drosha as an interferon-independent antiviral factor. Proc Natl Acad Sci USA 111:7108–7113

Sharma S, Anand R, Zhang X, Francia S, Michelini F, Galbiati A, Williams H, Ronato DA, Masson JY, Rothenberg E, Cejka P, d'Adda di Fagagna F (2021) MRE11-RAD50-NBS1 complex is sufficient to promote transcription by RNA polymerase II at double-Strand breaks by melting DNA ends. Cell Rep 34:108565

Shenoy A, Blelloch R (2009) Genomic analysis suggests that mRNA destabilization by the microprocessor is specialized for the auto-regulation of Dgcr8. PLoS One 4:e6971

Shi H, Tschudi C, Ullu E (2006) An unusual dicer-like1 protein fuels the RNA interference pathway in Trypanosoma brucei. RNA 12:2063–2072

Singleton MR, Dillingham MS, Wigley DB (2007) Structure and mechanism of helicases and nucleic acid translocases. Annu Rev Biochem 76:23–50

Sinha NK, Iwasa J, Shen PS, Bass BL (2018) Dicer uses distinct modules for recognizing dsRNA termini. Science 359:329–334

Sinkkonen L, Hugenschmidt T, Filipowicz W, Svoboda P (2010) Dicer is associated with ribosomal DNA chromatin in mammalian cells. PLoS One 5:e12175

Son S, Kim B, Yang J, Kim VN (2023) Role of the proline-rich disordered domain of DROSHA in intronic microRNA processing. Genes Dev 37:383–397

Song JJ, Liu J, Tolia NH, Schneiderman J, Smith SK, Martienssen RA, Hannon GJ, Joshua-Tor L (2003) The crystal structure of the Argonaute2 PAZ domain reveals an RNA binding motif in RNAi effector complexes. Nat Struct Biol 10:1026–1032

Song H, Fang X, Jin L, Shaw GX, Wang YX, Ji X (2017) The functional cycle of Rnt1p: five consecutive steps of double-stranded RNA processing by a eukaryotic RNase III. Structure 25:353–363

St Johnston D, Brown NH, Gall JG, Jantsch M (1992) A conserved double-stranded RNA-binding domain. Proc Natl Acad Sci USA 89:10979–10983

Su S, Wang J, Deng T, Yuan X, He J, Liu N, Li X, Huang Y, Wang HW, Ma J (2022a) Structural insights into dsRNA processing by drosophila Dicer-2-Loqs-PD. Nature 607:399–406
Su SC, Wang J, Deng T, Yuan X, He JQ, Liu N, Li XM, Huang Y, Wang HW, Ma JB (2022b) Structural insights into dsRNA processing by drosophila Dicer-2-logs-PD. Nature 607:399-+
Sun W, Li G, Nicholson AW (2004) Mutational analysis of the nuclease domain of Escherichia coli ribonuclease III. Identification of conserved acidic residues that are important for catalytic function in vitro. Biochemistry 43:13054–13062
Tahbaz N, Kolb FA, Zhang H, Jaronczyk K, Filipowicz W, Hobman TC (2004) Characterization of the interactions between mammalian PAZ PIWI domain proteins and dicer. EMBO Rep 5:189–194
Tang X, Zhang Y, Tucker L, Ramratnam B (2010) Phosphorylation of the RNase III enzyme Drosha at Serine300 or Serine302 is required for its nuclear localization. Nucleic Acids Res 38:6610–6619
Tang X, Li M, Tucker L, Ramratnam B (2011) Glycogen synthase kinase 3 beta (GSK3β) phosphorylates the RNAase III enzyme Drosha at S300 and S302. PLoS One 6:e20391
Tang X, Wen S, Zheng D, Tucker L, Cao L, Pantazatos D, Moss SF, Ramratnam B (2013) Acetylation of drosha on the N-terminus inhibits its degradation by ubiquitination. PLoS One 8:e72503
Taylor DW, Ma E, Shigematsu H, Cianfrocco MA, Noland CL, Nagayama K, Nogales E, Doudna JA, Wang HW (2013) Substrate-specific structural rearrangements of human dicer. Nat Struct Mol Biol 20:662–670
Tian Y, Simanshu DK, Ma JB, Park JE, Heo I, Kim VN, Patel DJ (2014) A phosphate-binding pocket within the platform-PAZ-connector helix cassette of human dicer. Mol Cell 53:606–616
Triboulet R, Chang HM, Lapierre RJ, Gregory RI (2009) Post-transcriptional control of DGCR8 expression by the microprocessor. RNA 15:1005–1011
Trotter JR, Ernst NL, Carnes J, Panicucci B, Stuart K (2005) A deletion site editing endonuclease in Trypanosoma brucei. Mol Cell 20:403–412
Tworak A, Urbanowicz A, Podkowinski J, Kurzynska-Kokorniak A, Koralewska N, Figlerowicz M (2016) Six Medicago truncatula dicer-like protein genes are expressed in plant cells and upregulated in nodules. Plant Cell Rep 35:1043–1052
Uebbing S, Kreiss M, Scholl F, Hafner AK, Surun D, Garscha U, Werz O, Basavarajappa D, Samuelsson B, Radmark O, Suess B, Steinhilber D (2021) Modulation of microRNA processing by 5-lipoxygenase. FASEB J 35:e21193
Valli AA, Santos BA, Hnatova S, Bassett AR, Molnar A, Chung BY, Baulcombe DC (2016) Most microRNAs in the single-cell alga Chlamydomonas reinhardtii are produced by dicer-like 3-mediated cleavage of introns and untranslated regions of coding RNAs. Genome Res 26:519–529
van Rij RP, Saleh MC, Berry B, Foo C, Houk A, Antoniewski C, Andino R (2006) The RNA silencing endonuclease Argonaute 2 mediates specific antiviral immunity in Drosophila melanogaster. Genes Dev 20:2985–2995
Verdel A, Jia S, Gerber S, Sugiyama T, Gygi S, Grewal SI, Moazed D (2004) RNAi-mediated targeting of heterochromatin by the RITS complex. Science 303:672–676
Volpe TA, Kidner C, Hall IM, Teng G, Grewal SI, Martienssen RA (2002) Regulation of heterochromatic silencing and histone H3 lysine-9 methylation by RNAi. Science 297:1833–1837
Wei W, Ba Z, Gao M, Wu Y, Ma Y, Amiard S, White CI, Rendtlew Danielsen JM, Yang YG, Qi Y (2012) A role for small RNAs in DNA double-strand break repair. Cell 149:101–112
Wei X, Ke H, Wen A, Gao B, Shi J, Feng Y (2021) Structural basis of microRNA processing by dicer-like 1. Nat Plants 7:1389–1396
Weinberg DE, Nakanishi K, Patel DJ, Bartel DP (2011) The inside-out mechanism of dicers from budding yeasts. Cell 146:262–276
Welker NC, Maity TS, Ye X, Aruscavage PJ, Krauchuk AA, Liu Q, Bass BL (2011) Dicer's helicase domain discriminates dsRNA termini to promote an altered reaction mode. Mol Cell 41:589–599

Wells SE, Neville M, Haynes M, Wang J, Igel H, Ares M Jr (1996) CUS1, a suppressor of cold-sensitive U2 snRNA mutations, is a novel yeast splicing factor homologous to human SAP 145. Genes Dev 10:220–232

Westra ER, Swarts DC, Staals RH, Jore MM, Brouns SJ, van der Oost J (2012) The CRISPRs, they are a-changin': how prokaryotes generate adaptive immunity. Annu Rev Genet 46:311–339

White E, Schlackow M, Kamieniarz-Gdula K, Proudfoot NJ, Gullerova M (2014) Human nuclear dicer restricts the deleterious accumulation of endogenous double-stranded RNA. Nat Struct Mol Biol 21:552–559

Wilson RC, Tambe A, Kidwell MA, Noland CL, Schneider CP, Doudna JA (2015) Dicer-TRBP complex formation ensures accurate mammalian microRNA biogenesis. Mol Cell 57:397–407

Wojcik K, Krzeminska P, Kurzynska-Kokorniak A (2025) Possible role of human ribonuclease dicer in the regulation of R loops. FEBS Open Bio. https://doi.org/10.1002/2211-5463.70026

Woolcock KJ, Stunnenberg R, Gaidatzis D, Hotz HR, Emmerth S, Barraud P, Bühler M (2012) RNAi keeps Atf1-bound stress response genes in check at nuclear pores. Genes Dev 26:683–692

Wu H, Xu H, Miraglia LJ, Crooke ST (2000) Human RNase III is a 160-kDa protein involved in preribosomal RNA processing. J Biol Chem 275:36957–36965

Xie Z, Johansen LK, Gustafson AM, Kasschau KD, Lellis AD, Zilberman D, Jacobsen SE, Carrington JC (2004) Genetic and functional diversification of small RNA pathways in plants. PLoS Biol 2:E104

Yadav RP, Mäkelä JA, Hyssälä H, Cisneros-Montalvo S, Kotaja N (2020) DICER regulates the expression of major satellite repeat transcripts and meiotic chromosome segregation during spermatogenesis. Nucleic Acids Res 48:7135–7153

Yoshida T, Asano Y, Ui-Tei K (2021) Modulation of MicroRNA Processing by Dicer via Its Associated dsRNA Binding Proteins. *Non-coding RNA*:7

Young RA, Steitz JA (1978) Complementary sequences 1700 nucleotides apart form a ribonuclease III cleavage site in Escherichia coli ribosomal precursor RNA. Proc Natl Acad Sci USA 75:3593–3597

Yu X, Li X, Zheng L, Ma J, Gan J (2017) Structural and functional studies of a noncanonical dicer from Entamoeba histolytica. Sci Rep 7:44832

Zamore PD, Bartel DP, Lehmann R, Williamson JR (1999) The PUMILIO-RNA interaction: a single RNA-binding domain monomer recognizes a bipartite target sequence. Biochemistry 38:596–604

Zapletal D, Taborska E, Pasulka J, Malik R, Kubicek K, Zanova M, Much C, Sebesta M, Buccheri V, Horvat F, Jenickova I, Prochazkova M, Prochazka J, Pinkas M, Novacek J, Joseph DF, Sedlacek R, Bernecky C, O'Carroll D, Stefl R, Svoboda P (2022) Structural and functional basis of mammalian microRNA biogenesis by dicer. Mol Cell 82(4064–4079):e13

Zer C, Chanfreau G (2005) Regulation and surveillance of normal and 3′-extended forms of the yeast aci-reductone dioxygenase mRNA by RNase III cleavage and exonucleolytic degradation. J Biol Chem 280:28997–29003

Zhang H, Kolb FA, Jaskiewicz L, Westhof E, Filipowicz W (2004) Single processing center models for human dicer and bacterial RNase III. Cell 118:57–68

Zhang X, Yang F, Liu F, Tian Q, Hu M, Li P, Zeng Y (2021) Conservation of differential animal MicroRNA processing by Drosha and dicer. Front Mol Biosci 8:730006

Zhang L, Xiang Y, Chen S, Shi M, Jiang X, He Z, Gao S (2022) Mechanisms of MicroRNA biogenesis and stability control in plants. Front Plant Sci 13:844149

Zhou R, Czech B, Brennecke J, Sachidanandam R, Wohlschlegel JA, Perrimon N, Hannon GJ (2009) Processing of drosophila endo-siRNAs depends on a specific loquacious isoform. RNA 15:1886–1895

Zhu H, Zhou Y, Castillo-González C, Lu A, Ge C, Zhao YT, Duan L, Li Z, Axtell MJ, Wang XJ, Zhang X (2013) Bidirectional processing of pri-miRNAs with branched terminal loops by Arabidopsis dicer-like1. Nat Struct Mol Biol 20:1106–1115

Zou J, Chang M, Nie P, Secombes CJ (2009) Origin and evolution of the RIG-I like RNA helicase gene family. BMC Evol Biol 9:85

Chapter 8
Conserved and Diverse Functions of Bacterial Small RNAs

Teppei Morita

Abstract Small regulatory RNAs are key to bacterial stress responses. The regulatory functions of small RNAs are exerted by base-pairing with mRNA targets, affecting the translation and stability of these targets. Small RNAs and their protein partners have been identified in a range of bacteria. This chapter summarizes a well-characterized class of small regulatory RNAs that bind to the partner protein Hfq, and discusses the commonalities and diversities among RNA/Hfq-mediated gene regulation.

Keywords Small RNA (sRNA) · Hfq · Gene expression regulation · Stress response · Bacteria

8.1 Introduction

Bacteria combat stress by plastically regulating their own gene expression. If the bacteria have difficulty coping with the stress, they may alter their own genome to overcome it. Consequently, stress-related gene regulation is key in determining the boundary between maintaining or mutating the bacterial genome. To promote evolution, the gene regulation mechanism is the primary target for disruption under severe stress conditions. For instance, some laboratory-based evolution studies using *Escherichia coli* have isolated multiple point mutations in the *rho* gene under exposure to various types of stress (Worthan et al. 2025), including sublethal ethanol stress, thermal stress, and alkaline pH stress (Haft et al. 2014;

T. Morita (✉)
Department of Infectious Diseases, Kyorin University School of Medicine, Mitaka, Tokyo, Japan

Faculty of Environment and Information studies, Keio University, Fujisawa, Kanagawa, Japan

Institute for Advanced Biosciences, Keio University, Tsuruoka, Yamagata, Japan
e-mail: teppei-morita@ks.kyorin-u.ac.jp

A. Kanai, A. W. Nicholson (eds.), *Molecular Evolution of RNA Regulatory Enzymes and Their Systems*, Nucleic Acids and Molecular Biology 38,
https://doi.org/10.1007/978-3-032-03300-0_8

Rodríguez-Verdugo et al. 2014; Worthan et al. 2024). The *rho* gene encodes Rho, a bacterial transcription factor that regulates gene transcription at the elongation step (Adams et al. 2021; Delaleau et al. 2024; Goodson and Winkler 2018).

Small noncoding RNAs function as post-transcriptional regulatory factors that bind to the target mRNAs by base-pairing, and alter their translation efficiency and stability. This type of functional RNA molecule has been identified in all kingdoms of life (McJunkin and Gottesman 2024; Wagner and Romby 2015), albeit with differences in their lengths and structures. In prokaryotes, since the discovery of MicF small regulatory RNA, which downregulates the expression of *ompF* encoding a major outer membrane porin in *E. coli* (Mizuno et al. 1984), numerous studies have uncovered a large class of small regulatory RNAs (Holmqvist and Wagner 2017; Hör et al. 2020; Kavita et al. 2018). These RNAs are generally induced in response to environmental stress signals and shape the transcriptome/proteome involved in combating stress (Nitzan et al. 2017; Papenfort and Melamed 2023). Furthermore, as next-generation sequencing technologies have advanced, the research on small regulatory RNAs has rapidly expanded beyond the model bacterial species, *E. coli* and *Salmonella*, to encompass pathogenic bacteria and environmental microorganisms (Adams and Storz 2020; Barquist and Vogel 2015; Westermann 2018). The variations in small regulatory RNAs suggest how these RNAs evolved or were distributed among the diverse range of bacterial species, and how they were integrated into the gene regulatory network in each bacterium (Peer and Margalit 2014; Updegrove et al. 2015). To deduce these common themes, this chapter shares the knowledge obtained by dissecting bacterial small regulatory RNAs.

This chapter focuses on a subset of regulatory RNAs that interact with the RNA binding protein Hfq in bacteria. Hfq-dependent regulatory RNAs are well understood and are generally called small RNAs (sRNAs). Hfq was initially identified as a host factor for bacteriophage Qβ RNA replication in *E. coli* (Franze de Fernandez et al. 1968). Hfq is now recognized as a pleiotropic RNA chaperone that helps sRNAs regulate their mRNA targets (Holmqvist and Vogel 2018; Santiago-Frangos and Woodson 2018). Hfq belongs to the large family of Sm-like RNA binding proteins and assembles into a stable homohexameric ring-like architecture (Dos Santos et al. 2019; Weichenrieder 2014).

In the first part of this chapter, the characteristics, biogenesis and regulatory mechanisms of sRNAs, almost all of which have been studied in *E. coli* and *Salmonella*, are documented. In the second part, the commonalities and divergences among sRNAs and the variations in their function are discussed, using several examples such as SgrS, which responds to glucose–phosphate stress.

8.2 Characteristics of sRNAs

sRNAs are typically transcripts with a size ranging from 50 to 200 nucleotides with three functional elements: (i) a region that base-pairs with target mRNAs, (ii) Hfq-binding sites, and (iii) an intrinsic transcription terminator (Fig. 8.1a). The base-pair

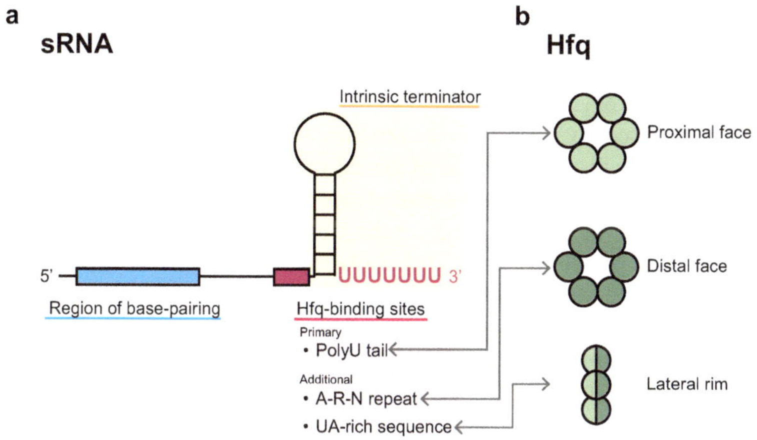

Fig. 8.1 Structures of an Hfq-dependent sRNA and the RNA-binding surfaces of Hfq homohexamer. (**a**) sRNAs consist of a base-pair region (blue), Hfq-binding sites (pink) and an intrinsic terminator (yellow). The base-pair region is typically in the 5′ portion relative to the other two elements. (**b**) N-terminal core of Hfq assembled in a homohexameric ring with three RNA-binding surfaces (proximal face, distal face, and rim). The C-terminal region of Hfq is unstructured and is less conserved (see Sect. 8.5.2 for details)

regions include a sequence of around eight nucleotides, referred to as the seed region (Kawamoto et al. 2006; Sharma et al. 2007), which allows sRNAs to bind multiple mRNA targets through short and imperfect base-pairing (Bobrovskyy et al. 2019; De Lay and Gottesman 2009; Guillier and Gottesman 2008; Miyakoshi et al. 2022b; Papenfort et al. 2010). This limited and noncontiguous complementarity and the nature of the target sites on the mRNAs are thought to cause a variety of regulatory outcomes.

The ability of sRNAs to bind Hfq is a critical feature that facilitates the base-pairing with mRNA targets (Møller et al. 2002; Zhang et al. 2002). There are at least two Hfq-binding sites in an sRNA, each of which binds to one of the three RNA binding surfaces on the ring-shaped Hfq hexamer (Fig. 8.1b) (Updegrove et al. 2016). One sequence motif of the Hfq-binding site is a polyU tail of seven or more Us that is the primary binding site preferred by the proximal face of the Hfq hexamer (Otaka et al. 2011; Sauer and Weichenrieder 2011). This polyU tail is contained in an intrinsic terminator (Morita et al. 2017), which has been found on all Hfq-binding sRNAs identified. Additionally, sRNAs encompass other sequence motifs for Hfq-binding; an A-R-N repeat (R = A/G, N = any residue) preferentially binds to the distal face of the Hfq hexamer (Mikulecky et al. 2004), whereas a UA-rich sequence interacts with the lateral rim of the ring (Ishikawa et al. 2012; Sauer et al. 2012). In general, sRNAs have one of these two motifs as an additional binding site; those containing an A-R-N repeat are classified as Class I sRNAs, and those containing an UA-rich sequence are classified as Class II sRNAs (Schu et al. 2015).

An intrinsic terminator is the main element for transcription termination in bacteria (Ray-Soni et al. 2016; Roberts 2019; You et al. 2023). The nucleotide sequence of intrinsic terminator consists of a GC-rich dyad repeat, which forms a hairpin structure in the RNA, followed by a T-rich stretch. A notable genetic feature of sRNAs is a consecutive T stretch longer than seven nucleotides in their intrinsic terminator (Holmqvist et al. 2016; Melamed et al. 2016; Otaka et al. 2011). The consecutive U residues in the RNA that are released after the termination are individually bound to the proximal face of the Hfq hexamer (Dimastrogiovanni et al. 2014; Sauer and Weichenrieder 2011). Although the nucleotide sequences of the hairpin structure do not appear to be conserved, the hairpin structure itself, particularly its thermodynamic stability, plays an important role in sRNA production by modulating the termination position within the T stretch (Balbontín et al. 2010; Morita et al. 2017; Morita et al. 2015; Sledjeski et al. 1996). In summary, the intrinsic terminators of sRNAs possess two features: a polyT stretch of seven or more T residues in DNA and a moderate-strength RNA hairpin, which ensures transcription is terminated after the T stretch.

8.3 Biogenesis of sRNAs

sRNAs typically have their own transcription units (a promoter and terminator), and the transcription products function as regulatory RNAs (Livny and Waldor 2007). This type of sRNA is mostly regulated by specific transcription factors in response to cognate environmental signals (Fig. 8.2a) (Gottesman and Storz 2011). For instance, the expression of sRNA RyhB is regulated by the Fe-responsive repressor Fur, which induces RyhB in response to iron depletion (Massé et al. 2003; Massé and Gottesman 2002). RyhB regulates the expression of several mRNAs, including those encoding iron storage proteins (Massé et al. 2005; Wang et al. 2015). Many transcription factors involved in sRNA production also regulate the initiation of mRNA transcription involved in cognate stress responses, forming a mixed network that overlays the transcriptional and post-transcriptional layers. The multilayered regulatory motifs are thought to ensure that the regulatory networks provide an appropriate outcome (Durand and Guillier 2021; Fröhlich and Gottesman 2018; Miyakoshi 2024; Papenfort and Melamed 2023; Wagner and Romby 2015).

Another type of sRNA originates from the processing of mRNAs. Among the processing products, those containing the 3′ portion of parental mRNAs act as regulatory RNAs and referred to as 3′-derived sRNAs (Fig. 8.2b) (Chao et al. 2012; Miyakoshi et al. 2015b; Ponath et al. 2022). The processing of parental mRNAs often relies on the major endoribonuclease RNase E (Chao and Vogel 2016; Miyakoshi et al. 2015a)(see also Chapter by Liu, et al. in this volume). In this biogenesis pathway, processing shortens the primary transcript, thereby separating the regulatory function as an sRNA from the parental mRNA. Additionally, endoribonucleolytic cleavage by RNase E converts the 5′ end of the downstream product

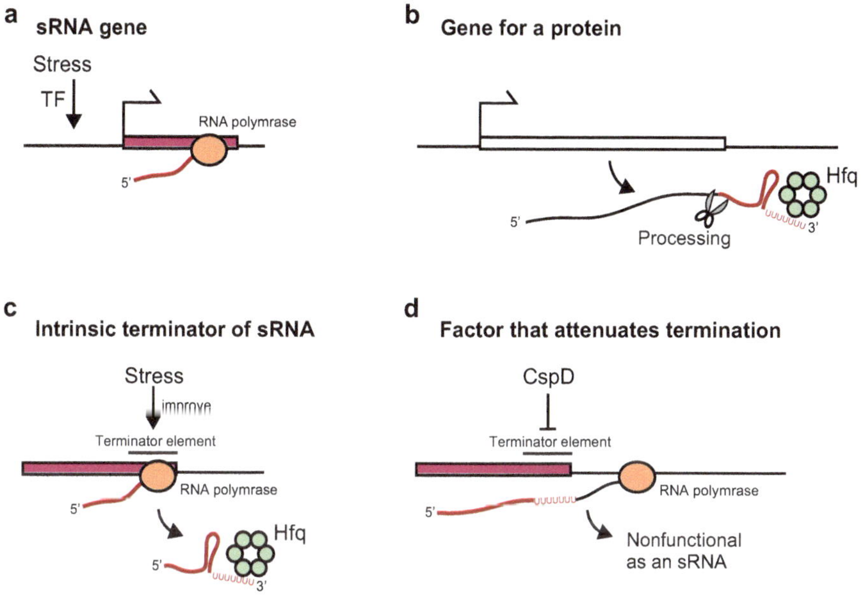

Fig. 8.2 Biosynthesis pathways of sRNAs. (**a**) Canonical sRNAs are transcribed from their own promoter by cognate transcription factors (TFs), and the transcription terminates at their intrinsic terminators. sRNA promoters are occasionally found inside an mRNA coding sequence [e.g., MicL sRNA (Guo et al. 2014)]. (**b**) Some sRNAs are produced by 3′ processing of an mRNA. Typically, RNA cleavage in 3′ processing involves RNase E. (**c**) Efficiency of sRNA transcription termination is increased during stress. Stressors, including low temperature (for DsrA sRNA) and glucose–phosphate and iron depletion (for both SgrS sRNA and RyhB sRNA), increase termination efficiency. (**d**) Protein factors affect sRNA transcription termination. CspD reduces SgrS production by attenuating transcription termination, thereby abrogating regulation by SgrS

from triphosphate to monophosphate (Mackie 1998; Misra and Apirion 1979), which may confer an additive role to 3′-derived sRNAs.

Regardless of whether the 5′ end of an sRNA is generated from its own promoter or by the mRNA from which it is processed, the 3′ end originates from an intrinsic terminator (Chen et al. 2019). Several studies have shown that transcription termination for sRNAs could also act as a regulatory layer to control their production (Fig. 8.2c). For example, the efficiency of the transcription termination for the sRNA DsrA, an activator for the expression of *rpoS*, is increased in cells growing at low temperature (Sledjeski et al. 1996). The termination efficiencies at the terminators of the sRNAs SgrS and RyhB are likewise increased under glucose–phosphate accumulation and iron depletion stress (Morita et al. 2015). An increase in the termination efficiency likely promotes sRNA production, contributing to the cellular response to the stressor. Additionally, when a protein-coding region is located immediately downstream of an sRNA gene, readthrough of the sRNA terminator enhances expression of the downstream gene, as the production of functional sRNA is reduced.

Furthermore, multicopy genetic screening identified the factors that attenuate the intrinsic terminators of SgrS and other sRNAs (Morita et al. 2022). One such factor is CspD, an RNA binding protein belonging to the cold-shock protein family. CspD accumulates under poor growth conditions and binds to the SgrS transcripts, attenuating the intrinsic termination for SgrS and stabilizing the transcripts (Fig. 8.2d) (Hasegawa et al. 2024; Morita et al. 2022). However, CspD did not appear to be involved in the mechanism of the increased termination efficiency during stress exposure (Morita et al. 2022). Accordingly, questions remain about how the efficiency of intrinsic termination for sRNAs is regulated (Chen et al. 2019).

8.4 Regulatory Mechanisms of sRNAs

sRNA transcripts are usually unstable. Once sRNAs bind to the Hfq hexamer, they are stabilized by Hfq (Gottesman 2004). Hfq bound to sRNAs is crucial in sRNA–target pairing. Similar to sRNAs, which contain functional elements for Hfq-mediated duplex formation (see Sect. 8.2), the target mRNAs have several structural features, particularly an Hfq-binding motif, with either an A-R-N repeat or a UA-rich sequence (Beisel et al. 2012; Schu et al. 2015). sRNAs carrying a UA-rich sequence regulate target mRNAs containing A-R-N repeats, whereas sRNAs carrying A-R-N repeats regulate target mRNAs containing a UA-rich sequence (Schu et al. 2015). These combinations allow single Hfq hexamers to bind with both an sRNA and its target mRNA. Duplex formation is then facilitated by the arginine-rich region in the lateral rim of the Hfq hexamer (Małecka and Woodson 2024; Panja et al. 2013; Sauer et al. 2012).

A typical regulatory outcome is target silencing that is attributed to translation repression (Vogel and Luisi 2011). In this mode, sRNAs generally base pair with the 5′ region of the target mRNA, impeding ribosome entry (Fig. 8.3a). An *in vitro* translation assay using the RNA pair of SgrS–*ptsG* mRNA demonstrated that efficient translation repression is still possible when base-pairing occurs in the absence of Hfq (Maki et al. 2010; Maki et al. 2008). This strongly suggests that the RNA molecule is the determinant of sRNA regulation. Additionally, some studies showed that the Hfq itself on mRNA could repress translation (Chen and Gottesman 2017; Desnoyers and Massé 2012; Morita and Aiba 2019).

Another mode of silencing involves the rapid degradation of coupled sRNA–target pair (Fig. 8.3b) (Lalaouna et al. 2013; Saramago et al. 2014). RNase E, which is recruited by Hfq to target mRNAs, is central to this sRNA-mediated decay pathway (Bruce et al. 2018; Massé et al. 2003; Morita et al. 2005). The target mRNAs degraded through this pathway contain the site favored by RNase E (Morita and Aiba 2011; Prévost et al. 2011). For the regulation mediated by both translation repression and mRNA decay, efficient repression of translation was observed, even in the RNase E variants lacking the decay pathway (Morita et al. 2006; Urban and Vogel 2007). This suggests that translation repression is sufficient for target silencing by the sRNAs.

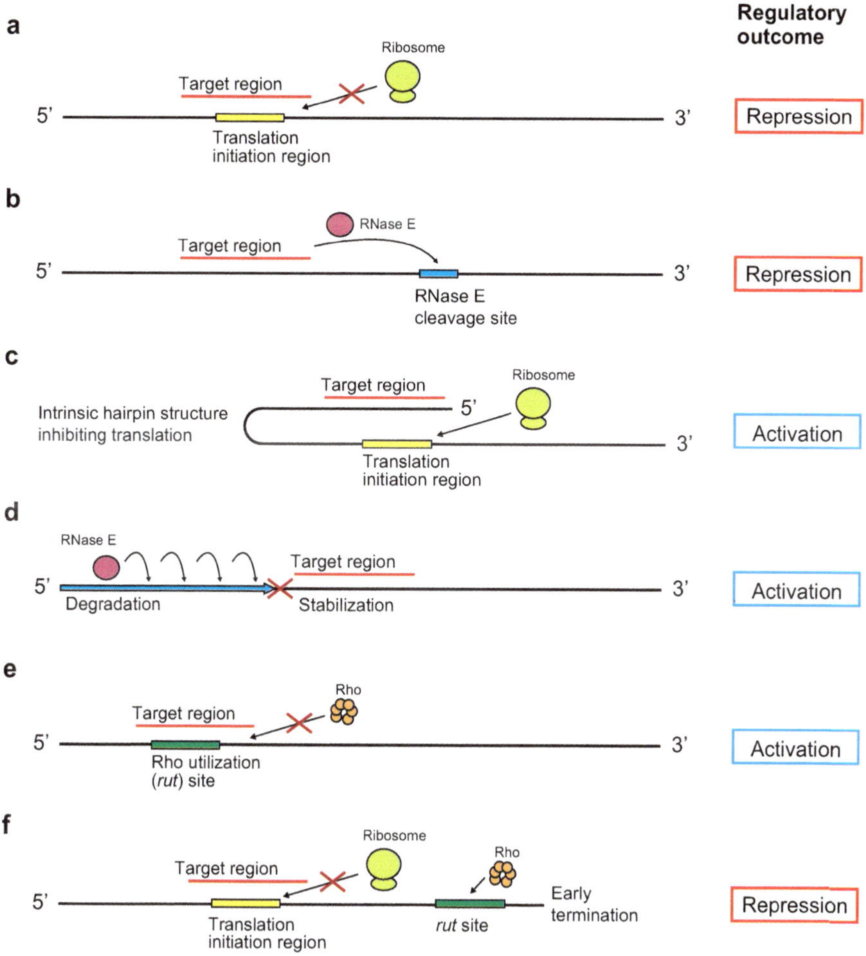

Fig. 8.3 Schematic overview of the regulatory mechanisms used by sRNAs. (**a–f**) In target mRNAs, the target regions that base pair with sRNAs are indicated by red lines. The translation initiation region (yellow), RNase E cleavage site (blue), and Rho utilization site (green) are shown as colored boxes. The regulatory outcome of each mechanism is shown to the right. All models are described in detail in Sect. 8.4

In addition to canonical regulation, some sRNAs promote target expression by either translation activation or mRNA stabilization (Papenfort and Vanderpool 2015). Target mRNAs regulated through translation activation intrinsically form a translation-inhibitory structure in the 5′ region (Majdalani et al. 2005). Some sRNAs base-pair with the 5′ region of target mRNAs, preventing the formation of the inhibitory structure (Fig. 8.3c). On the other hand, some target mRNAs are stabilized by sRNAs by interfering with ribonuclease access (Fig. 8.3d) (Fröhlich et al. 2013; Papenfort et al. 2013).

Several studies have also revealed that Rho-mediated transcription termination is regulated by sRNAs (Chen et al. 2019). Rho-dependent transcription termination is triggered by the binding of Rho to the nascent RNA transcripts at Rho-utilization sites, which are typically characterized by single-stranded C-rich/G-poor RNA (Adhya and Gottesman 1978). Some sRNAs modulate the accessibility of Rho to the Rho-utilization sites on target mRNAs, either directly by base-pairing (Fig. 8.3e) or indirectly by translational repression of the target (Fig. 8.3f) (Bossi et al. 2012; Sedlyarova et al. 2016; Silva et al. 2019).

8.5 Conserved and Divergent sRNA Functions Across Bacterial Species

Similar to many other areas of bacterial molecular biology, research on sRNAs began with the model organisms, *E. coli* and *Salmonella*. The accumulation of accurate genome sequences in databases has helped researchers find sRNAs in obscure bacterial species. This enables us to examine the roles of sRNAs in the physiology of each bacterium and to design bioengineering applications based on regulatory RNAs. These research directions are expected to provide clues to solving emerging problems, such as the development of drug resistance in pathogenic bacteria (Vogel et al. 2025; Westermann 2018). Furthermore, the identification of sRNAs in various bacteria has made it feasible to investigate the distribution of sRNAs. Unlike the coding sequences of proteins, which partly allow for changes even in functional domains owing to the presence of synonymous codons, a single nucleotide substitution in sRNA sequences is highly likely to have a large effect on their function. Therefore, the commonalities and divergences among sRNAs provide a simpler subject for research aiming to understand how the regulatory factors of gene expression vary among bacteria. The following sections highlight several examples of sRNA studies.

8.5.1 sRNA SgrS and Its Target ptsG *mRNA*

SgrS is a well-characterized sRNA in *E. coli* and *Salmonella* that is important in gene regulation when cells are exposed to glucose–phosphate stress (Bobrovskyy et al. 2019; Papenfort and Vanderpool 2015). The expression of *sgrS* is induced by the dedicated transcription factor SgrR in response to cellular accumulation of glucose–phosphate or analogs (Fig. 8.4a) (Morita et al. 2003; Richards et al. 2013; Vanderpool and Gottesman 2004). SgrS represses several mRNAs involved in glucose uptake/metabolism, including *ptsG* mRNA encoding the major glucose transporter $IICB^{Glc}$, while stabilizing the mRNA for *yigL* encoding a sugar phosphatase. These regulatory outcomes adjust the expression of the targets to the given levels,

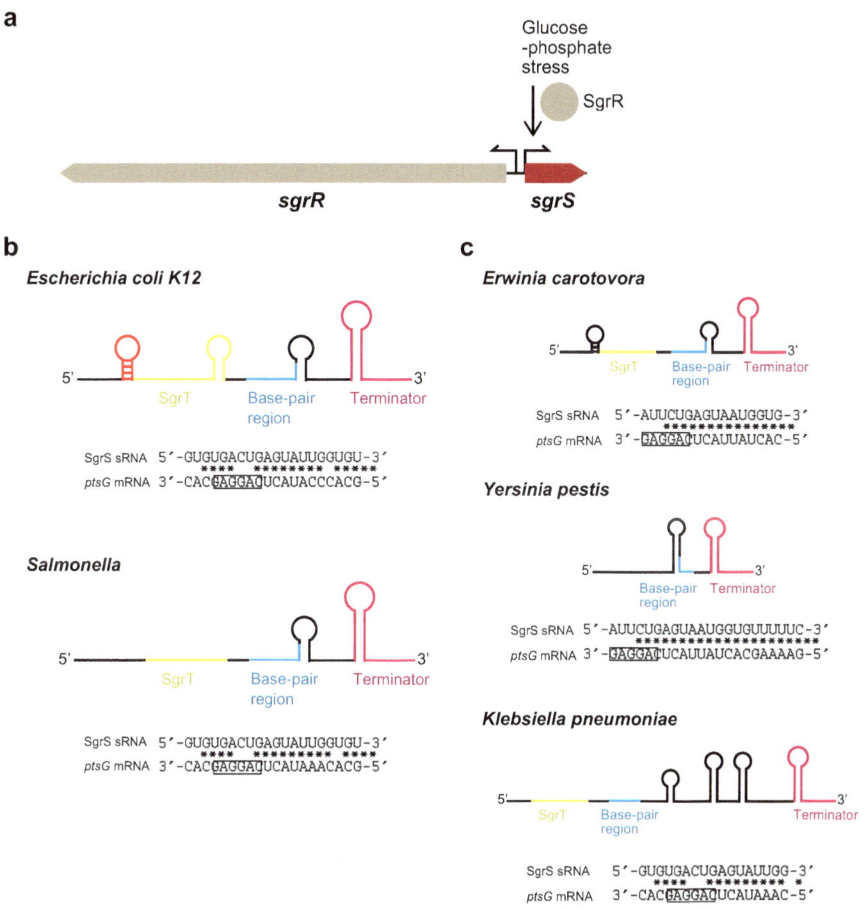

Fig. 8.4 SgrS is an sRNA that responds to glucose–phosphate stress. (**a**) Chromosomal *sgrR–sgrS* region with their promoters. SgrR protein induces SgrS expression in response to glucose–phosphate stress. (**b**, **c**) Examples of functional domains of SgrS in some bacteria. For each SgrS, the SgrT coding region (yellow), base-pair region (blue), and intrinsic terminator (pink) are shown. In *E. coli* SgrS, the hairpin structure that inhibits SgrT translation is shown in red. The predicted base-pairing of individual SgrS with the cognate *ptsG* mRNA is shown at the bottom of each panel. Accurate prediction of the secondary structures in a wider range of bacteria was previously described. (Horler and Vanderpool 2009)

thereby alleviating the accumulation of glucose–phosphate (Bobrovskyy et al. 2019; Papenfort et al. 2013). SgrS also regulates the expression of the virulence factor *sopD* in *Salmonella* (Papenfort et al. 2012), suggesting that SgrS monitors carbon metabolism and links the metabolic status with virulence to respond to environmental changes.

Studies on SgrS also led to the discovery of the small protein SgrT (around 40 amino acids long), which is translated from the *sgrS* transcript (Wadler and Vanderpool 2007). In the *E. coli sgrS* transcripts, SgrT was hardly translated owing to an RNA hairpin structure, but the production of SgrT was observed in *Salmonella* (Fig. 8.4b) (Balasubramanian and Vanderpool 2013). SgrT binds to $IICB^{Glc}$, inhibiting the transport of its substrates (Balasubramanian and Vanderpool 2013; Lloyd et al. 2017). Although the function of SgrT is exerted by a different mechanism to that of SgrS, SgrT contributes to essentially the same stress response as SgrS (denoted as a dual functional RNA). Other regulatory RNAs with two relevant functions (for example, SgrS sRNA/SgrT small protein) includes Spot42/SpfP (Aoyama et al. 2022) and AzuR/AzuC (Raina et al. 2022) identified in *E. coli*, and VcdR/VcdP (Venkat et al. 2021) identified in *Vibrio cholerae*.

In parallel with the functional characterization of SgrS, an early study used bioinformatics to investigate the distribution and conservation of SgrS (Horler and Vanderpool 2009). Because the functional elements of sRNAs are short (see Sect. 8.2), it is generally difficult to find sRNA homologs in the vast bacterial genomes. To search for homologs of SgrS, the SgrR protein, which is encoded by a gene located next to *sgrS* in the *E. coli* genome (Fig. 8.4a), provided a clue to narrow down the genomic loci. Subsequently, some features indicative of sRNAs, including the promoter and terminator sequences, were used for the search. This led to the identification of several SgrS homologs in bacterial genomes (Horler and Vanderpool 2009). This approach using a landmark gene identified homologs of GcvB sRNA, which is involved in amino acid metabolism and transport (Sharma et al. 2007; Urbanowski et al. 2000). The *sgrS* gene located next to *sgrR* was mostly identified in the genomes of enteric bacterial species. Sequence alignment of the *sgrS* locus across these species revealed limited overall conservation sequence of SgrS, with notable preservation of its functional elements for an sRNA (Fig. 8.4b,c) (Horler and Vanderpool 2009). This strongly suggests that these elements are the components required for sRNAs. Regarding SgrT, limited conservation was found in the sequence alignment of the *sgrS* locus, not only at the nucleotide sequence level, similar to that observed between *E. coli* and *Salmonella*, but also at the amino acid sequence level (Horler and Vanderpool 2009). The questions relating to how SgrT diverged within the *sgrS* loci and how the divergence of SgrS/SgrT is linked to the physiology of each bacterium are challenging and should be addressed in the future.

In studying the variation in an sRNA, it is important to compare the variation with the distribution and conservation of the target mRNAs of the sRNA. *ptsG* mRNA is a well-studied target of SgrS that encodes the membrane component of the major glucose transporter $IICB^{Glc}$ (Kawamoto et al. 2006; Kawamoto et al. 2005; Kimata et al. 2001). The *ptsG* gene is present in the genomes where SgrR was identified, and duplex formation is predictable between each identified SgrS–*ptsG* mRNA pair (Fig. 8.4b,c) (Horler and Vanderpool 2009). This implies that SgrS can regulate *ptsG* mRNA in the identified bacteria. Intriguingly, although the *ptsG* gene was found in a wider range of organisms than the *sgrR–sgrS* region, the 5′ target region of *ptsG* mRNA was only conserved in species containing SgrS homologs (Horler and Vanderpool 2009). This suggests that the cellular response to glucose–phosphate stress might be an evolutionary constraint in a subset of enteric bacteria, which could explain why base-pairing, and thus the regulation by SgrS, has remained conserved. Similarly, high degrees of conservation in the base-pairing region, but not in the targets in mRNAs, were also observed in various sRNAs (Peer and Margalit 2011, 2014; Skippington and Ragan 2012). Conservation of the base-pairing region in sRNAs probably became a selective pressure, favoring further accumulation of the target mRNAs and accelerating the complexity of the regulatory networks (Gottesman and Storz 2011; Richter and Backofen 2012).

8.5.2 RNA Chaperone Hfq

The distribution and structural diversity of Hfq homologs also appears to restrict the evolutionary spread and diversification of sRNAs across bacterial genomes (Peer and Margalit 2014). Phylogenetic analysis showed a broad distribution of Hfq in α-proteobacteria, β/γ-proteobacteria, and low GC Gram-positive bacteria (Sun et al. 2002). Regulation by sRNAs has been reported in studies on α- and β/γ-proteobacteria (Gottesman 2004; Moon et al. 2021; Vogt et al. 2024), and a marked accumulation of sRNAs was observed in *Enterobacteriales* (Peer and Margalit 2014). The Hfq homologs of *Enterobacteriales* contain an unstructured C-terminal region that is longer than that of most other species. An amino acid alignment of Hfq homologs showed that the C-terminal tail is less conserved than the N-terminal domain containing the Sm motifs (Sun et al. 2002). Furthermore, *in vivo* and *in vitro* experiments suggested that the C-terminal tail of *E. coli* Hfq cooperates with each of the three RNA binding surfaces, shaping the RNA selectivity of Hfq to achieve distinct regulatory outcomes for each sRNA (Kavita et al. 2022; Sarni et al. 2022). This extended function may contribute to sRNA accumulation in *Enterobacteriales*.

Although Hfq homologs have also been found in Gram-positive bacteria, their regulation by sRNAs has rarely been observed (Watkins and Arya 2023). The differing contribution of sRNA/Hfq between bacteria could be explained by the four amino acid residues on the lateral rim of the Hfq hexamer. The rim of *E. coli* (*Enterobacteriales*) consists of arginine-rich amino acid residues R-R-E-R, whereas the rims of *B. subtilis* and *S. aureus* (Gram-positive) consist of R-K-E-N and K-A-N-Q, respectively (Zheng et al. 2016). The Hfq proteins of these two species and the *E. coli* variants lacking an arginine residue in the rim facilitated to a lesser extent sRNA–target mRNA pairing *in vitro* (Panja et al. 2013; Zheng et al. 2016). The possibility that the replacement of amino acid residues in the rim might reflect the low GC content of Gram-positive bacteria was previously discussed (Watkins and Arya 2023).

8.6 Variations in 3′-Derived sRNAs

Variants of 3′-derived sRNAs (see Fig. 8.2b) can often be identified more readily because the originating parental mRNAs serve as genomic landmarks. One example is SroC sRNA, a 3′-derived sRNA produced by processing of *gltI* mRNA in

Salmonella and, as a sponge RNA, it hinders the function of GcvB sRNA through base-pairing (Miyakoshi et al. 2015a; Vogel et al. 2003). By using the *gltI* gene and the following *gltJKL* operon as landmarks, the gene locus of SroC homologs has been found in many enterobacterial species, including *E. coli*. Unsurprisingly, the genomes in which SroC was identified contain the gene for the target GcvB, and *E. coli* SroC retains the sponge activity for GcvB (Miyakoshi et al. 2015a). Interestingly, SroC sRNA found in *Yersinia pestis*, whose genome also contains the *gcvB* gene, lacks part of the base-pair region with GcvB, thus losing the sponge regulatory function (Miyakoshi et al. 2015a). This variation provides a hint for further research into the acquisition or loss of regulatory function in the divergent evolution of sRNAs.

Similar variations were observed in studies of other 3′-derived sRNAs. GlnZ sRNA is produced by processing of *glnA* mRNA and regulates multiple mRNAs, including *sucA* and *aceE*, which are involved in nitrogen and carbon metabolism (Kawano et al. 2005; Miyakoshi et al. 2022a; Sittka et al. 2009; Walling et al. 2022). For GlnZ homologs, the seed region and structural features of the intrinsic terminator are highly conserved in *Enterobacteriaceae*. However, detailed sequence comparison between *E. coli* and *Salmonella* revealed that an extra G nucleotide is inserted in the seed region of *Salmonella* GlnZ (Miyakoshi et al. 2022a). *Salmonella* GlnZ regulates *sucA* mRNA, similar to *E. coli* GlnZ, but it does not regulate *aceE* mRNA because of the extra G nucleotide (Fig. 8.5a). Likewise, several nucleotide differences shift the target repertoire of SdhX sRNA. SdhX is produced by processing of the *sdhCDAB–sucABCD* operon mRNA and regulates the expression of genes involved in acetate metabolism (Cronan Jr. and Laporte 2005; De Mets et al. 2019; Miyakoshi et al. 2019; Zhang et al. 2003). SdhX from *E. coli* and *Salmonella* efficiently represses the expression of *ackA*, which encodes acetate kinase (De Mets et al. 2019; Miyakoshi et al. 2019). *Salmonella* SdhX also regulates the *fumB* and *yfbV* mRNAs, whereas *E. coli* SdhX has lost the ability to regulate these targets and instead regulates *fdoG* and *katG* as specific targets (Fig. 8.5a) (Miyakoshi et al. 2019). Unlike the GlnZ–target pairs, these differences in SdhX recognition for the *fumB* and *yfbV* mRNAs arise from variations of the mRNA target sequence.

It is worth noting that the three sRNAs described above are involved in adjusting target expression according to available nutrient sources. It was proposed that the preferred levels of enzymes in metabolic pathways are conserved, even among completely different species, and that various types of post-transcriptional regulation have important roles in maintaining enzyme production at the required levels (Lalanne et al. 2018). Therefore, the differences in the sRNA target repertoires between *E. coli* and *Salmonella* may contribute to maintaining the preferred levels of enzymes, rather than regulating the expression of specific genes that help bacteria combat environmental stresses. If this is the case, the preferred enzyme levels defined in each metabolic pathway probably forced selective pressure on the evolution of these sRNAs.

The sequence variations are often found in the 3′ region. In each transcription unit of *Staphylococcus aureus* species, sequence variations are more frequently found in the 3′ untranslated region than in the coding or 5′ untranslated regions

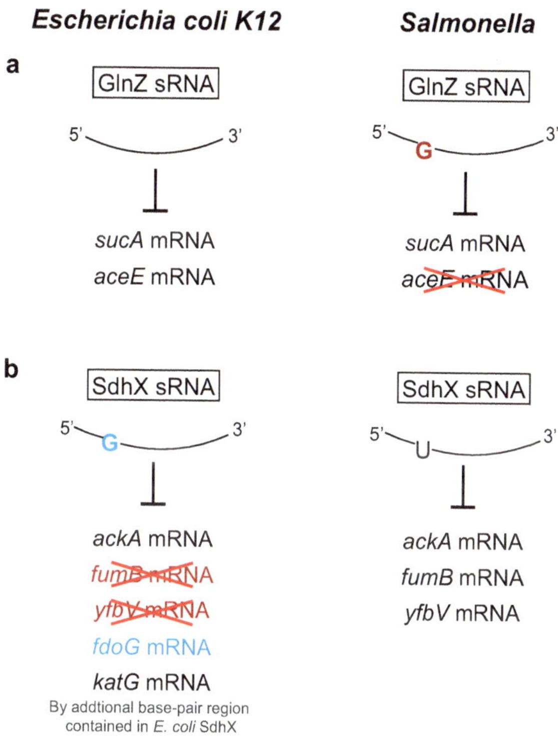

Fig. 8.5 Homologs of 3′-derived sRNAs in *E. coli* and *Salmonella* have different lists of mRNA targets due to nucleotide variations. (**a**) GlnZ, in which an extra G inserted within the seed region in *Salmonella* is shown in red. (**b**) SdhX, in which the target regions of *fumB* and *yfbV* vary between *E. coli* and *Salmonella*; these genes are not regulated by SdhX in *E. coli*. A single nucleotide variation in the seed regions appears as G (blue) in *E. coli* and U (black) in *Salmonella*, which affects the regulation of *fdoG* mRNA

(Menendez-Gil et al. 2020). Similarly, Group II introns, a type of mobile element, are frequently inserted in genetic loci near to the intrinsic terminators (Dai and Zimmerly 2002; Miura et al. 2022; Waldern et al. 2020). It is plausible that the genetic loci corresponding to the 3′ regions of transcripts may be an ideal playground where sRNAs can gain or lose function, or to rearrange their target lists (Miyakoshi et al. 2015b; Ponath et al. 2022).

8.7 Concluding Remarks

This chapter has summarized current insights into both the conserved and divergent features of bacterial sRNAs, with a particular focus on Hfq-binding sRNAs. Advancing sRNA research along these lines will greatly improve our ability to predict sRNA-mediated regulation in obscure bacterial genomes and help researchers design effective synthetic sRNAs for genetic engineering. There are several other important aspects of sRNA-mediated regulation that could not be covered fully here, including studies of proteins that modulate sRNA function/biogenesis, such as PNPase (De Lay and Gottesman 2011; Dendooven et al. 2021; Sinha and De Lay 2022), Crc (Malecka et al. 2021; Pei et al. 2019; Sonnleitner et al. 2018) and HqbA

(Chen et al. 2021; Luo et al. 2023). Other sRNA binding proteins, including ProQ (Liao and Smirnov 2023; Olejniczak et al. 2022) and CsrA (Pourciau et al. 2020; Romeo and Babitzke 2018), along with their partner RNAs, are also beyond the scope of this chapter. Further investigation of the species distribution and molecular divergence of these proteins —as well as key RNases such as RNase E — would reinforce our knowledge on how RNA-mediated regulation is integrated into the gene regulatory network in each bacterium. Another area not discussed here is the origin of sRNAs and the mechanisms by which ancestral sRNA in a bacterium transfer to other bacteria. Prior reviews provide broader perspectives on these areas (Dutcher and Raghavan 2018; Peer and Margalit 2014; Updegrove et al. 2015). Together, these research directions will contribute to a more comprehensive understanding of how RNA-based regulation shapes bacterial physiology and evolution.

Acknowledgements The author thanks Masatoshi Miyakoshi and Miu Naruki for comments on the manuscript. Preparation of this article was supported by the Japan Society for the Promotion of Science KAKENHI (grant number 23 K04996), and in part by funds from the Yamagata Prefectural Government and Tsuruoka City, Japan.

Conflict of Interest The author declared no conflict of interest.

References

Adams PP, Storz G (2020) Prevalence of small base-pairing RNAs derived from diverse genomic loci. Biochim Biophys Acta Gene Regul Mech 1863(7):194524. https://doi.org/10.1016/j.bbagrm.2020.194524

Adams PP, Baniulyte G, Esnault C, Chegireddy K, Singh N, Monge M, Dale RK, Storz G, Wade JT (2021) Regulatory roles of Escherichia coli 5′ UTR and ORF-internal RNAs detected by 3′ end mapping. elife 10. https://doi.org/10.7554/eLife.62438

Adhya S, Gottesman M (1978) Control of transcription termination. Annu Rev Biochem 47:967–996. https://doi.org/10.1146/annurev.bi.47.070178.004535

Aoyama JJ, Raina M, Zhong A, Storz G (2022) Dual-function spot 42 RNA encodes a 15-amino acid protein that regulates the CRP transcription factor. Proc Natl Acad Sci USA 119(10):e2119866119. https://doi.org/10.1073/pnas.2119866119

Balasubramanian D, Vanderpool CK (2013) Deciphering the interplay between two independent functions of the small RNA regulator SgrS in salmonella. J Bacteriol 195(20):4620–4630. https://doi.org/10.1128/jb.00586-13

Balbontín R, Fiorini F, Figueroa-Bossi N, Casadesús J, Bossi L (2010) Recognition of heptameric seed sequence underlies multi-target regulation by RybB small RNA in salmonella enterica. Mol Microbiol 78(2):380–394. https://doi.org/10.1111/j.1365-2958.2010.07342.x

Barquist L, Vogel J (2015) Accelerating discovery and functional analysis of small RNAs with new technologies. Annu Rev Genet 49:367–394. https://doi.org/10.1146/annurev-genet-112414-054804

Beisel CL, Updegrove TB, Janson BJ, Storz G (2012) Multiple factors dictate target selection by Hfq-binding small RNAs. EMBO J 31(8):1961–1974. https://doi.org/10.1038/emboj.2012.52

Bobrovskyy M, Azam MS, Frandsen JK, Zhang J, Poddar A, Ma X, Henkin TM, Ha T, Vanderpool CK (2019) Determinants of target prioritization and regulatory hierarchy for the bacterial small RNA SgrS. Mol Microbiol 112(4):1199–1218. https://doi.org/10.1111/mmi.14355

Bossi L, Schwartz A, Guillemardet B, Boudvillain M, Figueroa-Bossi N (2012) A role for rho-dependent polarity in gene regulation by a noncoding small RNA. Genes Dev 26(16):1864–1873. https://doi.org/10.1101/gad.195412.112

Bruce HA, Du D, Matak-Vinkovic D, Bandyra KJ, Broadhurst RW, Martin E, Sobott F, Shkumatov AV, Luisi BF (2018) Analysis of the natively unstructured RNA/protein-recognition core in the Escherichia coli RNA degradosome and its interactions with regulatory RNA/Hfq complexes. Nucleic Acids Res 46(1):387–402. https://doi.org/10.1093/nar/gkx1083

Chao Y, Vogel J (2016) A 3′ UTR-derived small RNA provides the regulatory noncoding arm of the inner membrane stress response. Mol Cell 61(3):352–363. https://doi.org/10.1016/j.molcel.2015.12.023

Chao Y, Papenfort K, Reinhardt R, Sharma CM, Vogel J (2012) An atlas of Hfq-bound transcripts reveals 3′ UTRs as a genomic reservoir of regulatory small RNAs. EMBO J 31(20):4005–4019. https://doi.org/10.1038/emboj.2012.229

Chen J, Gottesman S (2017) Hfq links translation repression to stress-induced mutagenesis in E. Coli. Genes Dev 31(13):1382–1395. https://doi.org/10.1101/gad.302547.117

Chen J, Morita T, Gottesman S (2019) Regulation of transcription termination of small RNAs and by small RNAs: molecular mechanisms and biological functions. Front Cell Infect Microbiol 9:201. https://doi.org/10.3389/fcimb.2019.00201

Chen J, To, L., de Mets F, Luo X, Majdalani N, Tai CH, Gottesman S (2021) A fluorescence-based genetic screen reveals diverse mechanisms silencing small RNA signaling in E. Coli. Proc Natl Acad Sci USA 118(27). https://doi.org/10.1073/pnas.2106964118

Cronan JE Jr, Laporte D (2005) Tricarboxylic acid cycle and Glyoxylate bypass. EcoSal Plus 1(2). https://doi.org/10.1128/ecosalplus.3.5.2

Dai L, Zimmerly S (2002) Compilation and analysis of group II intron insertions in bacterial genomes: evidence for retroelement behavior. Nucleic Acids Res 30(5):1091–1102. https://doi.org/10.1093/nar/30.5.1091

De Lay N, Gottesman S (2009) The Crp-activated small noncoding regulatory RNA CyaR (RyeE) links nutritional status to group behavior. J Bacteriol 191(2):461–476. https://doi.org/10.1128/jb.01157-08

De Lay N, Gottesman S (2011) Role of polynucleotide phosphorylase in sRNA function in Escherichia coli. RNA 17(6):1172–1189. https://doi.org/10.1261/rna.2531211

De Mets F, Van Melderen L, Gottesman S (2019) Regulation of acetate metabolism and coordination with the TCA cycle via a processed small RNA. Proc Natl Acad Sci USA 116(3):1043–1052. https://doi.org/10.1073/pnas.1815288116

Delaleau M, Figueroa-Bossi N, Do TD, Kerboriou P, Eveno E, Bossi L, Boudvillain M (2024) Rho-dependent transcriptional switches regulate the bacterial response to cold shock. Mol Cell 84(18):3482–3496. https://doi.org/10.1016/j.molcel.2024.07.034

Dendooven T, Sinha D, Roeselová A, Cameron TA, De Lay NR, Luisi BF, Bandyra KJ (2021) A cooperative PNPase-Hfq-RNA carrier complex facilitates bacterial riboregulation. Mol Cell 81(14):2901–2913. https://doi.org/10.1016/j.molcel.2021.05.032

Desnoyers G, Massé E (2012) Noncanonical repression of translation initiation through small RNA recruitment of the RNA chaperone Hfq. Genes Dev 26(7):726–739. https://doi.org/10.1101/gad.182493.111

Dimastrogiovanni D, Fröhlich KS, Bandyra KJ, Bruce HA, Hohensee S, Vogel J, Luisi BF (2014) Recognition of the small regulatory RNA RydC by the bacterial Hfq protein. elife 3. https://doi.org/10.7554/eLife.05375

Dos Santos RF, Arraiano CM, Andrade JM (2019) New molecular interactions broaden the functions of the RNA chaperone Hfq. Curr Genet 65(6):1313–1319. https://doi.org/10.1007/s00294-019-00990-y

Durand S, Guillier M (2021) Transcriptional and post-transcriptional control of the nitrate respiration in bacteria. Front Mol Biosci 8:667758. https://doi.org/10.3389/fmolb.2021.667758

Dutcher HA, Raghavan R (2018) Origin, evolution, and loss of bacterial small RNAs. Microbiol Spectr 6(2). https://doi.org/10.1128/microbiolspec.RWR-0004-2017

Franze de Fernandez MT, Eoyang L, August JT (1968) Factor fraction required for the synthesis of bacteriophage Qbeta-RNA. Nature 219(5154):588–590. https://doi.org/10.1038/219588a0
Fröhlich KS, Gottesman S (2018) Small regulatory RNAs in the Enterobacterial response to envelope damage and oxidative stress. Microbiol Spectr 6(4). https://doi.org/10.1128/microbiolspec.RWR-0022-2018
Fröhlich KS, Papenfort K, Fekete A, Vogel J (2013) A small RNA activates CFA synthase by isoform-specific mRNA stabilization. EMBO J 32(22):2963–2979. https://doi.org/10.1038/emboj.2013.222
Goodson JR, Winkler WC (2018) Processive Antitermination. Microbiol Spectr 6(5). https://doi.org/10.1128/microbiolspec.RWR-0031-2018
Gottesman S (2004) The small RNA regulators of Escherichia coli: roles and mechanisms*. Ann Rev Microbiol 58:303–328. https://doi.org/10.1146/annurev.micro.58.030603.123841
Gottesman S, Storz G (2011) Bacterial small RNA regulators: versatile roles and rapidly evolving variations. Cold Spring Harb Perspect Biol 3(12). https://doi.org/10.1101/cshperspect.a003798
Guillier M, Gottesman S (2008) The 5′ end of two redundant sRNAs is involved in the regulation of multiple targets, including their own regulator. Nucleic Acids Res 36(21):6781–6794. https://doi.org/10.1093/nar/gkn742
Guo MS, Updegrove TB, Gogol EB, Shabalina SA, Gross CA, Storz G (2014) MicL, a new σE-dependent sRNA, combats envelope stress by repressing synthesis of Lpp, the major outer membrane lipoprotein. Genes Dev 28(14):1620–1634. https://doi.org/10.1101/gad.243485.114
Haft RJ, Keating DH, Schwaegler T, Schwalbach MS, Vinokur J, Tremaine M, Peters JM, Kotlajich MV, Pohlmann EL, Ong IM, Grass JA, Kiley PJ, Landick R (2014) Correcting direct effects of ethanol on translation and transcription machinery confers ethanol tolerance in bacteria. Proc Natl Acad Sci USA 111(25):E2576–E2585. https://doi.org/10.1073/pnas.1401853111
Hasegawa S, Inose R, Igarashi M, Tsurumaki M, Saito M, Yanagisawa T, Kanai A, Morita T (2024) An internal loop region is responsible for inherent target specificity of bacterial cold-shock proteins. RNA 31(1):67–85. https://doi.org/10.1261/rna.080163.124
Holmqvist E, Vogel J (2018) RNA-binding proteins in bacteria. Nat Rev Microbiol 16(10):601–615. https://doi.org/10.1038/s41579-018-0049-5
Holmqvist E, Wagner EGH (2017) Impact of bacterial sRNAs in stress responses. Biochem Soc Trans 45(6):1203–1212. https://doi.org/10.1042/bst20160363
Holmqvist E, Wright PR, Li L, Bischler T, Barquist L, Reinhardt R, Backofen R, Vogel J (2016) Global RNA recognition patterns of post-transcriptional regulators Hfq and CsrA revealed by UV crosslinking in vivo. EMBO J 35(9):991–1011. https://doi.org/10.15252/embj.201593360
Hör J, Matera G, Vogel J, Gottesman S, Storz G (2020) Trans-acting small RNAs and their effects on gene expression in Escherichia coli and salmonella enterica. EcoSal Plus 9(1). https://doi.org/10.1128/ecosalplus.ESP-0030-2019
Horler RS, Vanderpool CK (2009) Homologs of the small RNA SgrS are broadly distributed in enteric bacteria but have diverged in size and sequence. Nucleic Acids Res 37(16):5465–5476. https://doi.org/10.1093/nar/gkp501
Ishikawa H, Otaka H, Maki K, Morita T, Aiba H (2012) The functional Hfq-binding module of bacterial sRNAs consists of a double or single hairpin preceded by a U-rich sequence and followed by a 3′ poly(U) tail. RNA 18(5):1062–1074. https://doi.org/10.1261/rna.031575.111
Kavita K, de Mets F, Gottesman S (2018) New aspects of RNA-based regulation by Hfq and its partner sRNAs. Curr Opin Microbiol 42:53–61. https://doi.org/10.1016/j.mib.2017.10.014
Kavita K, Zhang A, Tai CH, Majdalani N, Storz G, Gottesman S (2022) Multiple in vivo roles for the C-terminal domain of the RNA chaperone Hfq. Nucleic Acids Res 50(3):1718–1733. https://doi.org/10.1093/nar/gkac017
Kawamoto H, Morita T, Shimizu A, Inada T, Aiba H (2005) Implication of membrane localization of target mRNA in the action of a small RNA: mechanism of post-transcriptional regulation of glucose transporter in Escherichia coli. Genes Dev 19(3):328–338. https://doi.org/10.1101/gad.1270605

Kawamoto H, Koide Y, Morita T, Aiba H (2006) Base-pairing requirement for RNA silencing by a bacterial small RNA and acceleration of duplex formation by Hfq. Mol Microbiol 61(4):1013–1022. https://doi.org/10.1111/j.1365-2958.2006.05288.x

Kawano M, Reynolds AA, Miranda-Rios J, Storz G (2005) Detection of 5′- and 3′-UTR-derived small RNAs and cis-encoded antisense RNAs in Escherichia coli. Nucleic Acids Res 33(3):1040–1050. https://doi.org/10.1093/nar/gki256

Kimata K, Tanaka Y, Inada T, Aiba H (2001) Expression of the glucose transporter gene, ptsG, is regulated at the mRNA degradation step in response to glycolytic flux in Escherichia coli. EMBO J 20(13):3587–3595. https://doi.org/10.1093/emboj/20.13.3587

Lalanne JB, Taggart JC, Guo MS, Herzel L, Schieler A, Li GW (2018) Evolutionary convergence of pathway-specific enzyme expression stoichiometry. Cell 173(3):749–761. https://doi.org/10.1016/j.cell.2018.03.007

Lalaouna D, Simoneau-Roy M, Lafontaine D, Massé E (2013) Regulatory RNAs and target mRNA decay in prokaryotes. Biochim Biophys Acta 1829(6-7):742–747. https://doi.org/10.1016/j.bbagrm.2013.02.013

Liao Z, Smirnov A (2023) FinO/ProQ-family proteins: an evolutionary perspective. Biosci Rep 43(3). https://doi.org/10.1042/bsr20220313

Livny J, Waldor MK (2007) Identification of small RNAs in diverse bacterial species. Curr Opin Microbiol 10(2):96–101. https://doi.org/10.1016/j.mib.2007.03.005

Lloyd CR, Park S, Fei J, Vanderpool CK (2017) The small protein SgrT controls transport activity of the glucose-specific phosphotransferase system. J Bacteriol 199(11). https://doi.org/10.1128/jb.00869-16

Luo X, Zhang A, Tai CH, Chen J, Majdalani N, Storz G, Gottesman S (2023) An acetyltranferase moonlights as a regulator of the RNA binding repertoire of the RNA chaperone Hfq in Escherichia coli. Proc Natl Acad Sci USA 120(49):e2311509120. https://doi.org/10.1073/pnas.2311509120

Mackie GA (1998) Ribonuclease E is a 5′-end-dependent endonuclease. Nature 395(6703):720–723. https://doi.org/10.1038/27246

Majdalani N, Vanderpool CK, Gottesman S (2005) Bacterial small RNA regulators. Crit Rev Biochem Mol Biol 40(2):93–113. https://doi.org/10.1080/10409230590918702

Maki K, Uno K, Morita T, Aiba H (2008) RNA, but not protein partners, is directly responsible for translational silencing by a bacterial Hfq-binding small RNA. Proc Natl Acad Sci USA 105(30):10332–10337. https://doi.org/10.1073/pnas.0803106105

Maki K, Morita T, Otaka H, Aiba H (2010) A minimal base-pairing region of a bacterial small RNA SgrS required for translational repression of ptsG mRNA. Mol Microbiol 76(3):782–792. https://doi.org/10.1111/j.1365-2958.2010.07141.x

Małecka EM, Woodson SA (2024) RNA compaction and iterative scanning for small RNA targets by the Hfq chaperone. Nat Commun 15(1):2069. https://doi.org/10.1038/s41467-024-46316-6

Malecka EM, Bassani F, Dendooven T, Sonnleitner E, Rozner M, Albanese TG, Resch A, Luisi B, Woodson S, Bläsi U (2021) Stabilization of Hfq-mediated translational repression by the co-repressor Crc in Pseudomonas aeruginosa. Nucleic Acids Res 49(12):7075–7087. https://doi.org/10.1093/nar/gkab510

Massé E, Gottesman S (2002) A small RNA regulates the expression of genes involved in iron metabolism in Escherichia coli. Proc Natl Acad Sci USA 99(7):4620–4625. https://doi.org/10.1073/pnas.032066599

Massé E, Escorcia FE, Gottesman S (2003) Coupled degradation of a small regulatory RNA and its mRNA targets in Escherichia coli. Genes Dev 17(19):2374–2383. https://doi.org/10.1101/gad.1127103

Massé E, Vanderpool CK, Gottesman S (2005) Effect of RyhB small RNA on global iron use in Escherichia coli. J Bacteriol 187(20):6962–6971. https://doi.org/10.1128/jb.187.20.6962-6971.2005

McJunkin K, Gottesman S (2024) What goes up must come down: off switches for regulatory RNAs. Genes Dev 38(13-14):597–613. https://doi.org/10.1101/gad.351934.124

Melamed S, Peer A, Faigenbaum-Romm R, Gatt YE, Reiss N, Bar A, Altuvia Y, Argaman L, Margalit H (2016) Global mapping of small RNA-target interactions in bacteria. Mol Cell 63(5):884–897. https://doi.org/10.1016/j.molcel.2016.07.026

Menendez-Gil P, Caballero CJ, Catalan-Moreno A, Irurzun N, Barrio-Hernandez I, Caldelari I, Toledo-Arana A (2020) Differential evolution in 3′UTRs leads to specific gene expression in staphylococcus. Nucleic Acids Res 48(5):2544–2563. https://doi.org/10.1093/nar/gkaa047

Mikulecky PJ, Kaw MK, Brescia CC, Takach JC, Sledjeski DD, Feig AL (2004) Escherichia coli Hfq has distinct interaction surfaces for DsrA, rpoS and poly(a) RNAs. Nat Struct Mol Biol 11(12):1206–1214. https://doi.org/10.1038/nsmb858

Misra TK, Apirion D (1979) RNase E, an RNA processing enzyme from Escherichia coli. J Biol Chem 254(21):11154–11159

Miura MC, Nagata S, Tamaki S, Tomita M, Kanai A (2022) Distinct expansion of group II introns during evolution of prokaryotes and possible factors involved in its regulation. Front Microbiol 13:849080. https://doi.org/10.3389/fmicb.2022.849080

Miyakoshi M (2024) Multilayered regulation of amino acid metabolism in Escherichia coli. Curr Opin Microbiol 77:102406. https://doi.org/10.1016/j.mib.2023.102406

Miyakoshi M, Chao Y, Vogel J (2015a) Cross talk between ABC transporter mRNAs via a target mRNA-derived sponge of the GcvB small RNA. EMBO J 34(11):1478–1492. https://doi.org/10.15252/embj.201490546

Miyakoshi M, Chao Y, Vogel J (2015b) Regulatory small RNAs from the 3′ regions of bacterial mRNAs. Curr Opin Microbiol 24:132–139. https://doi.org/10.1016/j.mib.2015.01.013

Miyakoshi M, Matera G, Maki K, Sone Y, Vogel J (2019) Functional expansion of a TCA cycle operon mRNA by a 3′ end-derived small RNA. Nucleic Acids Res 47(4):2075–2088. https://doi.org/10.1093/nar/gky1243

Miyakoshi M, Morita T, Kobayashi A, Berger A, Takahashi H, Gotoh Y, Hayashi T, Tanaka K (2022a) Glutamine synthetase mRNA releases sRNA from its 3′UTR to regulate carbon/nitrogen metabolic balance in Enterobacteriaceae. elife 11. https://doi.org/10.7554/eLife.82411

Miyakoshi M, Okayama H, Lejars M, Kanda T, Tanaka Y, Itaya K, Okuno M, Itoh T, Iwai N, Wachi M (2022b) Mining RNA-seq data reveals the massive regulon of GcvB small RNA and its physiological significance in maintaining amino acid homeostasis in Escherichia coli. Mol Microbiol 117(1):160–178. https://doi.org/10.1111/mmi.14814

Mizuno T, Chou MY, Inouye M (1984) A unique mechanism regulating gene expression: translational inhibition by a complementary RNA transcript (micRNA). Proc Natl Acad Sci USA 81(7):1966–1970. https://doi.org/10.1073/pnas.81.7.1966

Møller T, Franch T, Højrup P, Keene DR, Bächinger HP, Brennan RG, Valentin-Hansen P (2002) Hfq: a bacterial Sm-like protein that mediates RNA-RNA interaction. Mol Cell 9(1):23–30. https://doi.org/10.1016/s1097-2765(01)00436-1

Moon K, Sim M, Tai CH, Yoo K, Merzbacher C, Yu SH, Kim DD, Lee J, Förstner KU, Chen Q, Stibitz S, Knipling LG, Hinton DM (2021) Identification of BvgA-dependent and BvgA-independent small RNAs (sRNAs) in Bordetella pertussis using the prokaryotic sRNA prediction toolkit ANNOgesic. Microbiol Spectr 9(2):e0004421. https://doi.org/10.1128/Spectrum.00044-21

Morita T, Aiba H (2011) RNase E action at a distance: degradation of target mRNAs mediated by an Hfq-binding small RNA in bacteria. Genes Dev 25(4):294–298. https://doi.org/10.1101/gad.2030311

Morita T, Aiba H (2019) Mechanism and physiological significance of autoregulation of the Escherichia coli hfq gene. RNA 25(2):264–276. https://doi.org/10.1261/rna.068106.118

Morita T, El-Kazzaz W, Tanaka Y, Inada T, Aiba H (2003) Accumulation of glucose 6-phosphate or fructose 6-phosphate is responsible for destabilization of glucose transporter mRNA in Escherichia coli. J Biol Chem 278(18):15608–15614. https://doi.org/10.1074/jbc.M300177200

Morita T, Maki K, Aiba H (2005) RNase E-based ribonucleoprotein complexes: mechanical basis of mRNA destabilization mediated by bacterial noncoding RNAs. Genes Dev 19(18):2176–2186. https://doi.org/10.1101/gad.1330405

Morita T, Mochizuki Y, Aiba H (2006) Translational repression is sufficient for gene silencing by bacterial small noncoding RNAs in the absence of mRNA destruction. Proc Natl Acad Sci USA 103(13):4858–4863. https://doi.org/10.1073/pnas.0509638103

Morita T, Ueda M, Kubo K, Aiba H (2015) Insights into transcription termination of Hfq-binding sRNAs of Escherichia coli and characterization of readthrough products. RNA 21(8):1490–1501. https://doi.org/10.1261/rna.051870.115

Morita T, Nishino R, Aiba H (2017) Role of the terminator hairpin in the biogenesis of functional Hfq-binding sRNAs. RNA 23(9):1419–1431. https://doi.org/10.1261/rna.060756.117

Morita T, Majdalani N, Miura MC, Inose R, Oshima T, Tomita M, Kanai A, Gottesman S (2022) Identification of attenuators of transcriptional termination: implications for RNA regulation in Escherichia coli. MBio 13(6):e0237122. https://doi.org/10.1128/mbio.02371-22

Nitzan M, Rehani R, Margalit H (2017) Integration of bacterial small RNAs in regulatory networks. Annu Rev Biophys 46:131–148. https://doi.org/10.1146/annurev-biophys-070816-034058

Olejniczak M, Jiang X, Basczok MM, Storz G (2022) KH domain proteins: another family of bacterial RNA matchmakers? Mol Microbiol 117(1):10–19. https://doi.org/10.1111/mmi.14842

Otaka H, Ishikawa H, Morita T, Aiba H (2011) PolyU tail of rho-independent terminator of bacterial small RNAs is essential for Hfq action. Proc Natl Acad Sci USA 108(32):13059–13064. https://doi.org/10.1073/pnas.1107050108

Panja S, Schu DJ, Woodson SA (2013) Conserved arginines on the rim of Hfq catalyze base pair formation and exchange. Nucleic Acids Res 41(15):7536–7546. https://doi.org/10.1093/nar/gkt521

Papenfort K, Melamed S (2023) Small RNAs, large networks: posttranscriptional regulons in gram-negative bacteria. Ann Rev Microbiol 77:23–43. https://doi.org/10.1146/annurev-micro-041320-025836

Papenfort K, Vanderpool CK (2015) Target activation by regulatory RNAs in bacteria. FEMS Microbiol Rev 39(3):362–378. https://doi.org/10.1093/femsre/fuv016

Papenfort K, Bouvier M, Mika F, Sharma CM, Vogel J (2010) Evidence for an autonomous 5′ target recognition domain in an Hfq-associated small RNA. Proc Natl Acad Sci USA 107(47):20435–20440. https://doi.org/10.1073/pnas.1009784107

Papenfort K, Podkaminski D, Hinton JC, Vogel J (2012) The ancestral SgrS RNA discriminates horizontally acquired salmonella mRNAs through a single G-U wobble pair. Proc Natl Acad Sci USA 109(13):E757–E764. https://doi.org/10.1073/pnas.1119414109

Papenfort K, Sun Y, Miyakoshi M, Vanderpool CK, Vogel J (2013) Small RNA-mediated activation of sugar phosphatase mRNA regulates glucose homeostasis. Cell 153(2):426–437. https://doi.org/10.1016/j.cell.2013.03.003

Peer A, Margalit H (2011) Accessibility and evolutionary conservation mark bacterial small-rna target-binding regions. J Bacteriol 193(7):1690–1701. https://doi.org/10.1128/jb.01419-10

Peer A, Margalit H (2014) Evolutionary patterns of Escherichia coli small RNAs and their regulatory interactions. RNA 20(7):994–1003. https://doi.org/10.1261/rna.043133.113

Pei XY, Dendooven T, Sonnleitner E, Chen S, Bläsi U, Luisi BF (2019) Architectural principles for Hfq/Crc-mediated regulation of gene expression. elife 8. https://doi.org/10.7554/eLife.43158

Ponath F, Hör J, Vogel J (2022) An overview of gene regulation in bacteria by small RNAs derived from mRNA 3′ ends. FEMS Microbiol Rev 46(5). https://doi.org/10.1093/femsre/fuac017

Pourciau C, Lai YJ, Gorelik M, Babitzke P, Romeo T (2020) Diverse mechanisms and circuitry for global regulation by the RNA-binding protein CsrA. Front Microbiol 11:601352. https://doi.org/10.3389/fmicb.2020.601352

Prévost K, Desnoyers G, Jacques JF, Lavoie F, Massé E (2011) Small RNA-induced mRNA degradation achieved through both translation block and activated cleavage. Genes Dev 25(4):385–396. https://doi.org/10.1101/gad.2001711

Raina M, Aoyama JJ, Bhatt S, Paul BJ, Zhang A, Updegrove TB, Miranda-Ríos J, Storz G (2022) Dual-function AzuCR RNA modulates carbon metabolism. Proc Natl Acad Sci USA 119(10):e2117930119. https://doi.org/10.1073/pnas.2117930119

Ray-Soni A, Bellecourt MJ, Landick R (2016) Mechanisms of bacterial transcription termination: all good things must end. Annu Rev Biochem 85:319–347. https://doi.org/10.1146/annurev-biochem-060815-014844
Richards GR, Patel MV, Lloyd CR, Vanderpool CK (2013) Depletion of glycolytic intermediates plays a key role in glucose-phosphate stress in Escherichia coli. J Bacteriol 195(21):4816–4825. https://doi.org/10.1128/jb.00705-13
Richter AS, Backofen R (2012) Accessibility and conservation: general features of bacterial small RNA-mRNA interactions? RNA Biol 9(7):954–965. https://doi.org/10.4161/rna.20294
Roberts JW (2019) Mechanisms of bacterial transcription termination. J Mol Biol 431(20):4030–4039. https://doi.org/10.1016/j.jmb.2019.04.003
Rodríguez-Verdugo A, Carrillo-Cisneros D, González-González A, Gaut BS, Bennett AF (2014) Different tradeoffs result from alternate genetic adaptations to a common environment. Proc Natl Acad Sci USA 111(33):12121–12126. https://doi.org/10.1073/pnas.1406886111
Romeo T, Babitzke P (2018) Global regulation by CsrA and its RNA antagonists. Microbiol Spectr 6(2). https://doi.org/10.1128/microbiolspec.RWR-0009-2017
Santiago-Frangos A, Woodson SA (2018) Hfq chaperone brings speed dating to bacterial sRNA. Wiley Interdiscip Rev RNA 9(4):e1475. https://doi.org/10.1002/wrna.1475
Saramago M, Bárria C, Dos Santos RF, Silva IJ, Pobre V, Domingues S, Andrade JM, Viegas SC, Arraiano CM (2014) The role of RNases in the regulation of small RNAs. Curr Opin Microbiol 18:105–115. https://doi.org/10.1016/j.mib.2014.02.009
Sarni SH, Roca J, Du C, Jia M, Li H, Damjanovic A, Małecka EM, Wysocki VH, Woodson SA (2022) Intrinsically disordered interaction network in an RNA chaperone revealed by native mass spectrometry. Proc Natl Acad Sci USA 119(47):e2208780119. https://doi.org/10.1073/pnas.2208780119
Sauer E, Weichenrieder O (2011) Structural basis for RNA 3′-end recognition by Hfq. Proc Natl Acad Sci USA 108(32):13065–13070. https://doi.org/10.1073/pnas.1103420108
Sauer E, Schmidt S, Weichenrieder O (2012) Small RNA binding to the lateral surface of Hfq hexamers and structural rearrangements upon mRNA target recognition. Proc Natl Acad Sci USA 109(24):9396–9401. https://doi.org/10.1073/pnas.1202521109
Schu DJ, Zhang A, Gottesman S, Storz G (2015) Alternative Hfq-sRNA interaction modes dictate alternative mRNA recognition. EMBO J 34(20):2557–2573. https://doi.org/10.15252/embj.201591569
Sedlyarova N, Shamovsky I, Bharati BK, Epshtein V, Chen J, Gottesman S, Schroeder R, Nudler E (2016) sRNA-mediated control of transcription termination in E. Coli. Cell 167(1):111–121.e113. https://doi.org/10.1016/j.cell.2016.09.004
Sharma CM, Darfeuille F, Plantinga TH, Vogel J (2007) A small RNA regulates multiple ABC transporter mRNAs by targeting C/A-rich elements inside and upstream of ribosome-binding sites. Genes Dev 21(21):2804–2817. https://doi.org/10.1101/gad.447207
Silva IJ, Barahona S, Eyraud A, Lalaouna D, Figueroa-Bossi N, Massé E, Arraiano CM (2019) SraL sRNA interaction regulates the terminator by preventing premature transcription termination of rho mRNA. Proc Natl Acad Sci USA 116(8):3042–3051. https://doi.org/10.1073/pnas.1811589116
Sinha D, De Lay NR (2022) Target recognition by RNase E RNA-binding domain AR2 drives sRNA decay in the absence of PNPase. Proc Natl Acad Sci USA 119(48):e2208022119. https://doi.org/10.1073/pnas.2208022119
Sittka A, Sharma CM, Rolle K, Vogel J (2009) Deep sequencing of salmonella RNA associated with heterologous Hfq proteins in vivo reveals small RNAs as a major target class and identifies RNA processing phenotypes. RNA Biol 6(3):266–275. https://doi.org/10.4161/rna.6.3.8332
Skippington E, Ragan MA (2012) Evolutionary dynamics of small RNAs in 27 Escherichia coli and Shigella genomes. Genome Biol Evol 4(3):330–345. https://doi.org/10.1093/gbe/evs001
Sledjeski DD, Gupta A, Gottesman S (1996) The small RNA, DsrA, is essential for the low temperature expression of RpoS during exponential growth in Escherichia coli. EMBO J 15(15):3993–4000

Sonnleitner E, Wulf A, Campagne S, Pei XY, Wolfinger MT, Forlani G, Prindl K, Abdou L, Resch A, Allain FH, Luisi BF, Urlaub H, Bläsi U (2018) Interplay between the catabolite repression control protein Crc, Hfq and RNA in Hfq-dependent translational regulation in Pseudomonas aeruginosa. Nucleic Acids Res 46(3):1470–1485. https://doi.org/10.1093/nar/gkx1245

Sun X, Zhulin I, Wartell RM (2002) Predicted structure and phyletic distribution of the RNA-binding protein Hfq. Nucleic Acids Res 30(17):3662–3671. https://doi.org/10.1093/nar/gkf508

Updegrove TB, Shabalina SA, Storz G (2015) How do base-pairing small RNAs evolve? FEMS Microbiol Rev 39(3):379–391. https://doi.org/10.1093/femsre/fuv014

Updegrove TB, Zhang A, Storz G (2016) Hfq: the flexible RNA matchmaker. Curr Opin Microbiol 30:133–138. https://doi.org/10.1016/j.mib.2016.02.003

Urban JH, Vogel J (2007) Translational control and target recognition by Escherichia coli small RNAs in vivo. Nucleic Acids Res 35(3):1018–1037. https://doi.org/10.1093/nar/gkl1040

Urbanowski ML, Stauffer LT, Stauffer GV (2000) The gcvB gene encodes a small untranslated RNA involved in expression of the dipeptide and oligopeptide transport systems in Escherichia coli. Mol Microbiol 37(4):856–868. https://doi.org/10.1046/j.1365-2958.2000.02051.x

Vanderpool CK, Gottesman S (2004) Involvement of a novel transcriptional activator and small RNA in post-transcriptional regulation of the glucose phosphoenolpyruvate phosphotransferase system. Mol Microbiol 54(4):1076–1089. https://doi.org/10.1111/j.1365-2958.2004.04348.x

Venkat K, Hoyos M, Haycocks JR, Cassidy L, Engelmann B, Rolle-Kampczyk U, von Bergen M, Tholey A, Grainger DC, Papenfort K (2021) A dual-function RNA balances carbon uptake and central metabolism in vibrio cholerae. EMBO J 40(24):e108542. 10.15252/embj.2021108542

Vogel J, Luisi BF (2011) Hfq and its constellation of RNA. Nat Rev Microbiol 9(8):578–589. https://doi.org/10.1038/nrmicro2615

Vogel J, Bartels V, Tang TH, Churakov G, Slagter-Jäger JG, Hüttenhofer A, Wagner EG (2003) RNomics in Escherichia coli detects new sRNA species and indicates parallel transcriptional output in bacteria. Nucleic Acids Res 31(22):6435–6443. https://doi.org/10.1093/nar/gkg867

Vogel J, Faber F, Barquist L, Sparmann A, Popella L, Ghosh C (2025) Meeting report ASOBIOTICS 2024: an interdisciplinary symposium on antisense-based programmable RNA antibiotics. RNA. https://doi.org/10.1261/rna.080347.124

Vogt LN, Panis G, Schäpers A, Peschek N, Huber M, Papenfort K, Viollier PH, Fröhlich KS (2024) Genome-wide profiling of Hfq-bound RNAs reveals the iron-responsive small RNA RusT in Caulobacter crescentus. MBio 15(4):e0315323. https://doi.org/10.1128/mbio.03153-23

Wadler CS, Vanderpool CK (2007) A dual function for a bacterial small RNA: SgrS performs base pairing-dependent regulation and encodes a functional polypeptide. Proc Natl Acad Sci USA 104(51):20454–20459. https://doi.org/10.1073/pnas.0708102104

Wagner EGH, Romby P (2015) Small RNAs in bacteria and archaea: who they are, what they do, and how they do it. Adv Genet 90:133–208. https://doi.org/10.1016/bs.adgen.2015.05.001

Waldern J, Schiraldi NJ, Belfort M, Novikova O (2020) Bacterial group II intron genomic Neighborhoods reflect survival strategies: hiding and hijacking. Mol Biol Evol 37(7):1942–1948. https://doi.org/10.1093/molbev/msaa055

Walling LR, Kouse AB, Shabalina SA, Zhang H, Storz G (2022) A 3′ UTR-derived small RNA connecting nitrogen and carbon metabolism in enteric bacteria. Nucleic Acids Res 50(17):10093–10109. https://doi.org/10.1093/nar/gkac748

Wang J, Rennie W, Liu C, Carmack CS, Prévost K, Caron MP, Massé E, Ding Y, Wade JT (2015) Identification of bacterial sRNA regulatory targets using ribosome profiling. Nucleic Acids Res 43(21):10308–10320. https://doi.org/10.1093/nar/gkv1158

Watkins D, Arya D (2023) Models of Hfq interactions with small non-coding RNA in gram-negative and gram-positive bacteria. Front Cell Infect Microbiol 13:1282258. https://doi.org/10.3389/fcimb.2023.1282258

Weichenrieder O (2014) RNA binding by Hfq and ring-forming (L)Sm proteins: a trade-off between optimal sequence readout and RNA backbone conformation. RNA Biol 11(5):537–549. https://doi.org/10.4161/rna.29144

Westermann AJ (2018) Regulatory RNAs in virulence and host-microbe interactions. Microbiol Spectr 6(4). https://doi.org/10.1128/microbiolspec.RWR-0002-2017

Worthan SB, McCarthy RDP, Delaleau M, Stikeleather R, Bratton BP, Boudvillain M, Behringer MG (2024) Evolution of pH-sensitive transcription termination in Escherichia coli during adaptation to repeated long-term starvation. Proc Natl Acad Sci USA 121(39):e2405546121. https://doi.org/10.1073/pnas.2405546121

Worthan SB, Grant MI, Behringer MG (2025) Rho-dependent termination: a bacterial evolutionary capacitor for stress resistance. Transcription 1-14. https://doi.org/10.1080/21541264.2025.2474367

You L, Omollo EO, Yu C, Mooney RA, Shi J, Shen L, Wu X, Wen A, He D, Zeng Y, Feng Y, Landick R, Zhang Y (2023) Structural basis for intrinsic transcription termination. Nature 613(7945):783–789. https://doi.org/10.1038/s41586-022-05604-1

Zhang A, Wassarman KM, Ortega J, Steven AC, Storz G (2002) The Sm-like Hfq protein increases OxyS RNA interaction with target mRNAs. Mol Cell 9(1):11–22. https://doi.org/10.1016/s1097-2765(01)00437-3

Zhang A, Wassarman KM, Rosenow C, Tjaden BC, Storz G, Gottesman S (2003) Global analysis of small RNA and mRNA targets of Hfq. Mol Microbiol 50(4):1111–1124. https://doi.org/10.1046/j.1365-2958.2003.03734.x

Zheng A, Panja S, Woodson SA (2016) Arginine patch predicts the RNA annealing activity of Hfq from gram-negative and gram-positive bacteria. J Mol Biol 428(11):2259–2264. https://doi.org/10.1016/j.jmb.2016.03.027

Chapter 9
Molecular Basis and Evolution of Bacterial RNase E and Its Systems

Fang Liu, Philippe Icyishaka, Sarah L. Svensson, Masatoshi Miyakoshi, and Yanjie Chao

Abstract RNase E is a widely-conserved bacterial endoribonuclease with diverse roles in cellular RNA metabolism, and is also the central component of the bacterial RNA degradosome. Essential for viability in many species, RNase E is involved in processing of housekeeping RNAs and in general RNA turnover, and impacts many aspects of general physiology, environmental adaptation, and regulation of virulence. In particular, the crucial role of RNase E in post-transcriptional regulation via biogenesis of small regulatory RNAs and regulated decay of coding transcripts has emerged. Beyond its widespread importance to bacteria, RNase E has also been a model for understanding how RNA metabolism is organized across kingdoms of life. Here, we review the molecular details of RNase E and its protein partners as revealed from studies in Gram-negative models such as *E. coli* and *Salmonella*, and contrast them with themes uncovered in other species. We also discuss shared aspects of the RNase E degradosome with other RNA decay complexes that appear to have evolved convergently, such as those based on RNase Y/J and the eukaryotic exosome. Finally, we present emerging themes in RNase E, such as its role in phase separation and RNA networks, and propose that application of global deep sequencing approaches (TIER-seq, CLASH/RIL-seq, and Grad-seq) to additional species will provide a broader understanding of its structure and function.

Keywords RNase E · RNase G · Endoribonucleases · Degradosome · sRNA · 3′UTR

F. Liu · P. Icyishaka · S. L. Svensson · Y. Chao (✉)
Shanghai Institute of Materia Medica, University of Chinese Academy of Sciences, Chinese Academy of Sciences, Shanghai, China
e-mail: yjchao@ips.ac.cn

M. Miyakoshi
Institute of Medicine, University of Tsukuba, Tsukuba, Japan

A. Kanai, A. W. Nicholson (eds.), *Molecular Evolution of RNA Regulatory Enzymes and Their Systems*, Nucleic Acids and Molecular Biology 38,
https://doi.org/10.1007/978-3-032-03300-0_9

9.1 Introduction

Ribonucleases (RNases) catalyze the processing and degradation of ribonucleic acid (RNA) molecules into smaller components via cleavage of phosphorus-oxygen bonds (Bechhofer and Deutscher 2019; Broglia et al. 2023a; Nicholson 1999). Based on their cleavage patterns, RNases are classified into two primary categories: exoribonucleases and enribodonucleases (Aït-Bara and Carpousis 2015). Exoribonucleases degrade RNA molecules by sequentially removing terminal nucleotides from either the 5′ or 3′ end to release mononucleotides. In contrast, endoribonucleases cleave RNA molecules at internal sites. They can target either single- or double-stranded RNA to generate RNA fragments that contain 3′-hydroxyl or 3′-phosphate residues, reflecting the diverse enzymatic mechanisms and specificities. (Bechhofer and Deutscher 2019; Deutscher 2021).

Most bacteria encode a core suite of RNases covering these different activities, with several of them being widely conserved across taxa. One of the most conserved and well-studied examples is RNase E, a crucial endoribonuclease governing RNA metabolism that is encoded by most Gram-negative bacteria such as *Escherichia coli* (*E. coli*), and also by a handful of Gram-positive species such as *Mycobacterium tuberculosis* (Fig. 9.1). RNase E is essential for viability in *E. coli* and specifically cleaves single-stranded RNAs at internal positions. RNase E primarily participates in the processing and biogenesis of ribosomal RNA and transfer RNA, for example in the maturation of 5S rRNA and the processing of long tRNA precursors (Mackie 2013), but it is also required for the turnover of mRNAs. Notably, the inactivation

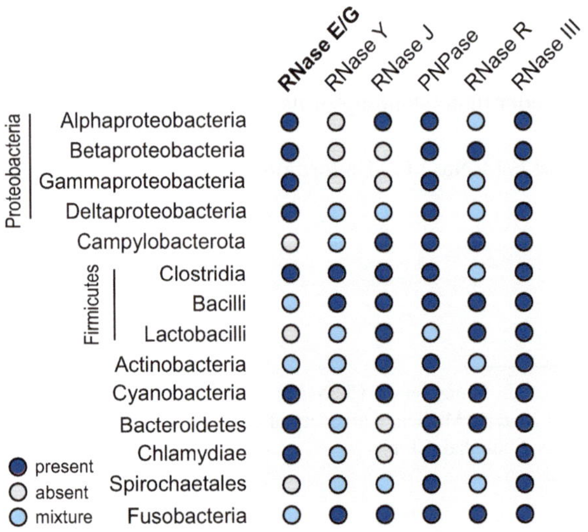

Fig. 9.1 The distribution of major RNase classes, including RNase E, across selected bacterial taxa (Based on (Hui et al. 2014). Present/absent: reported in 100%/0% of strains. Mixture: detected in only some strains)

of RNase E leads to a significant increase in the half-life of most mRNA transcripts in *E. coli*, demonstrating its crucial role in maintaining appropriate cellular levels of different RNA species (Baek et al. 2019; Hoffmann et al. 2024). Finally, RNase E also plays a central role in regulation of bacterial gene expression via its role in the biogenesis of small regulatory RNAs (sRNAs), as well as in the cleavage of many sRNA-mRNA duplexes to initiate decay. For example, CLASH (cross-linking and sequencing of hybrids) has recovered diverse sRNA-mRNA interactions associated with RNase E (Waters et al. 2016), while TIER-seq (transient inactivation of RNase E followed by sequencing) has identified many sRNAs with associated RNase E cleavage sites (Chao et al. 2017).

Most of the mechanistic studies of RNase E have been performed in model organisms such as *E. coli*. RNase E is a 5′-end-dependent endonuclease that senses the 5′-end phosphorylation state of target RNAs (Mackie 1998). Upon recognition of a 5′-monophosphate group, RNase E initiates rapid 5′ to 3′ cleavage of the transcript (Callaghan et al. 2005). Therefore, decay of primary transcripts (*e.g.*, mRNAs) by RNase E is usually initiated by a conversion of the 5′ terminus from a triphosphate to a monophosphate by an enzyme such as the pyrophosphohydrolase RppH (Deana et al. 2008). Nevertheless, RNase E can also bypass the 5′-end-dependence using a so-called "direct entry" mechanism to gain access to cleavage sites within primary transcripts (Mackie 2013; Baker and Mackie 2003; Kime et al. 2009). This initial internal cleavage generates RNA species with 5′-monophosphate ends, which are then subject to subsequent rapid processive degradation in part by the 5′-end dependent pathway (Bouvier and Carpousis 2011; Kime et al. 2009).

The essentiality of RNase E in *E. coli* and many other species indicates an evolutionarily conserved function that is crucial for bacterial viability, although the exact reason for this is not entirely clear. A temperature-sensitive RNase E mutant quickly loses colony-forming ability after heat shock treatment at 42 °C or above, which has been leveraged to study function of this essential enzyme. For example, RNase E cleavage sites have been mapped on a global scale by harnessing this mutant via the aforementioned TIER-seq method (Chao et al. 2017). The essentiality of RNase E might be related to its role in the biogenesis of housekeeping RNAs such as tRNA and rRNA, playing a crucial role in supporting general translation functions in the cell. However, since RNase E is also involved in the processing and function of sRNAs, which are crucial regulators of adaptation to diverse environmental conditions, its essentiality could also be attributed to this critical cellular role (Fu et al. 2019; Ponath et al. 2022).

9.2 Structural Organization of RNase E

RNase E is one of the largest characterized bacterial proteins, with a length of 1061 amino acids in *E. coli* (molecular mass ~ 118 kDa) (Baek et al. 2019). Structural analyses have revealed that *E. coli* RNase E possesses a bipartite architecture comprising two functionally distinct regions: the evolutionarily conserved N-terminal

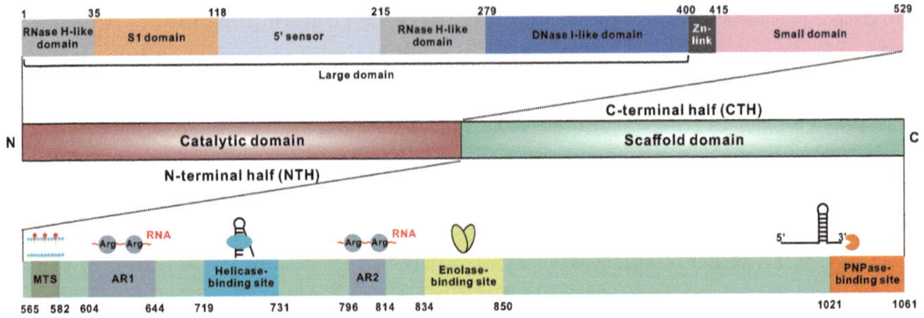

Fig. 9.2 Schematic representation of *E. coli* RNase E catalytic domains and binding sites for RNA degradosome components**.** *E. coli* RNase E is separated into an NTH (N-terminal half, the catalytic domain, residues 1–529) and CTH (C-terminal half, the scaffolding domain, residues 530–1061). The RNase H-like domain (residues 1–35, 215–279) is involved in RNA binding and organizing the active site. The S1 domain (residues 35–118) supports formation of the tetrameric quaternary structure. The 5′ sensor (residues 118–215) harbors the 5′-phosphate-sensing pocket, while the DNase I domain (residues 279–400) includes the catalytic center. Both the Zn-link (residues 400–415) and small domain (residues 415–529) are involved in tetramerization. The CTH contains several binding sites for RNA and various interacting proteins. The MTS (membrane targeting sequence, residues 565–582) is responsible for the interaction between RNase E and the inner membrane. AR1 and AR2 (Arginine-rich domains 1 & 2, residues 604–644 and 796–814, respectively) are RNA binding sites. The RNA helicase RhlB binds at residues 719–731, while enolase binds at residues 834–850 and polynucleotide phosphorylase (PNPase) binds at residues 1021–1061. The numbers of amino acids are indicated above or under the domains

half (NTH, residues 1–529) containing the catalytic domain, and the more variable C-terminal half (CTH, residues 530–1061) serving as a scaffold for the assembly of RNA degradosome (Mackie 2013; Koslover et al. 2008; Mardle et al. 2019; Liu et al. 2023b) (Fig. 9.2).

The RNase E NTH is highly conserved among prokaryotes, and enables specific interactions with RNA substrates as well as catalysis (Carpousis 2007). The NTH can be subdivided into a large domain (residues 1–400), Zn-link (residues 401–414), and a small domain (residues 415–529) (Callaghan et al. 2005). The large domain contains the catalytically-active core organized into four functionally specialized subdomains (Hammarlof et al. 2011): (i) an RNase H-like domain, which assumes the RNase H fold and serves a purely structural role; (ii) an S1 RNA-binding domain, which mediates the stabilization of the RNase E dimer, the assembly of tetrameric quaternary structure, and the binding of substrates; (iii) the 5′ sensor, harboring the 5′-phosphate-sensing pocket that recognizes and binds RNA substrates; and (iv) a DNase I-like domain, which serves as the catalytic core responsible for phosphodiester bond hydrolysis. The large domain is connected to a "Zn-link" interface, which is a metal-sharing surface for organizing the quaternary structure and catalytic site of RNase E (Liu et al. 2023b), and finally a small domain.

In contrast to the NTH, the CTH of RNase E is largely unstructured and disordered with low conservation at the sequence level. The CTH is organized with several fuctionally-conserved microdomains as binding sites for various interacting proteins (Erce et al. 2009; Hardwick et al. 2010), including the helicase RhlB,

Table 9.1 Comparison of major RNA metabolism complexes in bacteria and eukaryotes

Complex	RNase	RNA helicase	Exosome-like protein	Metabolic enzyme	Localization	Key references
E. coli degradosome	RNase E	RhlB	PNPase	Enolase	Inner membrane	Bandyra and Luisi (2018)
C. crescentus degradosome	RNase E	RhlB	PNPase	Aconitase	BR-bodies	Bayas et al. (2018)
B. subtilis degradosome	RNase Y	CshA	PNPase	Enolase, phosphofructokinase	Inner membrane	Lehnik-Habrink et al. (2010)
Pseudomonas aeruginosa degradosome	RNase E	DeaD	RNase R, PNPase	Unknown	Inner membrane	Geslain et al. (2025)
Helicobacter pylori degradosome	RNase J	RhpA	Not available	Unknown	Inner membrane	Tejada-Arranz et al. (2020)
Mycobacterium tuberculosis degradosome	RNase J, RNase E	RhlE	PNPase	Unknown	Not available	Płociński et al. (2019)
Anabaena degradosome	RNase E	CrhB	RNase II	Enolase	Cytoplasm	Zhou et al. (2020)
Eukaryotic exosome	Rrp44	Mtr4, Ski2	Rrp41, Rrp45, Rrp46, Rrp43, Mtr3, Rrp42, Rrp40, Rrp4, Csl4	Unknown	Nucleus, cytoplasm	Bandyra and Luisi (2018)

enolase, and polynucleotide phosphorylase (PNPase) in *E. coli*. The CTH of RNase E serves as a structural scaffold for the assembly of the RNA degradosome (Table 9.1), which is responsible for coordinated RNA processing and turnover (Bechhofer 2013). Within the degradosome complex, RNase E performs the primary endonucleolytic cleavage. RhlB functions as an ATP-dependent RNA helicase that resolves secondary structures through double-stranded RNA unwinding, while PNPase acts as a processive 3′ to 5′ exoribonuclease that degrades linearized RNA fragments. Finally, enolase is an enzyme involved in glycolysis, with an enigmatic role in the degradosome. In addition to the protein-protein interaction domain that scaffolds the degradosome, the CTH in *E. coli* also contains a membrane binding domain and two arginine-rich (AR) segments that play a part in RNA binding.

Structural studies have established that the *E. coli* RNase E enzyme forms a tetrameric architecture composed of four monomers that resemble two pairs of scissors connected at the handle region (Koslover et al. 2008; Schubert et al. 2004). The blades of the scissors are formed by the large domain of NTH, while the handles are composed of the small domain (Mackie 2013). A key feature of the tetrameric structure is the presence of a cysteine-zinc-cysteine linkage (Zn-link), which forms the bridge between adjacent monomers and enhances the stability (Moore et al. 2021). The tetrameric structure of RNase E is essential for both ribonucleolytic activity and degradosome organization.

9.3 Mechanisms of RNA Cleavage by RNase E

RNase E requires a divalent metal ion for its enzymatic activity, and *in vitro* studies have shown that both Mg^{2+} and Mn^{2+} can support cleavage, while it has a preference for Mn^{2+} over Mg^{2+} (Thompson et al. 2014). While Mg^{2+} is essential for activity *in vivo*, the enzyme can utilize Mn^{2+} when Mg^{2+} is absent.

RNase E has a strong preference for AU-rich regions, and in particular single-stranded RNA sequences with a purine base (adenine or guanine) located at the −2 position upstream of the scissile bond (Kaberdin 2003; Mackie 1992; McDowall et al. 1995). A typical RNase E cleavage site in *E. coli/Salmonella* is 5′-(A/G) N↓AU, where ↓ represents the site of RNase E cleavage and N stands for any nucleotide of the RNA substrate. Moreover, TIER-seq has revealed that a uridine at 2 nt downstream of the cleavage sites is crucial for RNase E cleavage in *Salmonella*, which appears to be a conserved structural mechanism involving several RNA-binding residues (Chao et al. 2017). Global studies of RNase E preference are so far rare, but some diversity might exist in evolutionarily-distant species. For example, *M. tuberculosis* RNase E seems to preferentially cleave cytosine-rich regions (Zhou et al. 2023).

During the RNA degradation process, RNase E cleaves the phosphodiester backbone of the RNA by hydrolysis. The initial endonucleolytic cleavage step generates two products. The first product is the 5′-end of the RNA substrate with 3′ hydroxyl group, which is then susceptible to cleavage by 3′ exonucleases such as RNase II or polynucleotide phosphorylase (PNPase). The second product of cleavage by RNase E is the 3′-half of the substrate, carrying a 5′-monophosphate and often the 3′-end termination stem loop from the RNA substrate. This product is in turn also highly susceptible to further RNase E-mediated cleavage via its 5′ monophosphate sensor (Carpousis 2007; Mackie 2013).

RNase E is reported to undertake either of two distinct cleavage mechanism pathways while degrading RNA substrates (Fig. 9.3). In one pathway (the 5′-end-dependent pathway), the 5′ terminal triphosphate needs first to be converted to a 5′-monophosphate, whereas in the other pathway (the direct entry or 5′-end-independent pathway), the cleavage begins with the direct attack of internal RNA sites by the endoribonuclease (Celesnik et al. 2007; Mackie 1998).

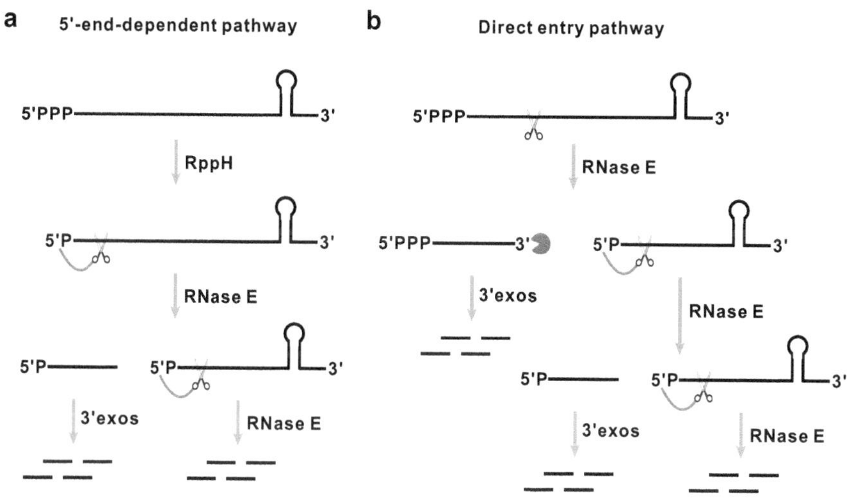

Fig. 9.3 Two RNA dtion pathways mediated by *E. coli* RNase E. (**a**) The 5′-end-dependent pathway. The 5′ triphosphate group (5′-PPP) is first converted to a 5′ monophosphate (5′-P) by a pyrophosphohydrolase such as RppH, or via processing by other RNases. The 5′ monophosphorylated RNA is then recognised by RNase E, leading to endonucleolytic cleavage. Subsequently, the RNA fragments are further degraded by RNase E or 3′-exonucleases (3′exos) such as PNPase. (**b**) The direct entry pathway. RNase E directly accesses internal sites within primary transcripts bearing 5′ triphosphorylated termini and cleaves them to produce two intermediate fragments. The 5′ triphosphorylated fragment that lacks a protective stem-loop at the 3′ end is rapidly degraded by 3′ to 5′ exoribonucleases such as RNase II, PNPase, and RNase R. The 5′ monophosphorylated fragment is further cleaved as described in (**a**)

Substrates that are resistant to RNase E cleavage appear to have one of two features. First, nascent transcripts are usually protected from RNase E degradation via their 5′-triphosphate (5′-PPP) and by a terminator stem-loop at their RNA 3′-end. For this reason, degradation by RNase E is often initiated either by the conversion of the 5′-PPP to 5′-monophosphate (5′-P) by the pyrophosphohydrolase RppH or by an internal cleavage within the RNA sequence (via the internal entry pathway) catalyzed by RNase E in Gram-negative bacteria (Celesnik et al. 2007; Deana et al. 2008).

9.3.1 5′-End-Dependent Cleavage Pathway

The 5′-end-dependent cleavage pathway is considered to be the major pathway used by RNase E to carry out degradation of RNA transcripts. In this mechanism, RNase E requires a 5′-monophosphate group on the RNA substrate for efficient cleavage (Fig. 9.3a). To achieve this, RppH first removes the 5′-terminal pyrophosphate from a triphosphorylated RNA to generate a 5′-monophosphorylated species, which can stimulate endonucleolytic cleavage by RNase E (Deana et al. 2008). The recognition of the 5′-monophosphate is facilitated by the 5′ sensor pocket residing within the catalytic domain of the *E. coli* RNase E NTH. Residues T170 and R169 within the 5′ sensor domain recognize the terminal phosphate via hydrogen bonding (Callaghan et al. 2005; Koslover et al. 2008). The 5′ sensor pocket binds the 5′-monophosphate, and residues in an RNA-binding channel contact the first nucleotides. As a consequence, this orients the single-stranded RNA with the active site of the enzyme to allow efficient cleavage (Bandyra et al. 2018). One transcript known to be cleaved by RNase E via the 5′-monophosphate pathway is the *yeiP* mRNA (Bechhofer 2013; Kime et al. 2009).

9.3.2 Direct Entry Pathway

The alternative approach bypasses the 5′ end monophosphate dependence, where RNase E can also cleave certain RNAs rapidly regardless the nature of the substrate 5′-end (Fig. 9.3b). This pathway has been referred to as 'direct' or 'internal' entry mechanism, which requires multiple single-stranded sites within the substrate with a conformation that facilitates their simultaneous interaction with RNase E (Kime et al. 2009). The enzyme's NTH harbors the catalytic domains required for this process. The 5′-monophosphate-independent cleavage activity of RNase E gives a strong explanation for why the degradation of several mRNAs is less affected by the loss of the RNA 5′ pyrophosphatase RppH compared to the disruption of RNase E itself (Kime et al. 2009). Some mRNAs, like *cspA* and *pgk* mRNA, are examples of transcripts that can be cleaved by RNase E without needing a 5′ monophosphate group, using a monophosphate-independent cleavage pathway (Clarke et al. 2014).

9.3.3 General Steps of RNase E Cleavage

Whether RNase E uses either the 5′ monophosphate dependent pathway or the direct entry pathway, the following series of six events occurs when the endoribonuclease generates the initial cleavage of RNA. (i) Initial substrate recognition, where RNase E recognizes and binds to its RNA substrate. This recognition is facilitated by specific RNA sequences or structural motifs, often found in single-stranded regions of the RNA. Binding may induce a conformational change in RNase E, positioning the RNA substrate appropriately for cleavage. (ii) Formation of the RNase E-RNA complex, where the interaction between RNase E and the RNA substrate is stabilized. This complex is crucial for the subsequent catalytic steps. (iii) Catalytic activation, where the divalent metal ion (such as Mg^{2+}) is positioned to stabilize the RNA backbone and facilitate the nucleophilic attack by water. (iv) Endonucleolytic cleavage, where a water molecule, activated by the metal ion and enzyme residues, attacks the phosphodiester bond within the RNA. This results in the cleavage of the RNA strand, generating two RNA fragments, one with a 5′ monophosphate group and another with a 3′ hydroxyl group. (v) Release of the RNA fragments following cleavage, where the two intermediate RNA fragments are released from the enzyme. This release can be facilitated by conformational changes in RNase E. RNase E then reverts to its original conformation, ready to bind to another RNA substrate. (vi) Subsequent processing and further degradation. The RNA fragments produced by RNase E may be further processed by other ribonucleases or cellular machinery, such as exonucleases, leading to complete degradation into oligonucleotides or maturation of the RNA. Many studies suggest that the initial RNA fragments carry a 5′-monophosphate end, which may make them preferable substrates for further degradation by RNase E (Durica-Mitic and Görke 2019).

9.4 Phylogenetic Conservation of RNase E and Functionally-Related Bacterial RNases

Orthologs of *E. coli* RNase E are found in approximately 80% of all sequenced bacterial genomes to date (Aït-Bara et al. 2014) (Fig. 9.1). Phylogenetic analysis has revealed that the sequence of the catalytic domain of RNase E family enzymes is highly conserved, whereas the noncatalytic regions are more variable in both sequence and length. While some Gram-positive bacterial species like *Bacillus subtilis* do not encode RNase E or its paralog RNase G, they express functionally equivalent enzymes such as RNase J1/J2 or RNase Y (Fig. 9.1). Therefore, homologs (functional or evolutionary) of these ribonucleases are ubiquitously distributed across most bacterial species, consistent with their crucial role in stable RNA processing and mRNA degradation. The general view is that RNase E and RNase G are widespread in Gram-negative bacteria, while RNase Y and RNase J are more commonly associated with Gram-positive bacteria. However, some mycobacteria strikingly encode both RNase E and RNase J (Taverniti et al. 2011). The phylogenetic distribution and conservation of RNase E across different bacterial species highlight its importance in bacterial physiology and RNA metabolism.

9.4.1 *Classification and Distribution of RNase E Homologs*

The high degree of conservation of the RNase E catalytic domain suggests that the core function of RNase E in RNA processing and degradation is maintained across different bacterial species. Homologs of RNase E have been classified into five groups (type I, type II, type III, type IV, and type V enzymes) based on their primary structure (Aït-Bara and Carpousis 2015) (Fig. 9.4). For type I enzymes, the catalytic domain is located near the N-terminus, while the scaffold is in the C-terminal region, as observed for *E. coli* RNase E. Type I RNase E homologs are mainly

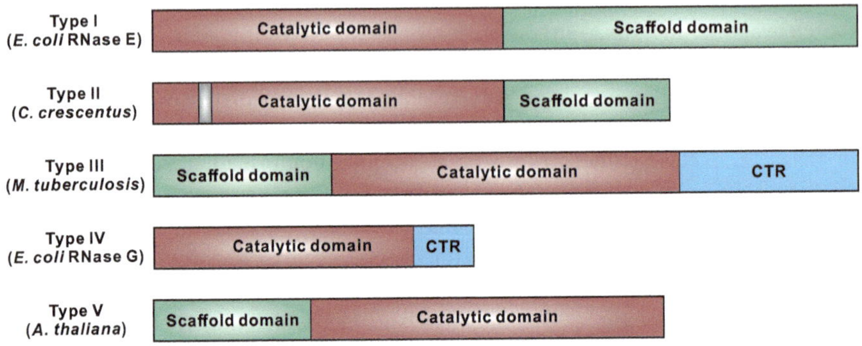

Fig. 9.4 Classification of RNase E homologues into five types. Type I RNase E, such as the well-studied homologue of *E. coli*, contains the catalytic domain at the N-terminus and a scaffold domain at the C-terminus (see also Fig. 9.2). Type II RNase E from *C. crescentus* and Alphaproteobacteria has a shorter scaffold domain, also at the C-terminus, and an arginine-proline-rich segment (small grey box) within the S1 RNA-binding domain within the N-terminal catalytic domain. The catalytic domain of type III RNase E encoded by *M. tuberculosis* and Actinobacteria is located near the center of the enzyme, with the scaffold instead at the N-terminus. Widely-distributed type IV RNase E is a short RNase G-like protein (see also Fig. 9.5). RNase E homologs encoded by chloroplasts of eukaryotes are classified as type V (*Arabidopsis thaliana*). The catalytic domain and scaffold domain are indicated in red and green, respectively. The C-terminal regions (CTR) are indicated in blue

distributed amongst γ-proteobacteria, including *Enterobacteriaceae* species, although type I homologs have also been identified and characterized in *Yersinia* (Yang et al. 2008), *Pseudoalteromonas* (Aït-Bara and Carpousis 2010; Purusharth et al. 2005), and *Vibrio* (Erce et al. 2010).

Type I RNase E is also commonly found in cyanobacteria. *Synechocystis* RNase E contains a relatively short C-terminal portion, which is thought to have diverged from *E. coli* over 3 billion years ago (Dickerson 1980; Kaberdin et al. 1998). TIER-seq revealed that *Synechocystis* RNase E has a cleavage signature consisting of an adenine at the −3 position and a uridine at the +2 position within a single-stranded segment of the RNA, similar to what is observed in *Salmonella* (Hoffmann et al. 2024; Chao et al. 2017). The C-terminal region of *Anabaena* RNase E has been shown to contain four conserved subregions (C1, C2, C3, and C4) and three highly variable subregions (V1, V2, and V3). Just as the CTH in *E. coli* RNase E serves as a scaffold, the C-terminal region in *Anabaena* also plays a similar role. The C4 subregion was found to interact with PNPase, forming a similar degradosome complex (Zhang et al. 2014). Currently, the functions of the other subregions in the C-terminal region of *Anabaena* RNase E remain unclear.

Type II RNase E is typically characterized by a relatively short C-terminal scaffold and an arginine-proline-rich segment inserted within the N-terminal S1 RNA-binding domain (Bycroft et al. 1997). Type II enzymes are extensively found among α-proteobacteria, such as *Caulobacter crescentus* and *Rhodobacter sphaeroides* (Hardwick et al. 2012; Spanka et al. 2021). The components of *C. crescentus* degradosome, like its *E. coli* counterpart, include RNase E, PNPase, and a DEAD-box

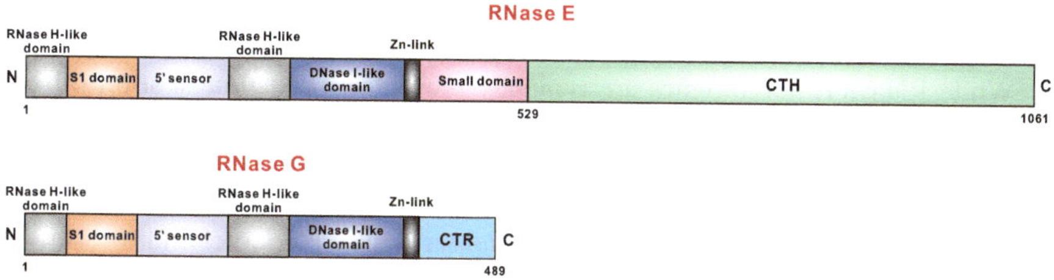

Fig. 9.5 Comparison of the domain structures of RNase E and RNase G from *E. coli*. The detailed structure of RNase E is shown in Fig. 9.2. RNase G contains a catalytic domain with similar components and organization as RNase E, but lacks the long disordered C-terminal half (CTH). In addition, RNase G possesses a short C-terminal region (CTR) downstream of the Zn-link which is not homologous to the small domain of RNase E

RNA helicase (Table 9.1). However, in the *C. crescentus* degradosome, aconitase substitutes for the enolase present in the *E. coli* degradosome (Hardwick et al. 2010). The RNA degradosome and mRNA are localized to ribonucleoprotein-bodies (BR-bodies), which are formed through liquid-liquid phase separation (Al-Husini et al. 2018; Carpousis et al. 2022). In *R. sphaeroides*, the inactivation of RNase E strongly impacts bacterial adaptation to phototrophic conditions (Börner et al. 2023; Förstner et al. 2018).

In type III RNase E, the catalytic domain is located near the center of enzyme, with extensions of varying lengths at the N- and C-termini. Typical organisms containing this type of RNase E are *Streptomyces coelicolor* and *M. tuberculosis*. In *S. coelicolor*, the RNase E homolog is also called RNase ES. Whereas RNase E is an essential gene in *E. coli*, RNase ES is not required for *S. coelicolor* survival, and compared to the wild-type strain, the growth rate of an RNase ES null mutant is decreased by only 10% (Lee and Cohen 2003). RNase ES can substitute for *E. coli rne* or *rng* null mutants (Inagawa et al. 2003; Lee and Cohen 2003), suggesting it has a similar activity. The NTH of *M. tuberculosis* type III RNase E contains two arginine-rich regions, located at positions 158–167 and 230–238, respectively (Jaiswal et al. 2024). In an *M. tuberculosis rne* knockdown strain, the degradation rate for almost 90% of cellular transcripts is affected. Unlike the *E. coli* homolog, *M. tuberculosis* RNase E preferentially cleaves at cytosine-rich regions (Zhou et al. 2023).

Type IV RNase E is more similar to RNase G of *E. coli*, lacking the small domain and the C-terminal non-catalytic region (Sect. 9.4.2 below). RNase G-like proteins are primarily found in archaea, some bacteria, and a few eukaryotes. Due to the distant phylogenetic relationship of some species carrying putative RNase G homologs with *E. coli*, it is not always easy to distinguish whether an RNase G-like protein is an RNase E or RNase G homolog, and thus the term "RNase E/G" has been adopted (Condon 2002). RNase E/G from *Aquifex aeolicus* exhibits temperature-dependent endoribonucleolytic activity, with the efficiency of cleavage increased as the temperature increases from 30 °C to 75 °C. The ability of *A. aeolicus* RNase

E/G to withstand high temperatures is consistent with the thermophilic properties of this organism (Kaberdin and Bizebard 2005).

While RNase E homologues are typically present in prokaryotes, several have also been detected in a handful of eukaryotes, where they are classified as type V RNase E homologs (Aït-Bara and Carpousis 2015). These enzymes have mainly been identified in plant chloroplasts, such as those of *Arabidopsis thaliana* and rice (Fang et al. 2025; Mudd et al. 2008). *Arabidopsis* RNase E is crucial for the development of chloroplasts and autotrophic growth (Mudd et al. 2008). Deletion of the gene encoding RNase E causes significant defects in RNA processing and chloroplast ribosome formation in *A. thaliana* (Walter et al. 2010). A recent study discovered that RNase E also plays a crucial role in chloroplast development, chlorophyll synthesis, and photosynthesis in rice (Fang et al. 2025).

9.4.2 Ribonuclease G

E. coli also encodes a much shorter paralog of RNase E called RNase G (originally named CafA), with a molecular weight of 55.4 kDa and length of 489 amino acids (Okada et al. 1994; Wachi et al. 1999). While its amino acid sequence shows ~30% identity and ~ 50% similarity with the N-terminal catalytic domain of RNase E, it lacks the small domain and the C-terminal non-catalytic region of RNase E (McDowall et al. 1993) (Fig. 9.5). RNase G primarily adopts a dimeric conformation under physiological conditions (Briant et al. 2003; Jiang and Belasco 2004). Since the short C-terminal region of RNase G lacks homology with the small domain of RNase E, RNase G is incapable of assembling an interface for tetramerization (Chung et al. 2010).

RNase G was initially identified as a ribonuclease involved in the processing and maturation of 16S rRNA precursors (Li et al. 1999; Wachi et al. 1999). It is now known that RNase G, similarly to RNase E, also participates in general RNA turnover. Although RNase G lacks the C-terminal scaffolding domain required for degradosome assembly, its conserved catalytic domain enables it to cleave several transcripts such as the *adhE* mRNA encoding alcohol dehydrogenase, albeit with reduced efficiency (Umitsuki et al. 2001). Analysis of global RNA-seq data revealed that RNase G may have an extensive role in the trimming of 5′-monophosphorylated ends from transcripts (Clarke et al. 2014). In contrast to RNase E, RNase G is a non-essential protein in *E. coli*. However, an *rng* mutant strain exhibits a defect in the 5′ end processing of 16S rRNA and accumulates abundant precursor forms of this RNA, and also shows significantly increased stability of the *adhE* mRNA (Li et al. 1999; Wachi et al. 1999; Wachi et al. 2001).

RNase G displays functional overlap with RNase E in several biological processes. These two paralogous RNases exhibit partial functional redundancy, as demonstrated by the ability of RNase G overexpression to partially compensate for the growth defect resulting from RNase E deficiency in specific contexts. For example, overexpression of RNase G can restore the defective colony-forming ability phenotype observed in cells lacking RNase E (Tamura et al. 2012). However, RNase G overproduction cannot fully replace RNase E function, such as its role in the maturation of tRNAs (Jiang et al. 2000; Lee et al. 2002; Ow et al. 2003). These observations not only demonstrate the evolutionary conservation of catalytic function between these two ribonucleases but also emphasize the non-redundant, system-level organization of bacterial RNA metabolism, where RNase E serves as the central regulatory hub.

9.4.3 *Ribonucleases Y and J in* Bacillus subtilis

Bacterial RNA processing and degradation in Gram-negative and Gram-positive species typically follows similar pathways, utilizing several evolutionarily distinct yet functionally analogous ribonucleases. Compared to *E. coli*, the model Gram-positive *Bacillus subtilis* displays substantial differences in its RNA degradation system, while maintaining analogous functions. Strikingly, *B. subtilis* lacks RNases E and G, but possesses two distinct ribonucleases (RNases Y and J) that are absent in *E. coli* (Bechhofer and Deutscher 2019) (Fig. 9.1). Although it exhibits no detectable similarity to RNase E in primary sequence and structure, RNase Y serves as the core endoribonuclease of RNA metabolism in Gram-positives such as *B. subtilis*. The enzyme is comprised of an N-terminal transmembrane domain, a coiled-coil domain important for oligomerization, a KH domain responsible for RNA binding, and a C-terminal hydrolytic domain that is essential for its catalytic activity (Shahbabian et al. 2009) (Fig. 9.6).

Despite structural differences, RNase Y shares several functional characteristics with RNase E. Both are initiating RNases and are localized at the cell membrane. The activity of RNase Y is also, like RNase E, stimulated by the presence of a 5′-monophosphate group in substrates (Lehnik-Habrink et al. 2011). Moreover, RNase Y is also part of a larger complex dedicated to RNA metabolism. Affinity

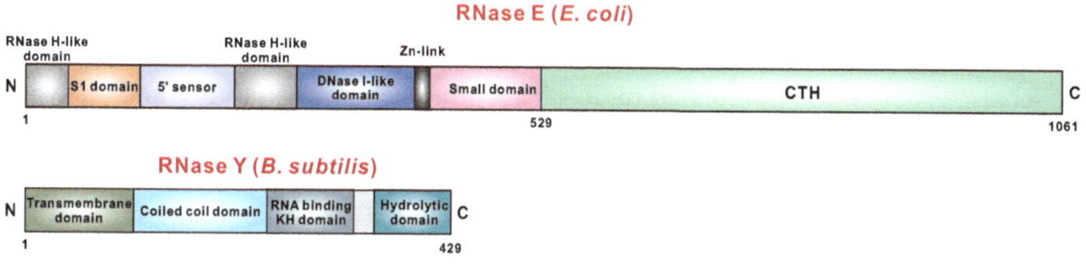

Fig. 9.6 Comparison of domains and organization of *E. coli* RNase E and *B. subtilis* RNase Y. The detailed structure of RNase E is shown in Fig. 9.2. *B. subtilis* RNase Y, despite being the main initiating endonuclease that also scaffolds a degradosome-like complex, includes different RNA binding and catalytic domains compared to RNase E. The enzyme contains an N-terminal transmembrane domain (TMD, residues 1–25) that mediates interaction with the inner membrane, followed by a large disordered coiled-coil domain (residues 25–210) which is important for oligomerization; an RNA-binding K homology (KH) domain (residues 211–270); and a hydrolytic domain (residues 336–429)

chromatography and two-hybrid analyses demonstrated that RNase Y can interact with several partner proteins (including RNase J, PNPase, the DEAD-box RNA helicase CshA, and the two glycolytic enzymes enolase and phosphofructokinase) to assemble a degradosome-like complex in *B. subtilis* (Commichau et al. 2009). Mutation of the gene encoding RNase Y, like RNase E, likewise has an impact on global RNA stability. Notably, *E. coli* RNase E can effectively replace RNase Y in *B. subtilis*. Studies demonstrated that wild type RNase E can rescue the poor growth phenotype of a mutant lacking RNase Y (Laalami et al. 2021).

RNase J is a unique dual-function bacterial RNase, exhibiting both 5′ to 3′ exonuclease and endonuclease activities (Figaro et al. 2013; Mathy et al. 2007). RNase J was originally identified in *B. subtilis,* which encodes two homologous isoforms, RNase J1 (encoded by *rnjA*) and RNase J2 (encoded by *rnjB*). RNases J1 and J2 associate into heterodimeric or heterotetrametric assemblies *in vivo* (Mathy et al. 2010). RNase J is the first identified 5′ to 3′ exonuclease in bacteria. Through collaboration with RNase Y, RNase J homologs support an RNA degradation pathway that is similar, but distinct from that in Gram-negative bacteria that may have evolved to suit their different cellular physiologies, information processing systems, and/or transcript/gene structures.

9.5 Regulation of RNase E Activity

The activity of RNase E must be tightly regulated to prevent spurious cleavage, while at the same time allowing selective maturation of stable RNAs and rapid degradation of the appropriate transcripts to completion. RNase E is controlled at several levels: transcriptional, post-transcriptional, and especially post-translational via its subcellular localization and interaction with other protein factors/sRNAs. The *rne* mRNA of *E. coli* is transcribed from three different promoters (Ow et al. 2002). An important level of control is negative autoregulation of its own mRNA to maintain steady-state levels. When RNase E levels are high, it binds a secondary structure in its long 5’UTR, leading to cleavage, transcript destabilization, and degradation (Diwa et al. 2000; Schuck et al. 2009).

RNase E activity is influenced by RNA substrate levels/availability and cleavage site availability, which can be influenced by RNA binding proteins, including Hfq, and sRNAs. Several accessory proteins dedicated to mediating RNase E activity or specificity have been identified, such as RapZ and CsrD in *E. coli*, which affect the secondary structure of target RNAs to increase accessibility to RNase E (Göpel et al. 2013). RraA and RraB of *Pseudomonas aeruginosa* bind the disordered C-terminus of RNase E to modulate degradosome composition (Gao et al. 2006; Górna et al. 2010). The disordered CTD of *P. aeruginosa* RNase E has even evolved to be targeted by a ϕKZ phage protein, Dip, which inhibits RNase E activity (Van den Bossche et al. 2016).

RNase E activity is also under tight control via its subcellular localization. In *E. coli*, the degradosome is tethered to the inner membrane via the RNase E MTS

(Fig. 9.2), and removal of this sequence impacts growth, global RNA metabolism, and rRNA processing (Hadjeras et al. 2019). Convergent evolution of membrane-association for other degradosome-like complexes (i.e., RNase Y or J-related complexes of in Gram-positives/*Helicobacter pylori*, respectively) demonstrates the importance of sequestering RNase activity in the cell. In eukaryotes, interaction of RNA-binding proteins (RBP) with RNA substrate can stimulate condensation into so-called liquid-liquid phase-separated membraneless organelles, which can increase the local concentration of residing species and impart distinct physico-chemical characteristics. Recent work has provided evidence that a similar phenomenon affects RNase E in bacteria. For example, condensation of the type II RNase E of *C. crescentus* with RNA into so-called BR-bodies is mediated by the intrinsically-disordered C-terminal domain (Bayas et al. 2018). A similar phenomenon was also reported for *E. coli* (Strahl et al. 2015). Future work will demonstrate if this phenomenon is mediated by other RNase E types, and how it impacts RNA metabolism and spatial localization in bacteria.

9.6 The RNA Degradosome: Convergent Evolution

E. coli RNase E is a central component of the so-called RNA degradosome, a multi-protein complex that mediates RNA decay and processing in bacteria. Therefore, a full comparison of the structure and function of RNase E between species also requires investigation of degradosome composition. Most of the degradosome complexes characterized so far include an endonucleolytic global RNase such as RNase E, which also scaffolds the complex and tethers it to the membrane, as well as an RNA helicase, PNPase, and curiously, a metabolic enzyme such as enolase (Table 9.1). Interestingly, although not all bacteria encode RNase E, many also assemble similar degradosome-like complexes that are instead scaffolded by RNase Y, which is unrelated to RNase E (Fig. 9.6). These complexes are also reminiscent of the exosomes of archaea and eukaryotes, which also include an RNase, a helicase, and a PNPase-like enzyme. Recently, even more diversity in RNA decay complexes has been revealed, such as the minimal membrane-associated RNase J-containing degradosome of *Helicobacter pylori*, a Gram-negative species that does not encode RNase E (Tejada-Arranz et al. 2020) or the blended degradosome of *M. tuberculosis*, which includes both RNase E and RNase J as well as additional KH domain proteins of currently unknown function (Płociński et al. 2019). While these examples suggest there is much more to uncover in terms of degradosome diversity, they also demonstrate, via convergent evolution of several characteristics, the universal importance of several aspects revealed primarily by the detailed study of the *E. coli* complex. Moreover, now that sequestration of the prokaryotic RNA decay machinery in membraneless organelles has been reported, this provides another example of how evolution has repeatedly evolved or maintained similar organizing principles for RNA decay machinery across kingdoms of life (Paris et al. 2025).

9.7 Role of RNase E in Post-transcriptional Regulation by Small RNAs

RNase E plays a central role in sRNA-mediated post-transcriptional regulation (Lalaouna et al. 2013). It prefers cleaving A/U rich regions in mRNAs that coincide with Hfq binding sites (Moll 2003); controls the half-lives of many sRNAs; and mediates the coupled degradation of sRNAs with target mRNAs (Massé et al. 2003; Viegas et al. 2007).

9.7.1 *Co-degradation of Target RNAs along with sRNAs*

At the post-transcriptional level, many sRNAs negatively regulate the stability of mRNAs through competition with 30S ribosome subunits at the translation initiation region, and/or by directly recruiting RNase E through interaction with Hfq (Lalaouna et al. 2013). For example, MicC sRNA induces the decay of *ompD* mRNA, encoding a *Salmonella*-specific outer membrane porin, by using its first 12 nt to base-pair with codons 23–26 of the *ompD* mRNA (Pfeiffer et al. 2009). The 5′-end phosphorylation status of MicC influences the *ompD* cleavage pattern *in vitro*: 5′-monophosphorylated MicC is more effective in *ompD* degradation by CTD-truncated RNase E, but is also less stable (Bandyra et al. 2012). When associated with the 30S translation pre-initiation complex (PIC), *ompD* mRNA is efficiently cleaved by the RNA degradosome but not by CTD-truncated RNase E (Bandyra et al. 2024). However, *in vivo*, MicC induces the decay of *ompD* mRNA by the 5′-sensor RNase E mutant, as well as by wild-type enzyme, but not by a CTD-truncated mutant protein, indicating that the RNA degradosome exerts sRNA-targeted mRNA decay independently of the 5′-end status of the targeting sRNA (Schilder and Görke 2023).

Alternatively, sRNAs can stabilize targets by preventing their cleavage by RNase E. For example, the sugar stress-responsive sRNA SgrS base-pairs with an RNase E cleavage site in the coding sequence (CDS) of *pldB* (Papenfort et al. 2013). This stabilizes the *pldB-yigL* bicistronic mRNA and activates the expression of YigL, a glucose-6-phosphate phosphatase, to detoxify phosphosugars. Binding sites for stabilizing sRNAs can also be located in either 5'UTRs or 3'UTRs. The *cfa* gene in *Salmonella enterica*, encoding cyclopropane fatty acid synthase, is transcribed from two promoters, and the mRNA transcribed from the distal promoter is activated by RydC sRNA via protection of an RNase E cleavage site in its longer 5'UTR (Bianco et al. 2019; Fröhlich et al. 2013; King et al. 2019). The *hilD* mRNA, encoding a *Salmonella* virulence transcriptional regulator, has an RNase E-susceptible region in its 310-nucleotide long 3'UTR, and SdsR and Spot 42 sRNAs protect *hilD* by binding at different sites in this region (Abdulla et al. 2023).

The targets of sRNAs are not always full-length mRNAs. So-called "sponge" RNAs can target or regulate other sRNAs (Figueroa-Bossi and Bossi 2018). The

SroC sponge RNA induces the decay of the GcvB sRNA by RNase E upon base-pairing at two distinct sequences (Miyakoshi et al. 2015a). Interestingly, RNase E generates SroC from the 3'UTR of *gltI* mRNA encoding a glutamate/aspartate ABC transporter through the 5′-dependent pathway, while it degrades GcvB targeted by SroC through the 5′-independent pathway. In *Vibrio cholerae*, the QrrX sponge RNA base-pairs with and destabilizes Qrr1–4 sRNAs via cleavage by RNase E (Huber et al. 2022).

9.7.2 Maturation of sRNAs by RNase E

RNase E plays an important role in maturation of functional sRNAs. TIER-seq analysis revealed ~22,000 RNase E cleavage sites in the *Salmonella* transcriptome, of which ~600 were assigned to 107 annotated sRNAs (Chao et al. 2017), demonstrating its central role in sRNA maturation and biogenesis. For example, the ArcZ sRNA is primarily transcribed as a 118-nt precursor and is processed into a 56-nt mature sRNA by RNase E (Papenfort et al. 2009). Precursor ArcZ is deficient in target regulation, as its seed region requires exposure via removal of the upstream fragment (Chao et al. 2017).

In addition to maturation of stand-alone sRNAs, RNase E also generates many sRNAs that have been identified in mRNA 3'UTRs. RNase E cleavage sites are enriched around mRNA stop codons (Chao et al. 2017), and given that intrinsic terminators in the 3'UTRs of mRNAs serve as the binding sites for RNA chaperone Hfq (Holmqvist et al. 2016), these regions are prime candidates for evolution of sRNAs. Indeed, such 3'UTR-derived sRNAs have seed sequences that are conserved among bacterial species and are complementary to specific mRNAs (Miyakoshi et al. 2015b; Ponath et al. 2022).

CpxQ sRNA serves as a typical example of this class of sRNAs (Chao and Vogel 2016). When bacteria perceive envelope stress, the CpxA/R two-component system rapidly activates the transcription of *cpxP* mRNA. CpxQ is processed from the *cpxP* transcript by RNase E at a uridine-rich region located 9-nt downstream of the stop codon. CpxQ acts in the same extracytoplasmic stress response pathway as the CpxP protein, which both regulates the CpxA/R two-component system and acts as an adaptor targeting misfolded proteins to DegP. By repressing mRNAs encoding envelope proteins (*e.g.*, *skp*, *nhaB*), CpxQ protects bacterial cells against inner membrane damage (Chao and Vogel 2016; Grabowicz et al. 2016). Notably, CpxQ demonstrates significantly higher sequence conservation than the *cpxP* gene, indicating its potentially more fundamental biological function. This evolutionary pattern highlights the indispensability of sRNAs.

This class of processed sRNAs appear to have similar functions to the parental RNAs (Chao and Vogel 2016; Chen et al. 2025; Liu et al. 2023a; Wang et al. 2020). NarS sRNA is derived from the 3′ UTR of nitrate transporter NarK whose gene is silent during standard aerobic growth (Wang et al. 2020). NarS represses the synthesis of the nitrite transporter NirC by Hfq-dependent base pairing. This regulatory role is to protect the bacterial cytoplasm from excessive nitrite toxicity during anaerobic respiration with abundant nitrate (Wang et al. 2020). A recent iRIL-seq (intracellular RNA interaction by ligation and sequencing) analysis reveals FadZ sRNA is processed from the conserved 3'UTR of *fadBA* mRNA encoding fatty acid oxidation enzymes (Liu et al. 2023a). In *Salmonella*, FadZ shuts down the expression of two abundant porins, OmpC and OmpD, in response to fatty acid metabolism.

The SdhX sRNA is encoded in the 3'UTR of the ~10,000-nt long *sdhCDAB-sucABCD* operon mRNA, which encodes three enzyme complexes of the TCA cycle and is highly conserved among *Enterobacteriaceae* (de Mets et al. 2019; Miyakoshi et al. 2019). *E. coli* expresses SdhX in several different lengths: 101, 99, 92, and ~ 40 nt (de Mets et al. 2019). In *Salmonella*, SdhX is sequentially processed by RNase E from the mRNA; SdhX1 (~90 nt) is cleaved 9-nt upstream of the *sucD* stop codon by RNase E via direct entry, and then is processed into SdhX2 (~40 nt) via the 5′-end dependent pathway (Miyakoshi et al. 2019). SdhX2 is conserved between *E. coli* and *Salmonella* and is the shortest of all Hfq-dependent sRNAs known so far. SdhX2 represses *ackA* encoding acetate kinase in *E. coli* and *Salmonella*.

GlnZ sRNA is embedded in the 3'UTR of *glnA* mRNA encoding glutamine synthetase (Miyakoshi et al. 2022; Walling et al. 2022). Although GlnZ is expressed as different isoforms among different clades of *E. coli* and *Salmonella*, its common targets are *sucA* and *glnP* mRNAs, which encode 2-oxoglutarate dehydrogenase and glutamine ABC transporter, respectively. The *E. coli glnA* mRNA harbors two RNase E cleavage sites ~20-nt upstream and ~ 5-nt downstream of the stop codon. In the absence of RNase E cleavage, *glnA* mRNA is unable to repress the expression of *sucA*, even though the GlnZ seed sequence exists downstream of the *glnA* stop codon. This indicates that the sRNA needs to be released from its parental mRNA by RNase E to regulate its target mRNAs *in trans* (Miyakoshi et al. 2022).

TIER-seq analysis in *V. cholerae* has also revealed many 3'UTR-derived sRNAs processed by RNase E (Hoyos et al. 2020). OppZ sRNA is produced from the 3'UTR of *oppABCDF* operon mRNA encoding an oligopeptide ABC transporter via RNase E cleavage just downstream of the *oppF* stop codon. Interestingly, the target of OppZ is *oppB*, constituting negative feedback autoregulation. CarZ sRNA, encoded in the 3'UTR of *carAB* mRNA encoding carbamoyl phosphate synthetase, also negatively autoregulates its parental mRNA through direct base-pairing at the *carA* translation initiation region (Hoyos et al. 2020). Finally, FarS sRNA is also processed by RNase E from the 3'UTR of *fabB* mRNA encoding β-ketoacyl-ACP synthase (Huber et al. 2020). FarS regulates two *fadE* paralogs, VC1740 and VC2231, encoding acyl-CoA dehydrogenases.

Another class of sRNAs is associated with RNA chaperone ProQ, which binds to stable secondary structures, and mRNA 3'UTRs are also enriched among ProQ ligands (Holmqvist et al. 2018). RaiZ sRNA is processed by RNase E from the 3'UTR of *raiA* mRNA encoding one of the ribosome hibernation factors and binds to ProQ (Smirnov et al. 2017). With assistance from ProQ, RaiZ base-pairs with a single mRNA, *hupA*, which encodes histone-like protein HU, and inhibits translation initiation. In *Salmonella*, STnc540 sRNA is also processed by RNase E from the 3'UTR of *ihfA* mRNA encoding integration host factor α and regulates *mgtB* within the *mgtCBR* operon, which encodes manganese transporter (Chao et al. 2017; Westermann et al. 2019). In *V. cholerae*, the ProQ-dependent sRNA FlaX is released from the 3'UTR of *flaA* mRNA, encoding the major flagellin, via RNase E cleavage and regulates the other genes involved in motility (Ghandour et al. 2025). In *E. coli*, 3'UTR-derived sRNAs generated by RNase E cleavage are also involved in the post-transcriptional regulatory network for flagellar biosynthesis, but instead depend on Hfq (Melamed et al. 2023).

9.8 Methods to Globally Study RNase E Mechanisms and Function across Taxa

As noted above, methods such as TIER-seq have been essential for defining the targetome of RNase E (Chao et al. 2017), for understanding its specificity and revealing its broad role in the biogenesis of sRNAs. In fact, several different RNA-seq-based methods have been developed and/or applied to study different RNases in different species. These methods have been recently reviewed in detail (Broglia et al. 2023b), and we summarize them in Table 9.2. For example, RNA-seq methods that leverage modified sample treatment or library preparation, such as differential RNA-seq, allow the identification of primary (5′-PPP) vs. processed (5′-P) 5′-ends (Bischler et al. 2017; Sharma et al. 2010), while comparative transcriptomics in WT vs. RNase mutant strains can reveal cleavage sites or measure global RNA stability (Khemici et al. 2015; Dugar et al. 2018; Altuvia et al. 2018). Other methods that were developed to capture and identify RBP binding partners on a global scale, such as RIP/CLIP-seq, have likewise been applied to WT or catalytically-dead RNases to map targets (Hör et al. 2020; Lioliou et al. 2012). The Grad-seq approach, where protein-RNA complexes in bacterial lysates are separated based on their size and shape and then characterized by RNA-seq and mass spectrometry, can interrogate degradosome composition (Smirnov et al. 2016). Finally, recent years have seen an increased appreciation for RNA-RNA interaction networks based on new methods to identify chimeric RNAs after proximity ligation on RBPs (Svensson and Chao 2022; Papenfort and Melamed 2023), including RNases E and III (McKellar et al. 2022; Papenfort and Melamed 2023). With the advent of *intracellular* RNA-interactome profiling (Liu et al. 2023a), this potentially allows more relevant and high-throughput characterization of the target landscape of RBPs such as RNases. We propose that the methods listed in Table 9.2 are well positioned, especially when applied comparatively and/or to species with structurally diverse RNase E homologues, to expand our understanding of the scope of RNase E mechanism and biological function, and how its structural diversity relates to this.

Table 9.2 Global deep sequencing methods for studying ribonuclease function

Method	Principle	Example applications (species/RNase)		References
TIER-seq	Transient inactivation of RNase E followed by sequencing, differential identification of 5′ ends after inactivation of temperature-sensitive RNase E	*Salmonella, V. cholerae, E. coli, Synechocystis*	RNase E	Chao et al. (2017), Hoffmann et al. (2021), Hoyos et al. (2020), and Vigoda et al. (2024)
RNA-seq (modified)	RNA-seq with different 5′-end treatments (e.g., dRNA-seq, EMOTE, REND-seq, PARE, and others), identification of, e.g., processing sites, primary transcripts, capping status	*H. pylori, S. aureus, B. subtilis, E. coli,* etc	RNase III, RppH, RNase E, RNase Y	Altuvia et al. (2018), Bischler et al. (2017), Clarke et al. (2014), DeLoughery et al. (2018), Khemici et al. (2015), and Sharma et al. (2010)
RIF-seq	Measuring transcript decay rates after halting transcription with rifampicin	*B. subtilis, Salmonella*	RNase Y	Jenniches et al. (2024), Khemici et al. (2015)
RIP-seq	RNA binding protein immunoprecipitation followed by cRNA sequencing, identification of RBP RNA binding partners	*S. aureus, Campylobacter jejuni,* etc.	RNase III, Cas9	Dugar et al. (2018), Lioliou et al. (2012)
CLIP-seq	Cross-linking and immunoprecipitation followed by sequencing, identification of RBP RNA binding partners and interacting nucleotides	*S. pneumoniae*	YhaM	Hör et al. (2020)
Grad-seq	Gradient profiling by sequencing, isolation and protein/RNA characterization of cellular complexes	*Salmonella, C. difficile, S. pneumoniae, Fusobacterium*	Degradosomes	Hör et al. (2020), Lamm-Schmidt et al. (2021), Smirnov et al. (2016), and Zhu et al. (2024)

(continued)

Table 9.2 (continued)

Method	Principle	Example applications (species/RNase)		References
CLASH	Cross-linking, ligation, and sequencing of hybrids, identification of RNA interaction partners bound to an RBP	*E. coli, S. aureus*	RNase E, RNase III	McKellar et al. (2022), Mediati et al. (2022), and Waters et al. (2016)
CRAC	UV cross-linking and analysis of cDNAs, identification of RNA interaction partners bound to an RBP	Enterohaemorrhagic *E. coli*	YbeY	McAteer et al. (2018)
RIL/ iRIL-seq	(*intracellular*) RNA interaction by ligation and sequencing, identification of RNA interaction partners bound to an RBP	*Salmonella, E. coli, V. cholerae, Klebsiella,* etc.	Hfq, ProQ	Ghandour et al. (2025), Huber et al. (2022), Liu et al. (2023a, b), Melamed et al. (2016), and Wu et al. (2024)

For further details and examples, see the reference Broglia et al. (2023b)

9.9 Conclusions and Future Perspectives

The broad impact of RNase E on RNA metabolism and physiology in *E. coli* and *Salmonella* is clear, and detailed studies in these organisms have established several key paradigms in RNA enzymology and post-transcriptional regulation. However, the focus on models in only a handful of taxa is a major barrier to gaining a full understanding of the scope and diversity of RNase E structure and function across bacterial lineages. Even in our understanding of *E. coli* RNase E, there are several gaps in our broader understanding of its structure and function. For example, while the overall structure of RNase E is known, but we cannot yet explain the dynamic conformational changes that occur during substrate recognition, binding, and cleavage.

A major gap is in understanding how insights gained from *E. coli* applies to other RNase E homologs, both in terms of mechanism and biological role. For example, we have reviewed that RNase E recognizes RNA substrates for cleavage using two modes (the major 5′ monophosphate-dependent pathway as well as a direct entry pathway), but it is so far unclear how far this extends phylogenetically. It is unclear how such functions of RNase E in *E. coli* have diverged across different bacterial lineages, and if additional mechanisms might be present in other species. For example, the recent focus on type II RNase E from *Caulobacter* has shed light on the possible role of RNA phase separation in bacteria.

The importance of RNase E in sRNA-mediated regulation in various species has been emerged. While we have reviewed the crucial role of RNase E in both sRNA biogenesis and activity on post-transcriptional regulation in enteric bacteria, it is not clear how far these observations extend to other RNase E homologs. Application of global approaches, such as TIER-seq and RNase E-CLASH, to more diverse bacteria will allow a broader understanding of how the targetome and specificity of the enzyme vary between species, and reveal new examples for molecular dissection to understand the full scope of mechanism and biological activity of RNase E.

We especially focused on the role of RNase E in the biogenesis of 3′UTR-derived sRNAs, a well characterized class of sRNAs. However, so far, less effort has been applied in understanding the role RNase E plays in generation of the many processed sRNAs that have been revealed by global sequencing approaches arising from 5'UTRs and coding sequences (Adams et al. 2021). We expect that addressing these gaps will require interdisciplinary approaches, combining structural biology, biochemistry, genetics, systems biology, and computational modeling using RNase E homologs from more diverse bacterial species to fully elucidate the roles and regulatory mechanisms of RNase E in bacterial cells across different lineages.

Studies of RNase E have also shed light on the molecular basis of bacterial virulence and pathogenesis. In several pathogenic bacteria, RNase E is involved in the regulation of virulence factors, including toxins and secretion systems. By controlling the stability and translation of mRNAs encoding these factors, RNase E plays a crucial role in modulating the infectivity and survival of pathogens within host environments. Given its essential role in bacterial physiology and gene regulation, RNase E represents a promising target for novel antimicrobial agents. Understanding the mechanisms of RNase E has important implications for the development of strategies to combat bacterial infections.

Acknowledgements This work was supported by National Key R&D Program of China (2022YFE0111800, 2022YFC2303200), Chinese Academy of Sciences (XDB0570000, 176002GJHZ2022022MI), JSPS-CAS Bilateral Joint Research Project (JPJSBP120237201), Natural Science Foundation of China (92478118, 32270064, 32300028), and Shanghai Municipal Science and Technology Commission (24ZR1493200).

References

Abdulla SZ, Kim K, Azam MS, Golubeva YA, Cakar F, Slauch JM, Vanderpool CK (2023) Small RNAs activate Salmonella Pathogenicity Island 1 by modulating mRNA stability through the hilD mRNA 3' untranslated region. J Bacteriol 205:e0033322. https://doi.org/10.1128/jb.00333-22

Adams PP, Baniulyte G, Esnault C, Chegireddy K, Singh N, Monge M, Dale RK, Storz G, Wade JT (2021) Regulatory roles of Escherichia coli 5'UTR and ORF-internal RNAs detected by 3'end mapping. elife 10:e62438

Aït-Bara S, Carpousis AJ (2010) Characterization of the RNA Degradosome ofPseudoalteromonas haloplanktis: conservation of the RNase E-RhlB interaction in the Gammaproteobacteria. J Bacteriol 192:5413–5423. https://doi.org/10.1128/jb.00592-10

Aït-Bara S, Carpousis AJ (2015) RNA degradosomes in bacteria and chloroplasts: classification, distribution and evolution of RNase E homologs. Mol Microbiol 97:1021–1135

Aït-Bara S, Carpousis AJ, Quentin Y (2014) RNase E in the γ-Proteobacteria: conservation of intrinsically disordered noncatalytic region and molecular evolution of microdomains. Mol Gen Genomics 290:847–862. https://doi.org/10.1007/s00438-014-0959-5

Al-Husini N, Tomares DT, Bitar O, Childers WS, Schrader JM (2018) proteobacterial RNA degradosomes assemble liquid-liquid phase-separated RNP bodies. Mol Cell 71:1027–1039. https://doi.org/10.1016/j.molcel.2018.08.003

Altuvia Y, Bar A, Reiss N, Karavani E, Argaman L, Margalit H (2018) In vivo cleavage rules and target repertoire of RNase III in Escherichia coli. Nucleic Acids Res 46:10380–10394

Baek YM, Jang K-J, Lee H, Yoon S, Baek A, Lee K, Kim D-E (2019) The bacterial endoribonuclease RNase E can cleave RNA in the absence of the RNA chaperone Hfq. J Biol Chem 294:16465–16478

Baker KE, Mackie GA (2003) Ectopic RNase E sites promote bypass of 5′-end-dependent mRNA decay in Escherichia coli. Mol Microbiol 47:75–88

Bandyra KJ, Luisi BF (2018) RNase E and the high-fidelity orchestration of RNA metabolism. Microbiology Spectrum 6. https://doi.org/10.1128/microbiolspec.rwr-0008-2017

Bandyra KJ, Said N, Pfeiffer V, Górna MW, Vogel J, Luisi BF (2012) The seed region of a small RNA drives the controlled destruction of the target mRNA by the endoribonuclease RNase E. Mol Cell 47:943–953. https://doi.org/10.1016/j.molcel.2012.07.015

Bandyra KJ, Wandzik JM, Luisi BF (2018) Substrate recognition and autoinhibition in the central ribonuclease RNase E. Mol Cell 72:275–285.e4. https://doi.org/10.1016/j.molcel.2018.08.039

Bandyra KJ, Fröhlich KS, Vogel J, Rodnina M, Goyal A, Luisi BF (2024) Cooperation of regulatory RNA and the RNA degradosome in transcript surveillance. Nucleic Acids Res 52:9161–9173. https://doi.org/10.1093/nar/gkae455

Bayas CA, Wang J, Lee MK, Schrader JM, Shapiro L, Moerner WE (2018) Spatial organization and dynamics of RNase E and ribosomes in Caulobacter crescentus. Proc Natl Acad Sci 115:E3712–E3721

Bechhofer DH (2013) Messenger RNA degradation in bacteria. Encyclop Biol Chem 52–58. https://doi.org/10.1016/b978-0-12-378630-2.00281-4

Bechhofer DH, Deutscher MP (2019) Bacterial ribonucleases and their roles in RNA metabolism. Crit Rev Biochem Mol Biol 54:242–300

Bianco CM, Fröhlich KSA, Vanderpool CK (2019) Bacterial cyclopropane fatty acid synthase mRNA is targeted by activating and repressing small RNAs. J Bacteriol 201:e00461–e00419. https://doi.org/10.1128/JB.00461-19

Bischler T, Hsieh P, Resch M, Liu Q, Tan HS, Foley PL, Hartleib A, Sharma CM, Belasco JG (2017) Identification of the RNA pyrophosphohydrolase RppH of helicobacter pylori and global analysis of its RNA targets. J Biol Chem 292:1934–1950

Börner J, Friedrich T, Bartkuhn M, Klug G (2023) Ribonuclease E strongly impacts bacterial adaptation to different growth conditions. RNA Biol 20:120–135. https://doi.org/10.1080/15476286.2023.2195733

Bouvier M, Carpousis AJ (2011) A tale of two mRNA degradation pathways mediated by RNase E. Mol Microbiol 82:1305–1310

Briant DJ, Hankins JS, Cook MA, Mackie GA (2003) The quaternary structure of RNase G from Escherichia coli. Mol Microbiol 50:1381–1390. https://doi.org/10.1046/j.1365-2958.2003.03775.x

Broglia L, Le Rhun A, Charpentier E (2023a) Methodologies for bacterial ribonuclease characterization using RNA-seq. FEMS Microbiol Rev 47:fuad049

Bycroft M, Hubbard TJ, Proctor M, Freund SM, Murzin AG (1997) The solution structure of the S1 RNA binding domain: a member of an ancient nucleic acid–binding fold. Cell 88:235–242

Callaghan AJ, Marcaida MJ, Stead JA, McDowall KJ, Scott WG, Luisi BF (2005) Structure of Escherichia coli RNase E catalytic domain and implications for RNA turnover. Nature 437:1187–1191. https://doi.org/10.1038/nature04084

Carpousis AJ (2007) The RNA Degradosome of Escherichia coli: an mRNA-degrading machine assembled on RNase E. Ann Rev Microbiol 61:71–87. https://doi.org/10.1146/annurev.micro.61.080706.093440

Carpousis AJ, Campo N, Hadjeras L, Hamouche L (2022) Compartmentalization of RNA Degradosomes in bacteria controls accessibility to substrates and ensures concerted degradation of mRNA to nucleotides. Ann Rev Microbiol 76:533–552. https://doi.org/10.1146/annurev-micro-041020-113308

Celesnik H, Deana A, Belasco JG (2007) Initiation of RNA decay in Escherichia coli by 5′ pyrophosphate removal. Mol Cell 27:79–90

Chao Y, Vogel J (2016) A 3′ UTR-derived small RNA provides the regulatory noncoding arm of the inner membrane stress response. Mol Cell 61:352–363. https://doi.org/10.1016/j.molcel.2015.12.023

Chao Y, Li L, Girodat D, Förstner KU, Said N, Corcoran C, Śmiga M, Papenfort K, Reinhardt R, Wieden H-J, Luisi BF, Vogel J (2017) In vivo cleavage map illuminates the central role of RNase E in coding and non-coding RNA pathways. Mol Cell 65:39–51. https://doi.org/10.1016/j.molcel.2016.11.002

Chen Z, Yang Y, Chen X, Bei C, Gao Q, Chao Y, Wang C (2025) An RNase III–processed sRNA coordinates sialic acid metabolism of Salmonella enterica during gut colonization. Proc Natl Acad Sci 122. https://doi.org/10.1073/pnas.2414563122

Chung D, Min Z, Wang B-C, Kushner SR (2010) Single amino acid changes in the predicted RNase H domain of Escherichia coli RNase G lead to complementation of RNase E deletion mutants. RNA 16:1371–1385. https://doi.org/10.1261/rna.2104810

Clarke JE, Kime L, D Romero A, McDowall KJ (2014) Direct entry by RNase E is a major pathway for the degradation and processing of RNA in Escherichia coli. Nucleic Acids Res 42:11733–11751. https://doi.org/10.1093/nar/gku808

Commichau FM, Rothe FM, Herzberg C, Wagner E, Hellwig D, Lehnik-Habrink M, Hammer E, Völker U, Stülke J (2009) Novel activities of glycolytic enzymes in Bacillus subtilis. Mol Cell Proteomics 8:1350–1360. https://doi.org/10.1074/mcp.m800546-mcp200

Condon C (2002) The phylogenetic distribution of bacterial ribonucleases. Nucleic Acids Res 30:5339–5346. https://doi.org/10.1093/nar/gkf691

de Mets F, van Melderen L, Gottesman S (2019) Regulation of acetate metabolism and coordination with the TCA cycle via a processed small RNA. Proc Natl Acad Sci 116:1043–1052. https://doi.org/10.1073/pnas.1815288116

Deana A, Celesnik H, Belasco JG (2008) The bacterial enzyme RppH triggers messenger RNA degradation by 5′ pyrophosphate removal. Nature 451:355–358. https://doi.org/10.1038/nature06475

DeLoughery A, Lalanne J-B, Losick R, Li G-W (2018) Maturation of polycistronic mRNAs by the endoribonuclease RNase Y and its associated Y-complex in Bacillus subtilis. Proc Natl Acad Sci 115:E5585–E5594

Deutscher MP (2021) Regulation of bacterial ribonucleases. Ann Rev Microbiol 75:71–86

Dickerson RE (1980) Cytochrome c and the evolution of energy metabolism. Sci Am 242:136–153. https://doi.org/10.1038/scientificamerican0380-136

Diwa A, Bricker AL, Jain C, Belasco JG (2000) An evolutionarily conserved RNA stem–loop functions as a sensor that directs feedback regulation of RNase E gene expression. Genes Dev 14:1249–1260

Dugar G, Leenay RT, Eisenbart SK, Bischler T, Aul BU, Beisel CL, Sharma CM (2018) CRISPR RNA-dependent binding and cleavage of endogenous RNAs by the Campylobacter jejuni Cas9. Mol Cell 69:893–905

Durica-Mitic S, Görke B (2019) Feedback regulation of small RNA processing by the cleavage product. RNA Biol 16:1055–1065. https://doi.org/10.1080/15476286.2019.1612693

Erce MA, Low JKK, March PE, Wilkins MR, Takayama KM (2009) Identification and functional analysis of RNase E of Vibrio angustum S14 and two-hybrid analysis of its interaction part-

ners. Biochim Biophy Acta – Proteins Proteom 1794:1107–1114. https://doi.org/10.1016/j.bbapap.2009.03.016
Erce MA, Low JKK, Wilkins MR (2010) Analysis of the RNA degradosome complex in Vibrio angustum S14. FEBS J 277:5161–5173. https://doi.org/10.1111/j.1742-4658.2010.07934.x
Fang H, Song L, Liu K, Gu Y, Guo Y, Zhang C, Zhang L (2025) OsRNE encodes an RNase E/G--type endoribonuclease required for chloroplast development and seedling growth in rice. Int J Mol Sci 26:2375. https://doi.org/10.3390/ijms26052375
Figaro S, Durand S, Gilet L, Cayet N, Sachse M, Condon C (2013) Bacillus subtilis mutants with knockouts of the genes encoding ribonucleases RNase Y and RNase J1 are viable, with major defects in cell morphology, sporulation, and competence. J Bacteriol 195:2340–2348. https://doi.org/10.1128/jb.00164-13
Figueroa-Bossi N, Bossi L (2018) Sponges and predators in the small RNA world. Microbiol Spectr 6:10-1128
Förstner KU, Reuscher CM, Haberzettl K, Weber L, Klug G (2018) RNase E cleavage shapes the transcriptome ofRhodobacter sphaeroidesand strongly impacts phototrophic growth. Life Sci Alliance 1:e201800080. https://doi.org/10.26508/lsa.201800080
Fröhlich KS, Papenfort K, Fekete A, Vogel J (2013) A small RNA activates CFA synthase by isoform-specific mRNA stabilization. EMBO J 32:2963–2979. https://doi.org/10.1038/emboj.2013.222
Fu H, Elena RC, Marquez PH (2019) The roles of small RNAs: insights from bacterial quorum sensing. ExRNA 1. https://doi.org/10.1186/s41544-019-0027-8
Gao J, Lee K, Zhao M, Qiu J, Zhan X, Saxena A, Moore CJ, Cohen SN, Georgiou G (2006) Differential modulation of E. Coli mRNA abundance by inhibitory proteins that alter the composition of the degradosome. Mol Microbiol 61:394–406
Geslain SAM, Hausmann S, Geiser J, Allen GE, Gonzalez D, Valentini M (2025) Critical functions and key interactions mediated by the RNase E scaffolding domain in Pseudomonas aeruginosa. PLoS Genet 21:e1011618
Ghandour R, Devlitsarov D, Popp P, Melamed S, Huber M, Siemers M, Krüger T, Kniemeyer O, Klingl A, Brakhage AA, Erhardt M, Papenfort K (2025) ProQ-associated small RNAs control motility in vibrio cholerae. Nucleic Acids Res 53. https://doi.org/10.1093/nar/gkae1283
Göpel Y, Papenfort K, Reichenbach B, Vogel J, Görke B (2013) Targeted decay of a regulatory small RNA by an adaptor protein for RNase E and counteraction by an anti-adaptor RNA. Genes Dev 27:552–564
Górna MW, Pietras Z, Tsai Y-C, Callaghan AJ, Hernández H, Robinson CV, Luisi BF (2010) The regulatory protein RraA modulates RNA-binding and helicase activities of the E. coli RNA degradosome. RNA 16:553–562
Grabowicz M, Koren D, Silhavy TJ (2016) The CpxQ sRNA negatively regulates Skp to prevent mistargeting of β-barrel outer membrane proteins into the cytoplasmic membrane. MBio 7:e00312–e00316. https://doi.org/10.1128/mBio.00312-16
Hadjeras L, Poljak L, Bouvier M, Morin-Ogier Q, Canal I, Cocaign-Bousquet M, Girbal L, Carpousis AJ (2019) Detachment of the RNA degradosome from the inner membrane of Escherichia coli results in a global slowdown of mRNA degradation, proteolysis of RNase E and increased turnover of ribosome-free transcripts. Mol Microbiol 111:1715–1731
Hammarlof DL, Liljas L, Hughes D (2011) Temperature-sensitive mutants of RNase E in Salmonella enterica. J Bacteriol 193:6639–6650. https://doi.org/10.1128/jb.05868-11
Hardwick SW, Chan VSY, Broadhurst RW, Luisi BF (2010) An RNA degradosome assembly in Caulobacter crescentus. Nucleic Acids Res 39:1449–1459. https://doi.org/10.1093/nar/gkq928
Hardwick SW, Gubbey T, Hug I, Jenal U, Luisi BF (2012) Crystal structure of Caulobacter crescentus polynucleotide phosphorylase reveals a mechanism of RNA substrate channelling and RNA degradosome assembly. Open Biol 2:120028. https://doi.org/10.1098/rsob.120028
Hoffmann UA, Heyl F, Rogh SN, Wallner T, Backofen R, Hess WR, Steglich C, Wilde A (2021) Transcriptome-wide in vivo mapping of cleavage sites for the compact cyanobacterial ribo-

nuclease E reveals insights into its function and substrate recognition. Nucleic Acids Res 49:13075–13091. https://doi.org/10.1093/nar/gkab1161
Hoffmann UA, Lichtenberg E, Rogh SN, Bilger R, Reimann V, Heyl F, Backofen R, Steglich C, Hess WR, Wilde A (2024) The role of the 5' sensing function of ribonuclease E in cyanobacteria. RNA Biol 21:373–390
Holmqvist E, Wright PR, Li L, Bischler T, Barquist L, Reinhardt R, Backofen R, Vogel J (2016) Global RNA recognition patterns of post-transcriptional regulators Hfq and CsrA revealed by UV crosslinking in vivo. EMBO J 35:991–1011. https://doi.org/10.15252/embj.201593360
Holmqvist E, Li L, Bischler T, Barquist L, Vogel J (2018) Global maps of ProQ binding in vivo reveal target recognition via RNA structure and stability control at mRNA 3′ ends. Mol Cell 70:971–982.e6. https://doi.org/10.1016/j.molcel.2018.04.017
Hör, J., Garriss, G., Di Giorgio, S., Hack, L.-M., Vanselow, J.T., Förstner, K.U., Schlosser, A., Henriques-Normark, B., Vogel, J., 2020. Grad-seq in a Gram-positive bacterium reveals exonucleolytic sRNA activation in competence control. EMBO J 39, e103852
Hoyos M, Huber M, Förstner KU, Papenfort K (2020) Gene autoregulation by 3' UTR-derived bacterial small RNAs. elife 9:e58836. https://doi.org/10.7554/ELIFE.58836
Huber M, Fröhlich KS, Radmer J, Papenfort K (2020) Switching fatty acid metabolism by an RNA-controlled feed forward loop. Proc Natl Acad Sci USA 117:8044–8054. https://doi.org/10.1073/pnas.1920753117
Huber M, Lippegaus A, Melamed S, Siemers M, Wucher BR, Hoyos M, Nadell C, Storz G, Papenfort K (2022) An RNA sponge controls quorum sensing dynamics and biofilm formation in Vibrio cholerae. Nat Commun 13:7585. https://doi.org/10.1038/s41467-022-35261-x
Hui MP, Foley PL, Belasco JG (2014) Messenger RNA degradation in bacterial cells. Annu Rev Genet 48:537–559
Inagawa T, Okamoto S, Wachi M, Ochi K (2003) RNase ES of Streptomyces coelicolor A3(2) can complement the rne and rng mutations in Escherichia coli. Biosci Biotechnol Biochem 67:1767–1771. https://doi.org/10.1271/bbb.67.1767
Jaiswal LK, Singh RK, Nayak T, Kakkar A, Kandwal G, Singh VS, Gupta A (2024) A comparative analysis of mycobacterial ribonucleases: towards a therapeutic novel drug target. Infect Genet Evol 123:105645. https://doi.org/10.1016/j.meegid.2024.105645
Jenniches L, Michaux C, Popella L, Reichardt S, Vogel J, Westermann AJ, Barquist L (2024) Improved RNA stability estimation through Bayesian modeling reveals most salmonella transcripts have subminute half-lives. Proc Natl Acad Sci 121:e2308814121
Jiang X, Belasco JG (2004) Catalytic activation of multimeric RNase E and RNase G by 5′-monophosphorylated RNA. Proc Natl Acad Sci 101:9211–9216. https://doi.org/10.1073/pnas.0401382101
Jiang X, Diwa A, Belasco JG (2000) Regions of RNase E important for 5′-end-dependent RNA cleavage and autoregulated synthesis. J Bacteriol 182:2468–2475. https://doi.org/10.1128/jb.182.9.2468-2475.2000
Kaberdin VR (2003) Probing the substrate specificity of Escherichia coli RNase E using a novel oligonucleotide-based assay. Nucleic Acids Res 31:4710–4716
Kaberdin VR, Bizebard T (2005) Characterization of Aquifex aeolicus RNase E/G. Biochem Biophys Res Commun 327:382–392. https://doi.org/10.1016/j.bbrc.2004.12.017
Kaberdin VR, Miczak A, Jakobsen JS, Lin-Chao S, McDowall KJ, von Gabain A (1998) The endoribonucleolytic N-terminal half of Escherichia coli RNase E is evolutionarily conserved in Synechocystis sp. and other bacteria but not the C-terminal half, which is sufficient for degradosome assembly. Proc Natl Acad Sci 95:11637–11642. https://doi.org/10.1073/pnas.95.20.11637
Khemici V, Prados J, Linder P, Redder P (2015) Decay-initiating endoribonucleolytic cleavage by RNase Y is kept under tight control via sequence preference and sub-cellular localisation. PLoS Genet 11:e1005577

Kime L, Jourdan SS, Stead JA, Hidalgo-Sastre A, McDowall KJ (2009) Rapid cleavage of RNA by RNase E in the absence of 5′ monophosphate stimulation. Mol Microbiol 76:590–604. https://doi.org/10.1111/j.1365-2958.2009.06935.x

King AM, Vanderpool CK, Degnan PH (2019) sRNA target prediction organizing tool (SPOT) integrates computational and experimental data to facilitate functional characterization of bacterial small RNAs. mSphere 4:e00561–e00518. https://doi.org/10.1128/mSphere.00561-18

Koslover DJ, Callaghan AJ, Marcaida MJ, Garman EF, Martick M, Scott WG, Luisi BF (2008) The crystal structure of the Escherichia coli RNase E Apoprotein and a mechanism for RNA degradation. Structure 16:1238–1244. https://doi.org/10.1016/j.str.2008.04.017

Laalami S, Cavaiuolo M, Roque S, Chagneau C, Putzer H (2021) Escherichia coliRNase E can efficiently replace RNase Y inBacillus subtilis. Nucleic Acids Res 49:4643–4654. https://doi.org/10.1093/nar/gkab216

Lalaouna D, Simoneau-Roy M, Lafontaine D, Massé E (2013) Regulatory RNAs and target mRNA decay in prokaryotes. Biochim Biophys Acta – Gene Regul Mech 1829:742–747. https://doi.org/10.1016/j.bbagrm.2013.02.013

Lamm-Schmidt V, Fuchs M, Sulzer J, Gerovac M, Hör J, Dersch P, Vogel J, Faber F (2021) Grad-seq identifies KhpB as a global RNA-binding protein in Clostridioides difficile that regulates toxin production. Microlife 2:uqab004

Lee K, Cohen SN (2003) A Streptomyces coelicolor functional orthologue of Escherichia coli RNase E shows shuffling of catalytic and PNPase-binding domains. Mol Microbiol 48:349–360. https://doi.org/10.1046/j.1365-2958.2003.03435.x

Lee K, Bernstein JA, Cohen SN (2002) RNase G complementation of rne null mutation identifies functional interrelationships with RNase E in Escherichia coli. Mol Microbiol 43:1445–1456. https://doi.org/10.1046/j.1365-2958.2002.02848.x

Lehnik-Habrink M, Newman J, Rothe FM, Solovyova AS, Rodrigues C, Herzberg C, Commichau FM, Lewis RJ, Stülke J (2011) RNase Y in Bacillus subtilis: a natively disordered protein that is the functional equivalent of RNase E from Escherichia coli. J Bacteriol 193:5431–5441. https://doi.org/10.1128/jb.05500-11

Lehnik-Habrink M, Pförtner H, Rempeters L, Pietack N, Herzberg C, Stülke J (2010) The RNA degradosome in Bacillus subtilis: identification of CshA as the major RNA helicase in the multiprotein complex. Mol Microbiol 77:958–971

Li Z, Pandit S, Deutscher MP (1999) RNase G (CafA protein) and RNase E are both required for the 5′ maturation of 16S ribosomal RNA. EMBO J 18:2878–2885. https://doi.org/10.1093/emboj/18.10.2878

Lioliou E, Sharma CM, Caldelari I, Helfer A-C, Fechter P, Vandenesch F, Vogel J, Romby P (2012) Global regulatory functions of the Staphylococcus aureus endoribonuclease III in gene expression. PLoS Genet 8:e1002782

Liu F, Chen Z, Zhang S, Wu K, Bei C, Wang C, Chao Y (2023a) In vivo RNA interactome profiling reveals 3'UTR-processed small RNA targeting a central regulatory hub. Nat Commun 14. https://doi.org/10.1038/s41467-023-43632-1

Liu S-J, Lin G-M, Yuan Y-Q, Chen W, Zhang J-Y, Zhang C-C (2023b) A conserved protein inhibitor brings under check the activity of RNase E in cyanobacteria. Nucleic Acids Res 52:404–419. https://doi.org/10.1093/nar/gkad1094

Mackie GA (1992) Secondary structure of the mRNA for ribosomal protein S20. Implications for cleavage by ribonuclease E. J Biol Chem 267:1054–1061

Mackie GA (1998) Ribonuclease E is a 5′-end-dependent endonuclease. Nature 395:720–724

Mackie GA (2013) RNase E: at the interface of bacterial RNA processing and decay. Nat Rev Microbiol 11:45–57

Mardle CE, Shakespeare TJ, Butt LE, Goddard LR, Gowers DM, Atkins HS, Vincent HA, Callaghan AJ (2019) A structural and biochemical comparison of ribonuclease E homologues from pathogenic bacteria highlights species-specific properties. Sci Rep 9. https://doi.org/10.1038/s41598-019-44385-y

Massé E, Escorcia FE, Gottesman S (2003) Coupled degradation of a small regulatory RNA and its mRNA targets in Escherichia coli. Genes Dev 17:2374–2383

Mathy N, Bénard L, Pellegrini O, Daou R, Wen T, Condon C (2007) 5′-to-3′ exoribonuclease activity in bacteria: role of RNase J1 in rRNA maturation and 5′ stability of mRNA. Cell 129:681–692. https://doi.org/10.1016/j.cell.2007.02.051

Mathy N, Hébert A, Mervelet P, Bénard L, Dorléans A, Li de la Sierra-Gallay I, Noirot P, Putzer H, Condon C (2010) Bacillus subtilis ribonucleases J1 and J2 form a complex with altered enzyme behaviour. Mol Microbiol 75:489–498. https://doi.org/10.1111/j.1365-2958.2009.07004.x

McAteer SP, Sy BM, Wong JL, Tollervey D, Gally DL, Tree JJ (2018) Ribosome maturation by the endoribonuclease YbeY stabilizes a type 3 secretion system transcript required for virulence of enterohemorrhagic Escherichia coli. J Biol Chem 293:9006–9016

McDowall KJ, Hernandez RG, Lin-Chao S, Cohen SN (1993) The ams-1 and rne-3071 temperature-sensitive mutations in the ams gene are in close proximity to each other and cause substitutions within a domain that resembles a product of the Escherichia coli mre locus. J Bacteriol 175:4245–4249. https://doi.org/10.1128/jb.175.13.4245-4249.1993

McDowall KJ, Kaberdin VR, Wu S-W, Cohen SN, Lin-Chao S (1995) Site-specific RNase E cleavage of oligonucleotides and inhibition by stem–loops. Nature 374:287–290

McKellar SW, Ivanova I, Arede P, Zapf RL, Mercier N, Chu L-C, Mediati DG, Pickering AC, Briaud P, Foster RG, Kudla G, Ross Fitzgerald J, Caldelari I, Carroll RK, Tree JJ, Granneman S (2022) RNase III CLASH in MRSA uncovers sRNA regulatory networks coupling metabolism to toxin expression. Nat Commun 13:3560. https://doi.org/10.1038/s41467-022-31173-y

Mediati DG, Wong JL, Gao W, McKellar S, Pang CNI, Wu S, Wu W, Sy B, Monk IR, Biazik JM, Wilkins MR, Howden BP, Stinear TP, Granneman S, Tree JJ (2022) RNase III-CLASH of multi-drug resistant Staphylococcus aureus reveals a regulatory mRNA 3′UTR required for intermediate vancomycin resistance. Nat Commun 13:3558. https://doi.org/10.1038/s41467-022-31177-8

Melamed S, Peer A, Faigenbaum-Romm R, Gatt YE, Reiss N, Bar A, Altuvia Y, Argaman L, Margalit H (2016) Global mapping of small RNA-target interactions in bacteria. Mol Cell 63:884–897. https://doi.org/10.1016/j.molcel.2016.07.026

Melamed S, Zhang A, Jarnik M, Mills J, Silverman A, Zhang H, Storz G (2023) σ(28)-dependent small RNA regulation of flagella biosynthesis. elife 12. https://doi.org/10.7554/eLife.87151

Miyakoshi M, Chao Y, Vogel J (2015a) Cross talk between ABC transporter mRNAs via a target mRNA-derived sponge of the GcvB small RNA. EMBO J 34:1478–1492. https://doi.org/10.15252/embj.201490546

Miyakoshi M, Chao Y, Vogel J (2015b) Regulatory small RNAs from the 3′ regions of bacterial mRNAs. Curr Opin Microbiol 24:132–139. https://doi.org/10.1016/j.mib.2015.01.013

Miyakoshi M, Matera G, Maki K, Sone Y, Vogel J (2019) Functional expansion of a TCA cycle operon mRNA by a 3′ end-derived small RNA. Nucleic Acids Res 47:2075–2088. https://doi.org/10.1093/nar/gky1243

Miyakoshi M, Morita T, Kobayashi A, Berger A, Takahashi H, Gotoh Y, Hayashi T, Tanaka K (2022) Glutamine synthetase mRNA releases sRNA from its 3'UTR to regulate carbon/nitrogen metabolic balance in Enterobacteriaceae. elife 11:e82411. https://doi.org/10.7554/eLife.82411

Moll I (2003) Coincident Hfq binding and RNase E cleavage sites on mRNA and small regulatory RNAs. RNA 9:1308–1314. https://doi.org/10.1261/rna.5850703

Moore CJ, Go H, Shin E, Ha H-J, Song S, Ha N-C, Kim Y-H, Cohen SN, Lee K (2021) Substrate-dependent effects of quaternary structure on RNase E activity. Genes Dev 35:286–299. https://doi.org/10.1101/gad.335828.119

Mudd EA, Sullivan S, Gisby MF, Mironov A, Kwon CS, Chung W-I, Day A (2008) A 125 kDa RNase E/G-like protein is present in plastids and is essential for chloroplast development and autotrophic growth in Arabidopsis*. J Exp Bot 59:2597–2610. https://doi.org/10.1093/jxb/ern126

Nicholson AW (1999) Function, mechanism and regulation of bacterial ribonucleases. FEMS Microbiol Rev 23:371–390

Okada Y, Wachi M, Hirata A, Suzuki K, Nagai K, Matsuhashi M (1994) Cytoplasmic axial filaments in Escherichia coli cells: possible function in the mechanism of chromosome segregation and cell division. J Bacteriol 176:917–922. https://doi.org/10.1128/jb.176.3.917-922.1994

Ow MC, Liu Q, Mohanty BK, Andrew ME, Maples VF, Kushner SR (2002) RNase E levels in Escherichia coli are controlled by a complex regulatory system that involves transcription of the rne gene from three promoters. Mol Microbiol 43:159–171

Ow MC, Perwez T, Kushner SR (2003) RNase G of Escherichia coli exhibits only limited functional overlap with its essential homologue, RNase E. Mol Microbiol 49:607–622. https://doi.org/10.1046/j.1365-2958.2003.03587.x

Papenfort K, Melamed S (2023) Small RNAs, large networks: posttranscriptional regulons in gram-negative bacteria. Ann Rev Microbiol 77:23–43. https://doi.org/10.1146/annurev-micro-041320-025836

Papenfort K, Said N, Welsink T, Lucchini S, Hinton JCD, Vogel J (2009) Specific and pleiotropic patterns of mRNA regulation by ArcZ, a conserved Hfq-dependent small RNA. Mol Microbiol 74:139–158. https://doi.org/10.1111/j.1365-2958.2009.06857.x

Papenfort K, Sun Y, Miyakoshi M, Vanderpool CK, Vogel J (2013) Small RNA-mediated activation of sugar phosphatase mRNA regulates glucose homeostasis. Cell 153:426–437. https://doi.org/10.1016/j.cell.2013.03.003

Paris G, Katsuya-Gaviria K, Luisi BF (2025) Machinery, mechanism, and information in post-transcription control of gene expression, from the perspective of unstable RNA. Q Rev Biophys 58:e10

Pfeiffer V, Papenfort K, Lucchini S, Hinton JCD, Vogel J (2009) Coding sequence targeting by MicC RNA reveals bacterial mRNA silencing downstream of translational initiation. Nat Struct Mol Biol 16:840–846. https://doi.org/10.1038/nsmb.1631

Płociński P, Macios M, Houghton J, Niemiec E, Płocińska R, Brzostek A, Słomka M, Dziadek J, Young D, Dziembowski A (2019) Proteomic and transcriptomic experiments reveal an essential role of RNA degradosome complexes in shaping the transcriptome of Mycobacterium tuberculosis. Nucleic Acids Res 47:5892–5905

Ponath F, Hör J, Vogel J (2022) An overview of gene regulation in bacteria by small RNAs derived from mRNA 3′ ends. FEMS Microbiol Rev 46:fuac017. https://doi.org/10.1093/femsre/fuac017

Purusharth RI, Klein F, Sulthana S, Jäger S, Jagannadham MV, Evguenieva-Hackenberg E, Ray MK, Klug G (2005) Exoribonuclease R interacts with endoribonuclease E and an RNA helicase in the psychrotrophic bacterium Pseudomonas syringae Lz4W. J Biol Chem 280:14572–14578. https://doi.org/10.1074/jbc.m413507200

Schilder A, Görke B (2023) Role of the 5′ end phosphorylation state for small RNA stability and target RNA regulation in bacteria. Nucleic Acids Res 51:5125–5143. https://doi.org/10.1093/nar/gkad226

Schubert M, Edge RE, Lario P, Cook MA, Strynadka NCJ, Mackie GA, McIntosh LP (2004) Structural characterization of the RNase E S1 domain and identification of its oligonucleotide-binding and dimerization interfaces. J Mol Biol 341:37–54. https://doi.org/10.1016/j.jmb.2004.05.061

Schuck A, Diwa A, Belasco JG (2009) RNase E autoregulates its synthesis in Escherichia coli by binding directly to a stem-loop in the rne 5′ untranslated region. Mol Microbiol 72:470–478

Shahbabian K, Jamalli A, Zig L, Putzer H (2009) RNase Y, a novel endoribonuclease, initiates riboswitch turnover in Bacillus subtilis. EMBO J 28:3523–3533. https://doi.org/10.1038/emboj.2009.283

Sharma CM, Hoffmann S, Darfeuille F, Reignier J, Findeiß S, Sittka A, Chabas S, Reiche K, Hackermüller J, Reinhardt R (2010) The primary transcriptome of the major human pathogen Helicobacter pylori. Nature 464:250–255

Smirnov A, Förstner KU, Holmqvist E, Otto A, Günster R, Becher D, Reinhardt R, Vogel J (2016) Grad-seq guides the discovery of ProQ as a major small RNA-binding protein. Proc Natl Acad Sci 113:11591–11596

Smirnov A, Wang C, Drewry LL, Vogel J (2017) Molecular mechanism of mRNA repression in trans by a ProQ-dependent small RNA. EMBO J 36:1029–1045. https://doi.org/10.15252/embj.201696127

Spanka D-T, Reuscher CM, Klug G (2021) Impact of PNPase on the transcriptome of Rhodobacter sphaeroides and its cooperation with RNase III and RNase E. BMC Genomics 22. https://doi.org/10.1186/s12864-021-07409-4

Strahl H, Turlan C, Khalid S, Bond PJ, Kebalo J-M, Peyron P, Poljak L, Bouvier M, Hamoen L, Luisi BF (2015) Membrane recognition and dynamics of the RNA degradosome. PLoS Genet 11:e1004961

Svensson SL, Chao Y (2022) RNase III-CLASH brings bacterial RNA networks into focus. Trends Microbiol 30:1125–1127. https://doi.org/10.1016/j.tim.2022.09.012

Tamura M, Moore CJ, Cohen SN (2012) Nutrient dependence of RNase E essentiality in Escherichia coli. J Bacteriol 195:1133–1141. https://doi.org/10.1128/jb.01558-12

Taverniti V, Forti F, Ghisotti D, Putzer H (2011) Mycobacterium smegmatis RNase J is a 5′-3′ exo−/endoribonuclease and both RNase J and RNase E are involved in ribosomal RNA maturation. Mol Microbiol 82:1260–1276. https://doi.org/10.1111/j.1365-2958.2011.07888.x

Tejada-Arranz A, Galtier E, El Mortaji L, Turlin E, Ershov D, De Reuse H (2020) The RNase J-based RNA degradosome is compartmentalized in the gastric pathogen Helicobacter pylori. MBio 11. https://doi.org/10.1128/mbio.01173-20

Thompson KJ, Zong J, Mackie GA (2014) Altering the divalent metal ion preference of RNase E. J Bacteriol 197:477–482. https://doi.org/10.1128/jb.02372-14

Umitsuki G, Wachi M, Takada A, Hikichi T, Nagai K (2001) Involvement of RNase G in in vivo mRNA metabolism in Escherichia coli. Genes Cells 6:403–410. https://doi.org/10.1046/j.1365-2443.2001.00430.x

Van den Bossche A, Hardwick SW, Ceyssens P-J, Hendrix H, Voet M, Dendooven T, Bandyra KJ, De Maeyer M, Aertsen A, Noben J-P (2016) Structural elucidation of a novel mechanism for the bacteriophage-based inhibition of the RNA degradosome. elife 5:e16413

Viegas SC, Pfeiffer V, Sittka A, Silva IJ, Vogel J, Arraiano CM (2007) Characterization of the role of ribonucleases in Salmonella small RNA decay. Nucleic Acids Res 35:7651–7664

Vigoda MB, Argaman L, Kournos M, Margalit H (2024) Unraveling the interplay between a small RNA and RNase E in bacteria. Nucleic Acids Res 52:8947–8966. https://doi.org/10.1093/nar/gkae621

Wachi M, Umitsuki G, Shimizu M, Takada A, Nagai K (1999) Escherichia coli cafA gene encodes a novel RNase, designated as RNase G, involved in processing of the 5′ end of 16S rRNA. Biochem Biophys Res Commun 259:483–488. https://doi.org/10.1006/bbrc.1999.0806

Wachi M, Kaga N, Umitsuki G, Clark DP, Nagai K (2001) A novel RNase G mutant that is defective in degradation of adhE mRNA but proficient in the processing of 16S rRNA precursor. Biochem Biophys Res Commun 289:1301–1306. https://doi.org/10.1006/bbrc.2001.6115

Walling LR, Kouse AB, Shabalina SA, Zhang H, Storz G (2022) A 3' UTR-derived small RNA connecting nitrogen and carbon metabolism in enteric bacteria. Nucleic Acids Res 50:10093–10109. https://doi.org/10.1093/nar/gkac748

Walter M, Piepenburg K, Schöttler MA, Petersen K, Kahlau S, Tiller N, Drechsel O, Weingartner M, Kudla J, Bock R (2010) Knockout of the plastid RNase E leads to defective RNA processing and chloroplast ribosome deficiency. Plant J 64:851–863. https://doi.org/10.1111/j.1365-313x.2010.04377.x

Wang C, Chao Y, Matera G, Gao Q, Vogel J (2020) The conserved 3′ UTR-derived small RNA NarS mediates mRNA crossregulation during nitrate respiration. Nucleic Acids Res 48:2126–2143. https://doi.org/10.1093/nar/gkz1168

Waters SA, McAteer SP, Kudla G, Pang I, Deshpande NP, Amos TG, Leong KW, Wilkins MR, Strugnell R, Gally DL, Tollervey D, Tree JJ (2016) Small RNA interactome of pathogenic E. coli revealed through crosslinking of RNase E. EMBO J 36:374–387. https://doi.org/10.15252/embj.201694639

Westermann AJ, Venturini E, Sellin ME, Förstner KU, Hardt W-D, Vogel J (2019) The major RNA-binding protein ProQ impacts virulence gene expression in Salmonella enterica Serovar Typhimurium. MBio 10. https://doi.org/10.1128/mBio.02504-18

Wu K, Lin X, Lu Y, Dong R, Jiang H, Svensson SL, Zheng J, Shen N, Camilli A, Chao Y (2024) RNA interactome of hypervirulent Klebsiella pneumoniae reveals a small RNA inhibitor of capsular mucoviscosity and virulence. Nat Commun 15:6946

Yang J, Jain C, Schesser K (2008) RNase E regulates the Yersinia type 3 secretion system. J Bacteriol 190:3774–3778. https://doi.org/10.1128/jb.00147-08

Zhang J-Y, Deng X-M, Li F-P, Wang L, Huang Q-Y, Zhang C-C, Chen W-L (2014) RNase E forms a complex with polynucleotide phosphorylase in cyanobacteria via a cyanobacterial-specific nonapeptide in the noncatalytic region. RNA 20:568–579. https://doi.org/10.1261/rna.043513.113

Zhou C, Zhang J, Hu X, Li C, Wang L, Huang Q, Chen W (2020) RNase II binds to RNase E and modulates its endoribonucleolytic activity in the cyanobacterium anabaena PCC 7120. Nucleic Acids Res 48:3922–3934

Zhou Y, Sun H, Rapiejko AR, Vargas-Blanco DA, Martini MC, Chase MR, Joubran SR, Davis AB, Dainis JP, Kelly JM, Ioerger TR, Roberts LA, Fortune SM, Shell SS (2023) Mycobacterial RNase E cleaves with a distinct sequence preference and controls the degradation rates of most Mycolicibacterium smegmatis mRNAs. J Biol Chem 299:105312. https://doi.org/10.1016/j.jbc.2023.105312

Zhu Y, Ponath F, Cosi V, Vogel J (2024) A global survey of small RNA interactors identifies KhpA and KhpB as major RNA-binding proteins in *Fusobacterium nucleatum*. Nucleic Acids Res. https://doi.org/10.1093/nar/gkae010

Chapter 10
CRISPR–Cas Systems ~ Diversity and Evolution ~

Yoshizumi Ishino

Abstract CRISPR (Clustered Regularly Interspaced Short Palindromic Repeats) was discovered when I was studying molecular genetics of *Escherichia coli,* and it has been about 40 years since I announced it in 1987. Nowadays, CRISPR is known to everyone in the field of life sciences, but I could not imagine what it meant physiologically when I discovered it. A sequence with remarkable characteristics was found downstream of a gene involved in conversion process for alkaline phosphatase to produce its isozyme molecules, which I was studying at that time. The characteristic of a constant length of sequence including a palindromic sequence repeated at equal intervals was so impressive that I could not help but write about it in our paper. It was later found to be responsible for the adaptive immune function of prokaryotes, and the mechanism was used to develop a practical genome editing technology. This chapter summarizes what is currently known about CRISPR and the group of genes encoding the associated Cas proteins, and also provides an analysis of their evolution and phylogeny.

Keywords RNA-programable nuclease · Acquired immunity · Genome editing · crRNA · ωRNA

Y. Ishino (✉)
Department of Bioscience and Biotechnology, Graduate School of Bioresource and Bioenvironmental Sciences, Kyushu University, Fukuoka, Japan

Genome Editing Research Institute, Nagahama Institute of Bio-Science and Technology, Nagahama, Shiga, Japan
e-mail: ishino@agr.kyushu-u.ac.jp

A. Kanai, A. W. Nicholson (eds.), *Molecular Evolution of RNA Regulatory Enzymes and Their Systems*, Nucleic Acids and Molecular Biology 38,
https://doi.org/10.1007/978-3-032-03300-0_10

10.1 Introduction

The characteristic repeated sequence derived from *E. coli* was reported in 1987 (Ishino et al. 1987). However, its physiological function was unclear. In 1993, it was reported that a similar sequence exists not only in bacteria but also in archaea (Mojica et al. 1993). Therefore, it was expected that CRISPR may exist widely in prokaryotes and play some universal function. The genomic era began in the middle 1990s with the advancement of DNA sequencing technology, and the complete sequences of particularly small microbial genomes were reported one after another. CRISPR-like sequences were identified in the sequencing studies. This sequence had come to be called by various names by different authors, such as Large Clusters of Tandem Repeat (LCTR), Direct Variable Repeat (DVR), SPacers Interspersed Direct Repeats (SPIDR), and Short Regularly Spaced Repeat (SRSR). Then, to unify the name of this genetic element, CRISPR (Clustered Regularly Interspaced Short Palindromic Repeats) was proposed, and the name, *cas* (CRISPR-*as*sociated) was proposed for the adjacent protein-coding genes (Jansen et al. 2002).

In the mid-2000s, it was discovered that some foreign genes derived from phages or plasmids were inserted into the spacer region between the repeating units of CRISPR, and this led to the speculation that CRISPR might function as a defense mechanism to protect the host cells against foreign genes (Mojica et al. 2005; Pourcel et al. 2005). A model was quickly proposed, in which the mechanism of action was similar to that of RNAi, a method of suppression for a specific gene expression, already known in eukaryotes that inhibits protein synthesis by degrading specific target mRNAs (Makarova et al. 2006). In the proposed mechanism, a short strand of RNA with a complementary sequence to the mRNA of the target gene destroys the foreign gene in a manner similar to the mechanism by which RNA-induced silencing complexes (RISCs) form and degrade the targeted RNA.

The hypothesis that CRISPR–Cas plays a role in the host defense was first proven by studies on lactic acid bacteria, to protect themselves from phages that infect them. Subsequently, detailed molecular mechanism was elucidated using the *streptococcal* CRISPR–Cas9. Shortly thereafter it was demonstrated that the purified Cas9–crRNA complexes can cleave target DNA *in vitro*. It was also proposed that this mechanism could be used as a technology to cut specific target genes in the genome, i.e., genome editing. This proposal was successfully realized, in which a target gene in the genome of human cells was artificially cut and edited using CRISPR–Cas9, and it has attracted attention as a practical genome editing technology (Jinek et al. 2012; Gasiunas et al. 2012).

CRISPR–Cas attracted two major points of interest: its biological defense function and its application to genome editing technology. Many researchers have entered this field, and research is being carried out around the world, with continuing rapid development. Following CRISPR–Cas9, Cas proteins with different properties have been reported, such as Cas12, Cas13, and Cas14, and the technical application of these proteins is also expanding. History of the research on CRISPR-Cas was summarized in the various review articles, including ours (Ishino et al.

2018) as representative. In this chapter, we summarize the current knowledge about the origin, evolution, and diversity of this system, which has been understood through the biochemistry and structural analysis of CRISPR–Cas.

10.2 Expanding Our Knowledge of CRISPR

After we published the first CRISPR sequence from *E. coli* K12 in 1987, other bacteria were examined using genomic Southern blotting analyses, involving the repeated sequence as a probe, to investigate whether this unique repeat sequence was unique to *E. coli* and bacteria. It was then reported that positive signals were detected in the genomes of other *E. coli* and bacteria, including *Shigella* and *Salmonella* (Nakata et al. 1989). *Mycobacteria* also contained a similar repeated sequence (Groenen et al. 1993). In the early 1990s, this repeat sequence was also found in the hyperhalophilic archaeon, *Haloferax mediterranei*, thus confirming its presence in two of the three domains of life. With the advent of the genomic era in the late 1990s, the repeated sequences with a similar characteristic feature were discovered continuously, and the genomic context of regions including the repeated sequences were compared in many organisms. In addition, four conserved genes were discovered that regularly exist adjacent to the repeat sequence regions. Then, to unify the name of this genetic element, CRISPR (Clustered Regularly Interspaced Short Palindromic Repeats) was proposed, and the name, *cas* (CRISPR-associated) was proposed for the adjacent protein-coding genes (Jansen et al. 2002). Cas1 and Cas2 proteins showed no similarity to the functional domains of known proteins, whereas Cas3 contains seven motifs characteristic of superfamily 2 helicases, and Cas4 is associated with RecB exonuclease, which functions as part of the RecBCD complex and plays an important role in initiating homologous recombination by end resection of double-stranded breaks. Thus, Cas3 and Cas4 were predicted to be involved in DNA metabolism, including DNA repair and recombination, transcriptional regulation, and chromosome segregation. The presence of Cas proteins in association with CRISPR suggested that they may be involved in the function of the CRISPR loci (Jansen et al. 2002).

From the beginning of the genomic era, thermophiles and hyperthermophiles were actively analyzed. Makarova, Koonin and colleagues systematically analyzed the conserved gene context in all prokaryotic genomes available at the time, and found multiple gene clusters in the hyperthermophilic archaea and bacterial genomes that were putatively DNA polymerases, helicases, and RecB-like nucleases (Makarova et al. 2002). Since these conserved genes were not present in mesophilic and moderately thermophilic archaea and bacteria at the time, they named these proteins RAMPs (Repeat-Associated Mysterious Proteins) and predicted that they may be part of a "mysterious" and unexplored DNA repair system unique to thermophiles. Many hyperthermophilic archaea, as well as hyperthermophilic bacteria such as *A. aeolicus* and *T. maritima,* have larger CRISPRs than mesophiles (Jansen et al. 2002), suggesting that the function of CRISPR may be related to the adaptation of organisms to high temperatures. However, as more sequences became available, it became clear that this correlation was not strong and that RAMPs are also present in many mesophilic bacteria and that they include Cas proteins.

The function of CRISPR was revealed when the research groups of Francisco Mojica of Alicante University and Christine Pourcel of University of Paris-Sud, Orsay independently noticed that the spacer regions between the repeats were homologous to sequences of bacteriophages, prophages, and plasmids (Mojica et al. 2005; Pourcel et al. 2005). After a detailed review of published papers, they noted that host strains, which contained sequences homologous to phages or plasmids in the spacer region of CRISPR in their genome, were not infected by them. These findings led them to propose that CRISPR–Cas functions similarly to the eukaryotic RNAi system to protect cells from the invasion of mobile genetic elements such as phages and plasmids. Both groups also suggested that CRISPR somehow induces the capture of foreign invading DNA fragments, thereby constituting a mechanism for remembering past genetic invasions. A paper in the same year was the third influential paper in elucidating the function of CRISPR–Cas (Bolotin et al. 2005). It confirmed these observations, and noted a correlation between the number of phage-derived spacers and the degree of resistance to phage infection. They suggested that it may be possible to generate antisense RNA using CRISPR. In these papers, the authors acknowledged the previous discovery of *cas* genes and suggested that the proteins encoded by these genes should be involved in the function of this putative prokaryotic immune system. The predicted role of Cas proteins was published a year later by the Koonin group (Makarova et al. 2006). They performed a detailed analysis of the Cas protein sequence and predicted its function with a mechanism similar to the eukaryotic RNAi system. These mechanistic predictions were experimentally confirmed a few years later and have helped as a guideline for the experiments of CRISPR–Cas system to date. Importantly, they pointed out that the CRISPR–Cas system with its memory component (acquisition process) is very similar to the adaptive immune system of vertebrates, with the crucial difference that the adaptive immune system of animals is not inherited. Considering the diversity of CRISPR–Cas systems, its high mobility suggested by irregular distribution, and some universality in archaea, the authors hypothesized that the CRISPR–Cas system emerged from an ancient ancestor of archaea and spread to bacteria by horizontal transfer. They also pointed out from a practical standpoint that the CRISPR-Cas system could be used for gene silencing in organisms that encode Cas proteins.

The function of the CRISPR–Cas system as an adaptive immune system in prokaryotes was experimentally demonstrated in 2007 by using the lactic acid bacterium *Streptococcus thermophilus*, which is important for the production of dairy products such as yogurt, to show that the CRISPR system protects itself from phages that infect them (Barrangou et al. 2007). Insertion of a phage sequence into the spacer region of the CRISPR in the *S. thermophilus* genome conferred resistance to the corresponding phage, whereas the deletion of the corresponding protospacer sequence from the spacer region of the host genome eliminated the resistance of the bacterium to phage infection. Furthermore, transformation of the host bacterium with a plasmid containing a sequence matching the spacer in CRISPR was inhibited

(Marraffini and Sontheimer 2008). Shortly thereafter, reconstitution of the immune system using the *E. coli* CRISPR, which I discovered in 1987, was presented. RNA molecules processed from the transcripts of the CRISPR region functioned as a complex with Cas proteins produced from the adjacent genes in the CRISPR region (Brouns et al. 2008). Concurrently, metagenomic analysis of archaea demonstrated that sequences at CRISPR loci dynamically change over a period of months, with new spacer sequences emerging that correspond to phages from the same population (Andersson and Banfield 2008).

The reconstituted *S. thermophilus* CRISPR–Cas9 system using the materials produced in *E. coli* protected the host against plasmid transformation and phage infection (Sapranauskas et al. 2011). It was also demonstrated that the *cas9* is the only *cas* gene encoded in the CRISPR–Cas system required for the targeted cleavage of DNA. Finally, the detailed molecular mechanism of the cleavage of DNA. target gene cleavage reaction was elucidated using CRISPR–Cas9 from *S. thermophilus* (Gasiunas et al. 2012) and also from *S. pyogenes* (Jinek et al. 2012). Soon after, the *S. pyogenes* CRISPR–Cas system was applied to genome editing in human neuronal and mouse kidney cells (Cong et al. 2013; Mali et al. 2013).

The reaction mechanism that supports the functioning of this system is schematically shown in Fig. 10.1. A complete CRISPR–*cas* locus consists of a CRISPR array, ranging from two to several hundred direct repeats of 25–35 bp each, and which are separated by spacers with a constant length (usually 30–40 bp). The contiguous cluster of *cas* genes are organized into one or more operons that often encode both "adaptation modules" and "effector modules" together with accessory genes. The CRISPR–Cas immune response involves three distinct, often intertwined, stages: (i) adaptation, where the invading DNA is fragmented and incorporated into the CRISPR spacer region, (ii) expression, where the CRISPR region is transcribed and pre-crRNA is made, and crRNA is made by processing the pre-crRNA, and (iii) interference, where the Effector–crRNA complex is recruted to the target gene to specifically cleave it (Fig. 10.1). During the adaptation phase, a complex of Cas proteins binds to a target DNA molecule and introduces two double-stranded (ds) breaks in the target DNA, usually after encountering a short (2–4 bp) sequence motif, termed called PAM (Protospacer-Adjacent Motif). The released segment, termed the protospacer, is inserted between two repeats in the CRISPR array (most often into the proximal repeat unit immediately following the leader sequence) to become a new spacer. The CRISPR array is then repaired by the cellular repair machinery, leading to duplication of the proximal repeat. Some CRISPR–Cas systems employ another adaptation mechanism, namely, obtaining the spacer from RNA (a transcript of the DNA genome of the mobile genetic elements (MGE)) via reverse transcription by reverse transcriptase (RT), which is encoded in the CRISPR–*cas* locus and is most often fused to the Cas1 protein (Silas et al. 2016; Silas et al. 2017; Li et al. 2014).

In the processing stage, the CRISPR array is expressed as a single long transcript, and the produced pre–crRNA is processed to generate mature crRNA by different complexes of Cas proteins; a dedicated processing nuclease (Cas6), a single large Cas protein; or an external RNase (Andersson and Banfield 2008;

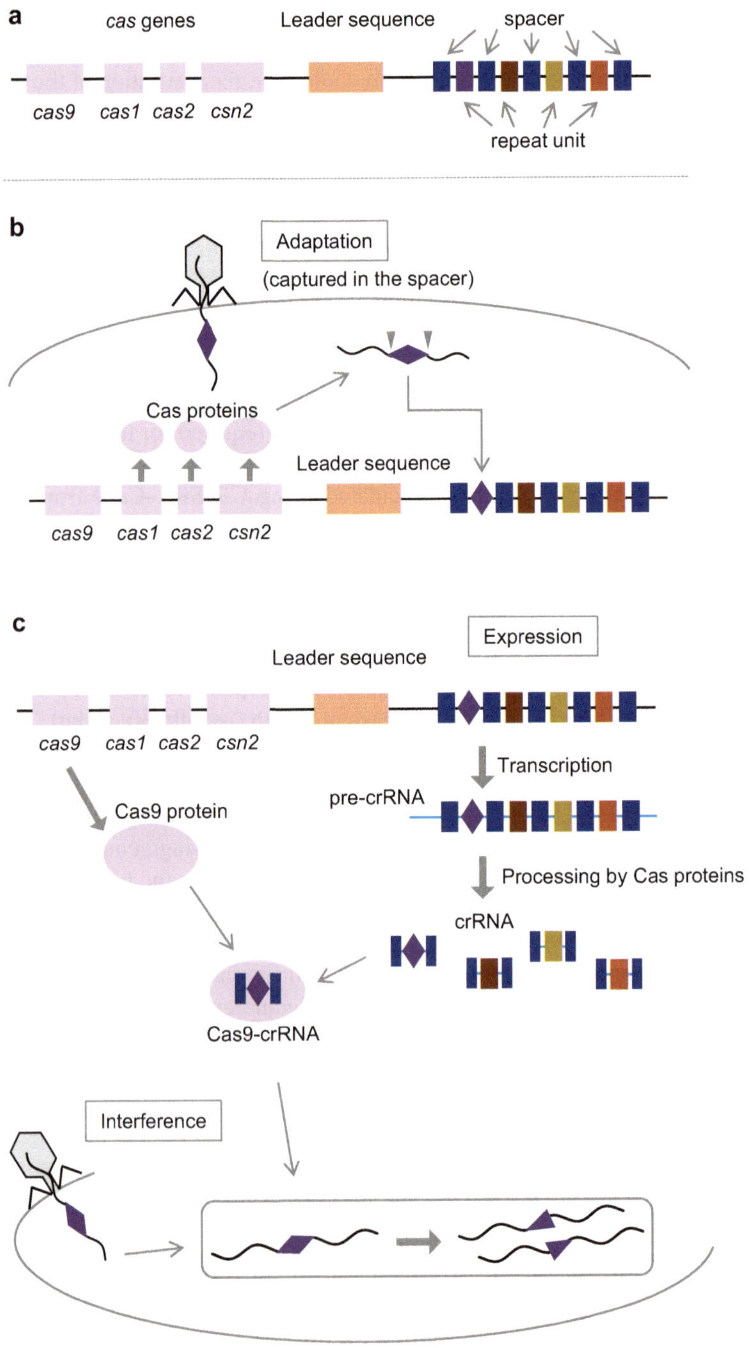

Fig. 10.1 Basic features of CRISPR–*cas*. (**a**) General gene organization of the CRISPR–*cas* region in the genome DNA. (**b**, **c**) Mechanism of action of CRISPR–*cas* adaptive immunity. (**b**) Adaptation process: Invading DNA is fragmented and incorporated into the CRISPR spacer region in the host cell genome for storage. (**c**) Expression and Cleavage processes: Transcription of the CRISPR region produces pre-crRNA, which is then processed into crRNA. Cleavage process: Foreign DNA is captured using homology in the crRNA sequence, and a complex with the Cas protein, which has nuclease activity, cleaves the DNA

Sapranauskas et al. 2011). In the final interference stage, the crRNA that remains bound to the processing complex is used as a guide (gRNA) to recognize a protospacer or a very similar sequence in the genome of the virus or plasmid. This sequence is then cleaved and inactivated by a Cas nuclease, which is either a component of the same effector complex or a separate Cas protein (Gasiunas et al. 2012; Jinek et al. 2012). This description is highly simplified, and more details can be found in numerous recent reviews on various aspects of CRISPR–Cas biology (Westra et al. 2016; Ishino et al. 2018; Makarova et al. 2020; Koonin et al. 2023; Pacesa et al. 2024).

10.3 Diversity and Classification of CRISPR-Cas

10.3.1 Distribution of CRISPR–Cas

When looking at the distribution of CRISPR, the content and distribution of CRISPR varies greatly even among closely related strains. For example, in the genus *Mycobacterium*, CRISPR is present in *M. tuberculosis* but not in *M. leprae*. On the other hand, phylogenetically distant *E. coli* and *M. avium*, as well as two archaea, *Methanothermobacter thermautotrophicus* and *Archaeoglobus fulgidus*, have nearly identical CRISPR repeat sequences (Jansen et al. 2002). The number of CRISPR arrays found in a single genome varies from 1 to 18, and the number of repeats in a single CRISPR array varies from 2 to more than 350 (Marraffini and Sontheimer 2010). In a previous CRISPR database (CRISPRdb), as of May 2017, CRISPR was predicted in the 202 whole genomes (87%) among 232 archaeal species and 3059 (45%) among 6782 bacterial species. However, a subsequent survey of 1724 draft genomes (Burstein et al. 2016) suggested that the prevalence of CRISPR–Cas systems in environmental microbial communities was very low, at 10.4% in bacteria and 10.1% in archaea. This large difference between the prevalence estimated from the whole genomes of cultured microorganisms and that of uncultured ones has been argued to be due to the lack of CRISPR–Cas systems in major representatives of the uncultured bacterial lineages (Burstein et al. 2016). CRISPRdb was subsequently closed, and its successor database, CRISPRCasdb, is now operative (https://crisprcas.i2bc.paris-saclay.fr/) (Pourcel et al. 2020). According to the report, as of 2022, the genomes of 72,104 species of bacteria and 1106 species of archaea have been sequenced, and 20,950 species (29.06%) of bacteria and 180 species of archaea (16.27%) are predicted to have CRISPR system.

10.3.2 Classification of CRISP-Cas

At both the structural and functional levels, CRISPR–Cas systems exhibit a defined modular organization. The two main components of a CRISPR–Cas system can be described as the "adaptation module" and the "effector module". In most CRISPR–Cas systems, the adaptation module is composed of the Cas1 and Cas2 proteins, which form a complex. Cas1 is the enzymatically active subunit, i.e., an endonuclease (integrase) responsible for cleaving DNA including the protospacer and the CRISPR array, while Cas2 forms the structural scaffold of the complex (Amitai and Sorek 2016; Wang et al. 2015; Nunez et al. 2014; Nunez et al. 2015). Reverse Transcriptase (RT), also contribute to the adaptation step, in some cases fused to Cas1 or Cas2, as described above (Silas et al. 2017; Li et al. 2014). In contrast to the relatively simple and uniform structure of the adaptation modules, effector modules are highly diverse among CRISPR–Cas systems, and this variation forms the basis of current CRISPR-Cas classification (Makarova et al. 2011, 2015; Shmakov et al. 2015; Koonin et al. 2017). In this section, we will look at the diversity of CRISPR–Cas.

Exploring the origin and evolutionary diversity of CRISPR–Cas molecules is an extremely interesting topic, but the rapid evolutionary diversity of CRISPR–Cas systems makes comprehensive classification difficult. Koonin et al. pointed out (2017) that because there is no universal *cas* gene, and that module recombination occurs frequently, it is not possible to draw an evolutionary tree with a single classification criterion. Therefore, a polythetic classification based on a systematic method for CRISPR–Cas was proposed, using a multifaceted approach that takes into account (i) signature *cas* genes specific to individual types and subtypes of CRISPR–Cas; (ii) sequence similarity between multiple common Cas proteins; (iii) the phylogeny of Cas1, the most conserved Cas protein; (iv) the arrangement of genes in the CRISPR–*cas* locus; and (v) the structure of CRISPR itself (Makarova et al. 2011, 2015).

In the classification method proposed by Koonin et al. (Makarova et al. 2015), CRISPR-Cas systems are first classified into two classes (class 1 and class 2) with different "effector module" configurations (Fig. 10.2). Approximately 90% of the identified CRISPR–*cas* loci are classified as a class 1 system, which functions as a multimeric effector complex composed of many different Cas proteins. The remaining 10% belong to the class 2 system, which uses a single multidomain effector protein and is found almost exclusively in the bacteria. Each class is then divided into three types: class 1 includes types I, III, and IV, and class 2 includes types II, V, and VI. Each type is further divided into multiple subtypes (type I, I-A–F and U; type III, III-A–D; type II, II-A–C, type V, V-A–E, and U; and type VI, VI-A–C) based on additional signature genes and characteristic gene sequences (Fig. 10.2).

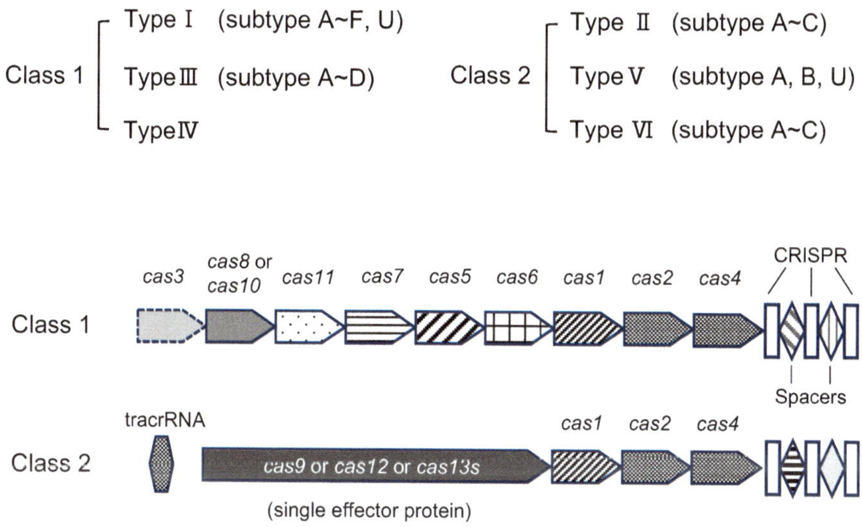

Fig. 10.2 Classification of CRISPR–Cas system. Based on detailed sequence analysis and genetic organization of the Cas proteins, CRISPR–Cas are classified into two classes. The major difference is the different organization of the effector Cas molecules. The effectors in class 1 are a complex composed of multiple Cas proteins. In contrast, the class 2 effectors are a single protein with multiple domains. Class 1 consists of the types I, III and type IV, while types II, V, VI belongs to class 2. The typical genetic organization of class 1 and 2 CRISPR loci is schematically shown

10.3.3 Class 1 Effectors

Within the Class 1 system, Type I is the most common with many examples and high diversity. On the other hand, Type III is present in many archaea but not in many bacteria. The multimeric effector complexes of Type I and Type III systems, known as Cascade (CRISPR–associated complex for antiviral defense) and the Csm/Cmr complex, respectively (Fig. 10.3), are structurally similar to each other, and although they are called by different names, they are expected to be evolutionarily related (Makarova et al. 2011, 2015 Rouillon et al. 2013; Jackson and Wiedenheft 2015; Venclovas 2016; Koonin et al. 2017). These effector complexes exhibit elaborate structures, with a backbone composed of Cas7 and Cas5 paralogs, which belong to RAMPs and contains an RNA Recognition Motif (RRM) fold and additional subunits. These effector complexes contain one Cas5 subunit and multiple Cas7 subunits. The complex accommodates a guide RNA (gRNA) that consists of a spacer and a portion of the repeat. The smaller subunit is usually present in multiple copies and interacts with the crRNA backbone bound to Cas7. Notably, the length of the bound spacer correlates with the number of Cas7 subunits in the backbone of the complex (Kuznedelov et al. 2016; Luo et al. 2016). Although the sequences of the protein subunits of type I and type III effector complexes show little sequence similarity to each other, the presence of homologous RAMPs in both types of complexes and the overall structural similarities revealed by cryo-electron

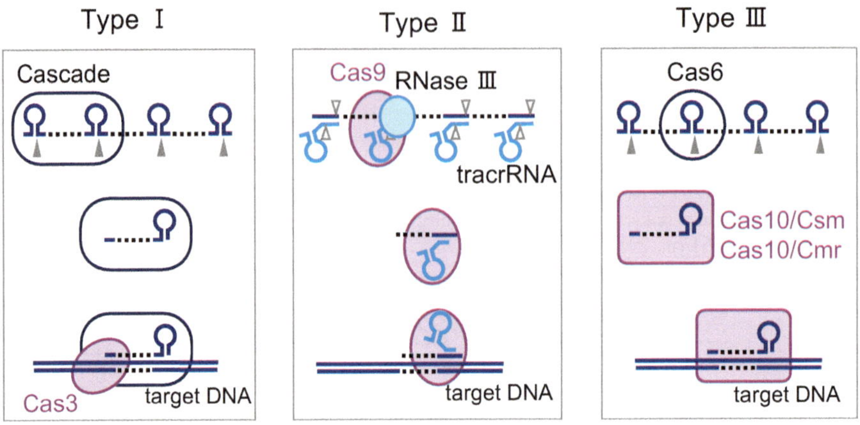

Fig. 10.3 Major differences in CRISPR mechanism of action in the Type I, II and III systems. CRISPR–Cas functions by the different mechanism of pre-crRNA processing and interference. In type I systems, Cascade, a complex of Cas proteins including Cas5, 6, 8 and Cas11, performs its function to process pre-crRNA, while Cas3 is needed to cleave target DNA. In type II system, RNase III is required to process pre-crRNA, and Cas9, a single protein effector is needed to cleave target DNA. In type III system, Cas6 acts in pre-crRNA processing, and Cas10/Csm or Cas10/Cmr multiprotein complexes are required for cleavage of target DNA and RNA, respectively

microscopy suggest that these complexes evolved from a common ancestor complex. An additional RAMP, Cas6, is loosely associated with effector complexes and normally functions as a repeat-specific RNase in pre-crRNA processing (Charpentier et al. 2015; Niewoehner and Jinek 2016) (Fig. 10.3).

Unlike other known CRISPR–Cas systems, the functionally uncharacterized type IV system does not contain an adaptive module consisting of nucleases Cas1 and Cas2, and furthermore, many of the genomes encoding type IV systems do not have identifiable CRISPR loci (Rouillon et al. 2013; Makarova et al. 2015). However, the effector module of the subtype III-B system utilizes a spacer generated by the type I system, which is located in a different position (Garrett et al. 2011). Type IV systems can also be considered as active CRISPR–Cas systems if they can utilize crRNAs produced from different CRISPR arrays (Makarova et al. 2015).

10.3.4 Class 2 Effectors

The Class 2 effector module is composed of a single large multidomain protein, so that each CRISPR–*cas* locus is much simpler and more uniform than those of class 1 (Fig. 10.2). In the current CRISPR–Cas classification (Makarova et al. 2015), class 2 included type II, which contains the effector endonuclease Cas9, a widely used genome editing enzyme, and type V, which contains the Cpf1 endonuclease (later called Cas12, a new name for the type V effector). Type V was experimentally uncharacterized at the time of its classification, but it was later proved that Cpf1 is in fact an active RNA-guided endonuclease with several different properties as compared with Cas9 (Zetsche et al. 2015).

To comprehensively analyze the diversity of class 2 CRISPR–Cas systems by genome and metagenomic mining, CRISPR–*cas* loci were identified either by the highly conserved *cas1* gene or by CRISPR arrays, and then the loci that could be classified into one of the known subtypes were subtracted. Selected loci encoding large proteins that are putative class 2 effectors were analyzed in detail (Shmakov et al. 2015). Three new type V subtypes containing a RuvC-like endonuclease domain and three type VI subtypes with two HEPN (higher eukaryotes and prokaryotes nucleotide-binding) domains predicted to have RNase activity were discovered (Shmakov et al. 2017). The interference activity of type V-B (Cas12b) (Shmakov et al. 2015), type VI-A (Cas13a) (Abudayyeh et al. 2016), and type VI-B (Cas13b) (Smargon et al. 2017) effectors was then experimentally verified.

In type II, pre-crRNA cleavage is catalyzed by RNase III, but as an effector, Cas9 requires a different RNA species, tracrRNA, which is encoded outside the CRISPR–*cas* locus. The sequence of tracrRNA is partially complementary to CRISPR and form a complex with crRNA (Charpentier et al. 2015). On the other hand, type V-A and type VI systems do not require tracrRNA, and the corresponding effectors Cas12 and Cas13 themselves have been shown to be fully involved in pre-crRNA processing (Liu et al. 2017; Fonfara et al. 2016; East-Seletsky et al. 2016; Zetsche et al. 2017) (Fig. 10.3).

A common feature of all type II and type V subtype effector molecules is the presence of a RuvC-like endonuclease domain with an RNase H fold. However, other parts of the Cas9 and Cas12 proteins of different subtypes do not show similarity to each other at the sequence or structural level (Fig. 10.4). Moreover, the sequence similarity between the RuvC-like domains of different subtypes is low. Therefore, they can only be recognized as homologous genes by comparison of conserved catalytic motifs and structures (Shmakov et al. 2015, 2017). Comparison of the structures of several Cas9 proteins (Jinek et al. 2014; Nishimasu et al. 2014, 2015; Hirano et al. 2016), three bacterial Cas12a (Cpf1) structures (Dong et al. 2016; Yamano et al. 2016; Gao et al. 2016; Swarts et al. 2017), and two Cas12b structures (originally called C2c1) in complex with guide RNA and target DNA (Liu et al. 2017; Yang et al. 2016) reveal a striking interplay of convergence and diversity (Fig. 10.4). All three effectors, Cas9, Cas12a, and Cas12b, are approximately the same size, but their structures are not superimposable except for the RuvC-like domain (Lewis and Ke 2017). The nuclease activity of the RuvC-like domain of Cas9 is complemented by the activity of RuvC-independent HNH family nucleases that insert into the RuvC-like domain (Fig. 10.4) and cleave the target (complementary to the gRNA) strand of the substrate DNA (Gasiunas et al. 2012; Jinek et al. 2012). Structural analysis of the CRISPR–Cas9 R-loop-containing target DNA complex shows that Cas9 induces a structural distortion of the DNA helix required for R-loop formation, which triggers a conformational change in Cas9 that

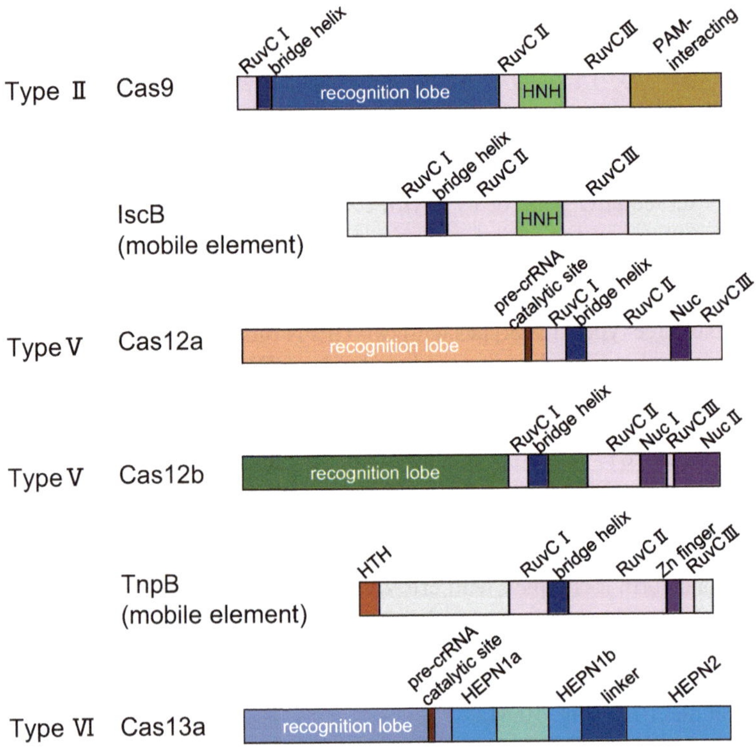

Fig. 10.4 General organization and domain architectures of class 2 effectors. The domain architectures of the effector proteins in each type of class 2 are schematically shown. Also shown for comparative purposes are the mobile genetic elements IscB and TnpB

positions the RuvC and HNH catalytic sites near the displaced and target strands, respectively (Jiang et al. 2016).

Compared with Cas9, the Cas12a protein lacks the HNH domain but contains a distinct domain with a unique fold inserted at a similar, but not identical, position in the RuvC-like domain (Fig. 10.4) (Lewis and Ke 2017). This domain (designated the Nuc domain of Cas12a) has been shown to be involved in target strand cleavage by Cas12a of the bacterium *Acidaminococcus* sp. (Yamano et al. 2016). The RuvC-like domain of Cas12b (C2c1) also contains an insertion domain with a unique fold, named Nuc, due to its similarity to Cas12a. However, the mechanism of action of Cas12b is different, as both target and non-target DNA strands have been shown to be accommodated and cleaved in the catalytic site of the RuvC-like domain (Yang et al. 2016). Later analysis of the structure and activity of Cas12a from *Francisella novicida* showed that mutations in the Nuc domain only partially inactivated target strand cleavage, suggesting that, like Cas12b, Cas12a cleaves both strands of the target DNA via the RuvC-like domain, and that the Nuc domain may only contribute to guide-target binding (Swarts et al. 2017). The details of target recognition and cleavage by different V-type effectors and the contributions of different domains are of great interest and are expected to progress.

Type VI effectors have come to be called Cas13, but the only recognizable common feature is the presence of two HEPN domains (Fig. 10.4). However, this domain shows even more significant sequence diversity among Cas13 proteins than the RuvC-like domains of Cas9 and Cas12, and therefore identification required extensive manual comparisons (Shmakov et al. 2015; Smargon et al. 2017). These observations are consistent with previous analyses showing that the HEPN superfamily is composed of highly diverse RNases involved primarily in various defense-related functions in both prokaryotes and eukaryotes (Anantharaman et al. 2013). Notably, HEPN RNases are present in the Toxin domain of numerous bacterial and archaeal Toxin-Antitoxin modules. The discovery that type VI putative effectors contain HEPN domains led to the prediction that this type would include dedicated CRISPR–Cas systems that target RNA (Shmakov et al. 2015) (RNA targeting has also been demonstrated in certain type III systems, but they also target DNA (Hale et al. 2012, 2014; Samai et al. 2015). This prediction has been verified by both type VI-A and type VI-B effectors (Cas13a and Cas13b, respectively), which have been characterized as programmable RNA-guided RNases that efficiently protect *E. coli* against infection by the RNA bacteriophage MS2 (Abudayyeh et al. 2016; Smargon et al. 2017). Structural analysis of Cas13a confirmed the presence of two HEPN domains and revealed another bilobed structure that is broadly similar to type II and type V effectors (Liu et al. 2017). In particular, in the case of type VI-B, the activity of the Cas13b protein was shown to be up- or down-regulated by small auxiliary Cas proteins encoded by the respective type VI-B variants (Smargon et al. 2017). Even more intriguing, Cas13a and Cas13b, when activated by target RNA, become promiscuous RNases that cleave RNA in a sequence-nonspecific manner (Abudayyeh et al. 2016; Smargon et al. 2017). Moreover, expression of Cas13 protein in concert with guide and target RNAs induces cytotoxicity in bacteria suggesting induction of dormancy or programmed cell death (Abudayyeh et al. 2016). This phenomenon seems to be consistent with previously proposed hypotheses on the correlation between immunity and dormancy induction and/or programmed cell death (PCD) in prokaryotes (Koonin and Zhang 2017; Makarova et al. 2012). Considering that RNA viruses have a narrow host range and occupy only a small portion of the prokaryotic virosphere, it seems likely that the type VI system targets DNA viruses, cleaves viral transcripts, and induces dormancy or PCD in response to strong expression of viral genes.

10.4 Evolution of CRISPR–Cas Molecules

10.4.1 Evolutionary Insights into CRISPR–Cas Molecules

As described above, CRISPR–Cas is an adaptive immune system that stores unique spacer sequences by inserting DNA sequences derived from MGEs into the CRISPR array as memories of encounters with foreign DNA such as MGEs. The spacer-containing CRISPR transcription RNA products can recognize and hybridize with nucleic acid strands with homologous sequences, so that Cas nucleases can be guided to specific target sites with homologous sequences upon new encounters with known MGEs. Thus, like all defense mechanisms, the CRISPR–Cas system has evolved in a perpetual arms race with MGEs, resulting in the rapid evolution of parts of the *cas* gene sequences, mainly the effector module components, and the remarkable diversification of the genetic composition and organization of the CRISPR-*cas* locus. This molecular diversity underlies the diversification of molecular mechanisms of CRISPR–mediated defense.

10.4.2 Origin of CRISPR–Cas

Compared to other defense systems in prokaryotes, the extraordinary complexity and diversity of the CRISPR–Cas system implies a complex evolutionary history of this molecule. Analysis of poorly characterized, narrowly distributed, and rapidly evolving gene clusters in archaeal genomes (called "dark matter islands" (Makarova et al. 2014) revealed the presence of multiple islands encoding Cas1 proteins (Cas1-solo) that are not associated with CRISPR *loci* (Makarova et al. 2013). A comprehensive survey of dark matter islands on the genome revealed that *cas1-solo* genes are always located in close proximity to genes encoding family B DNA polymerases and other conserved genes (Krupovic et al. 2014). Furthermore, these gene clusters were found to be surrounded by long inverted repeats and flanked by short direct repeats. These direct repeats resembled terminal inverted repeats (TIRs) and target site duplications (TSDs) characteristic of various transposable elements. However, none of the identified Cas1-solo-encoding genomic *loci* contained any known transposase or integrase genes. It was therefore hypothesized that Cas1 is the primary enzyme responsible for the transfer of these novel genetic elements, and these elements were named casposons (Krupovic et al. 2014). Casposons have been found to be widespread in the genomes of methanogenic archaea and *Thaumarchaeots* phyrum, as well as in various bacterial groups. A comparison of sequences from over 60 strains of *Methanosarcina mazei* showed that casposons have been inserted in multiple different sites in various ways, with gains and losses occurring, providing strong evidence that casposons have been transferred recently (Krupovic et al. 2016). Based on the gene content, taxonomic distribution, and phylogeny of the Cas1 protein, casposons can be classified into four families (Krupovic and Koonin 2016). Biochemical characterization of the casposon Cas1 (casposase) encoded in the genome of the thermophilic archaea *Aciduliprofundum boonei* confirmed the predicted integrase activity (Hickman and Dyda 2015; Béguin et al. 2016). Integration showed strong target site selectivity, and target site duplication regenerated the TSD observed in the *A. boonei* genome (Béguin et al. 2016). The latter feature is similar to the phenomenon of duplication of leader sequence-proximal CRISPR units upon protospacer integration catalyzed by the Cas1-Cas2 adaptation mechanism of CRISPR–Cas (Yosef et al. 2012; Rollie et al. 2015). Notably, the sequence features of the casposon target sites are functionally similar to those required for directional insertion of new protospacers into CRISPR arrays. In both systems, the functional target site consists of two elements: (1) a sequence that is duplicated upon integration of the introduced DNA duplex (TSD segment for casposon and CRISPR unit for protospacer integration), and (2) an upstream region that further determines the precise location of integration (leader sequence located upstream of the CRISPR array and TSD-proximal segment of the *A. boonei* genome) (Béguin et al. 2016).

Collectively, comparative genomics and experimental findings have strengthened the mechanistic similarities and evolutionary connections between casposons and the adaptive modules of prokaryotic adaptive immune systems, leading to an evolutionary scenario for the origin of the CRISPR–Cas system. It has been proposed that casposon insertion near "solo effector" innate immune loci and subsequent fixation of the ancestral casposon by inactivation of the TIR gave rise to the adaptive and effector modules, respectively, whereas CRISPR–repeats and leader sequences evolved directly from pre-existing casposon target sites (Koonin and Krupovic 2015; Krupovic et al. 2017). An important question in this scenario is how the substrate specificity of the ancestral casposase switched from the integration of defined casposon TIRs to the integration of essentially random and short (compared to the length of the casposon) protospacer sequences. It has been suggested that the coupling of Cas1 and Cas2 was crucial for this evolutionary transition (Krupovic et al. 2017). Similar target site specificity for casposon integration and CRISPR spacer integration has been demonstrated (Makarova et al. 2013). Currently identified Casposons do not code for Cas2, but some code for Cas4 and other nucleases (Murugan et al. 2017). It is likely that the entire adaptation module, and even additional Cas proteins, were contributed to the emerging CRISPR-Cas system by the ancestral Casposon (Krupovic et al. 2014). Furthermore, the prototypic CRISPR repeats and leader sequences may have been derived from the duplicated target sites of Casposon (Krupovic et al. 2016).

The ancestry of the effector module is unclear. Given that class 1 CRISPR–Cas systems are widespread in archaea and bacteria, whereas class 2 systems are much less abundant, it is likely that class 1 multisubunit effector complexes are ancestral (Shmakov et al. 2015). The core subunits of class 1 effector complexes contain divergent versions of RNA recognition motifs (RRMs). This fact may suggest that the effector module evolved by successive duplications of ancestral RRM proteins, followed by diversification driven by the host–enemy arms race (Makarova et al. 2013). The ancestral effector complexes before Casposon recruitment may have functioned as an RNA-guided innate immune system (Koonin and Krupovic 2015), such as the Argonaute-dependent immunity (Swarts et al. 2014).

The effector proteins of Type II and Type V evolved independently from a different group of small transposons that provided the corresponding RuvC-like nuclease domains (Shmakov et al. 2015, 2017). To think about the origin of the Class 2 effector modules, it is important to see that the sequences of the RuvC-like domains of Cas9 and Cas12 are more similar to the sequences of the transposon-encoded TnpB proteins than to each other (Fig. 10.4). The *tnpB* gene is highly abundant in bacterial and archaeal genomes. TnpB is not essential for transposition, and its role in the transposon life cycle is unclear (Pasternak et al. 2013), but the conservation of the catalytic site features of the RuvC-like endonuclease is conserved in most TnpB sequences suggests that these proteins are active nucleases. Cas9 in Type II and Cas12a, 12b, and 12c in Type V show the highest similarity to different groups of TnpB proteins, respectively, suggesting an independent origin of the different types and subtypes of class 2 effectors. The ancestry of Cas9 could be easily traced to the corresponding transposon family (denoted ISC for Insertion Sequences Cas9-related) thanks to the shared domain structure of IscB and Cas9 proteins, in which the HNH endonuclease domain is inserted into the RuvC-like domain (Chylinski et al. 2014; Kapitonov et al. 2015). On the other hand, assigning a specific ancestry of Cas12a–c to which transposon family remains unclear due to low sequence conservation (Shmakov et al. 2015). The descent of type V effectors from TnpB was revealed by identification of subtype V-U (U for uncharacterized) loci using CRISPR arrays as anchors in a computational search for the class 2 CRISPR– Cas (Shmakov et al. 2017).

While other class 2 effectors are large proteins that show limited similarity to TnpB within the RuvC-like domain, type V-U loci encode TnpB homologs that are the same size or slightly larger than transposon-encoded TnpB. In addition, the sequence showed high similarity. These high similarity of putative V-U effectors to TnpB allows a reliable phylogenetic analysis and supports the independent origin of class 2 effectors into distinct TnpB subfamilies. Although the functionality of type V-U systems remains to be demonstrated, they are likely to be at least partially active, since each CRISPR array contains a spacer homologous to phage genome sequences. The current evolutionary scenario for type II and type V effectors involves multiple insertions of nonautonomous transposons encoding TnpB adjacent to CRISPR arrays, followed by convergent "domestication" of TnpB and the acquisition of additional domains (Shmakov et al. 2017). The domains of the different subtypes are unrelated, but in all cases, they allow the effector protein to accommodate the crRNA guide complexed with the target DNA.

As with type V effectors, it is not easy to pinpoint the exact ancestor of type VI effectors. The Cas proteins containing HEPN domain, such as Csm6 and Csx1 from class 1 seem to be candidates (Shmakov et al. 2017; Jiang et al. 2016). There is no doubt that these Cas proteins evolved from HEPN-containing RNase toxins. Thus, the evolutionary history of class 2 systems is primarily the second story of the contribution of mobile elements to the evolution of CRISPR–Cas adaptive immunity, after casposons. Of all the seemingly independent cases of nuclease recruitment from mobile elements giving rise to class 2 systems, the capture of IscB giving rise to type II was by far the most successful one as compared with other types (Shmakov et al. 2017). Whether type II systems confer any distinct advantage to bacteria or are somehow more prone to horizontal gene transfer than other class 2 systems is an intriguing question for future research into CRISPR biology.

A protein containing a bridge helix (BH), connecting the REC lobe and NUC lobe, and two nuclease domains (RuvC and HNH), was found to be encoded by the IS200/605 transposon. This protein, called IscB (insertion sequences Cas9-like ORF B), was presumed to be the ancestor of Cas9 (Kapitonov et al. 2015). IsrB

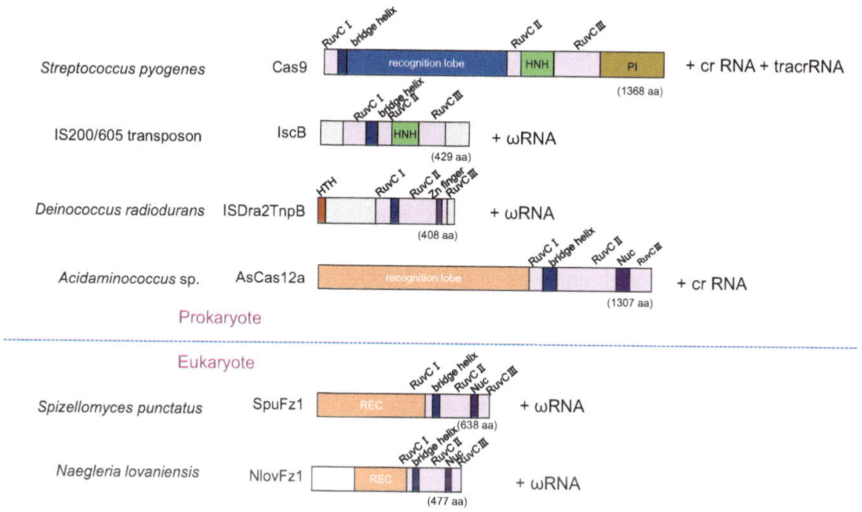

Fig. 10.5 OMEGA is a CRISPR–Cas-like RNA-programmable nucleases. Three transposon-encoded proteins, IscB, IsrB, and TnpB, are naturally RNA-programmable nucleases, and are named OMEGA (Obligate Mobile Element Guided Activity). ωRNA encoded in the adjacent region as a ncRNA, has guide RNA functions. This CRISPR–Cas-like system also exists in eukaryotic algae, in which the nuclease is called Fanzor

(insertion sequences RuvC-like ORF B), which has a RuvC domain and BH, was also discovered from this transposon. The relationship and function of IscB and IsrB with non-coding RNAs have been elucidated (Altae-Tran et al. 2021) (Fig. 10.5). The *iscB* gene was discovered with a CRISPR array, in which crRNA is encoded nearby, and this gene product was named CasIscB (CRISPR-associated IscB). CasIscB associates with non-coding RNA expressed from a site between the CRISPR array and the Isc-coding gene and this non-coding RNA-IscB complex cleaves target DNA just like crRNA-Cas9. In the *iscB* locus without a CRISPR array, an RNA associated with IscB is encoded in the terminal region of the transposon in the direction to *iscB*, and this can be expressed in microorganisms that have the *iscB* gene. This non-coding RNA was named ωRNA. CasIscB RNA and ωRNA have similar secondary structures, and the double-stranded region near the guide sequence is particularly conserved (Weinberg et al. 2009). IscB can cleave any DNA by adding a guide sequence to ωRNA (Altae-Tran et al. 2021). Such an RNA-dependent endonuclease encoded by the transposon together with ωRNA was named the OMEGA (obligate mobile element guided activity) system. To investigate the evolutionary relationship between the CRISPR–Cas and OMEGA systems, phylogenetic analysis of IscB, IsrB, and Cas9 has been done and found that the three nucleases arose from a common evolutionary lineage (Altae-Tran et al. 2021). The OMEGA protein is smaller than the Cas effector, and it can be expected to be easier easier to incorporate into gene therapy vectors such as adeno-associated viruses.

Although a wide variety of genome editing enzymes have been found, all of which are derived from prokaryotes, and it was a major mystery whether eukaryotes also have programmable RNA-guided DNA cleavage enzymes. In 2013, Bao et al. reported that a protein homologous to TnpB exists on the transposons of eukaryotic unicellular organisms such as fungi and amoebas, and it was named Fanzor (Fz) (Bao and Jurka 2013). At that time, the RNA-guided DNA cleavage activity of TnpB was not known, and it was discussed whether Fz controls the transposition of transposons by an epigenetic enzyme activity such as DNA methyltransferase (Bao and Jurka 2013). Through analysis of the OMEGA system, it was hypothesized that Fz may also be involved in OMEGA group, and it was confirmed that Fz is indeed an RNA-guided DNA cleavage enzyme (Saito et al. 2023; Jiang et al. 2023; Yoon et al. 2023) (Fig. 10.5). Ten years after discovery, Fz suddenly attracted attention as the first CRISPR–Cas-like enzyme in eukaryotes. Fz has been found in mollusks and insects, and it is interesting to see if more higher eukaryotes also have OMEGA system drived by Fz.

10.4.3 Functions Other Than the Adaptive Immune System

CRISPR–Cas systems are strongly suggested to be derived from analogues of MGEs (transposons), toxin and antitoxin systems, and host repair, recombination, and sensing systems (nucleases, helicases, and polymerases) (Makarova et al. 2020; Mohanraju et al. 2016; Koonin and Makarova 2019), as described above. In the adaptive immune system driven by CRISPR–Cas system, CRISPR array adaptation occurs when virus- and plasmid-derived sequences are incorporated as new CRISPR spacers. Cas proteins use CRISPR–derived RNA guides to specifically recognize and cleave the nucleic acids of invading mobile genetic elements. In addition to this most recognized function, other roles of CRISPR–Cas have been reported (Mohanraju et al. 2022). Viruses use some CRISPR-associated nucleases to attack host cells. Transposons are induced to transpose by CRISPR systems. In addition, some CRISPR–Cas systems control the gene expressions involved in bacterial physiology and virulence. Furthermore, pathogenic bacteria may indirectly use Cas nuclease activity to evade the human immune system or directly enter the nucleus to damage chromosomal DNA in infected human cells. Thus, the arms race between host and enemy has led to an exciting expansion of CRISPR function and mechanistic variation. It has led to the creation of new CRISPR–Cas systems with large variations in CRISPR-Cas composition, mechanism, and function (Makarova et al. 2020).

Although CRISPR–Cas systems share the ability of cleaving nucleic acid strands in a target sequence-specific manner using an RNA-guided protein complex, many CRISPR–Cas types involve an accompanying RNase or DNase activity without apparent sequence specificity, either by the effector Cas nuclease itself or by activation of auxiliary Cas nucleases. This accessory activity does not distinguish between self and non-self, and may cause the host cell to become growth-arrested, dormant, or even exert a lethal effect (Fig. 10.6). These functions run counter to the main task

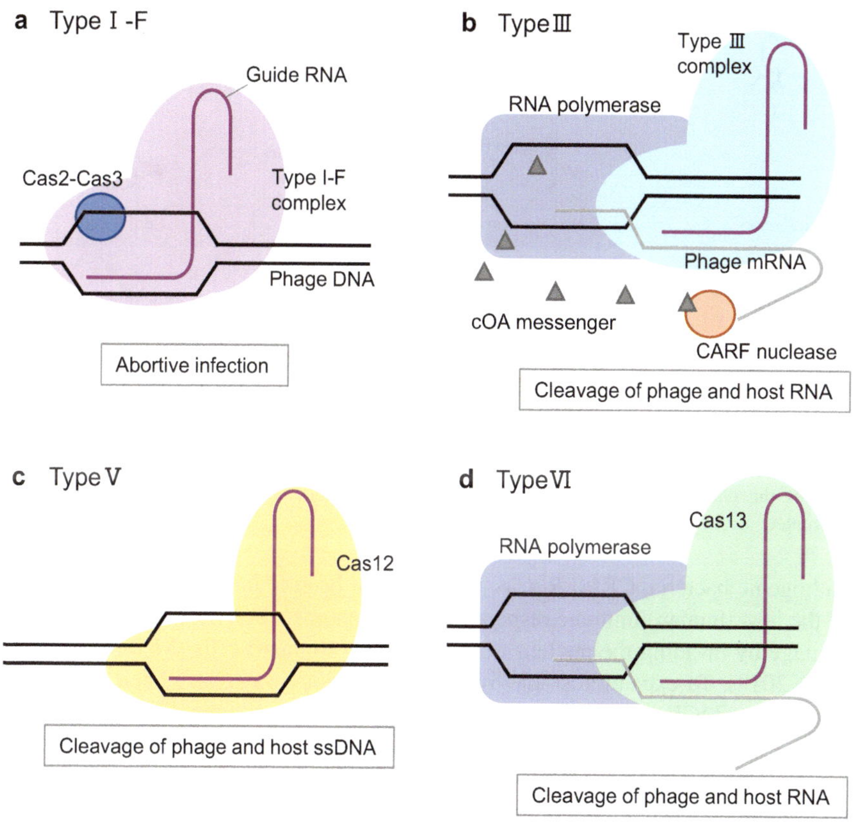

Fig. 10.6 Collateral nucleases associated with CRISPR–Cas systems. (**a**) Type I-F effector complex, bound to its target DNA, recruits the nuclease Cas2–Cas3 to degrade viral DNA and triggers abortive infection. The detailed mechanism has not been elucidated. (**b**) Type III effector complex, bound to the viral RNA produced by transcription of the viral DNA produces cyclic oligoadenylate (cOA) messenger molecules that allosterically activate CARF (CRISPR-associated Rossman fold) nuclease, which cleaves both phage and host RNAs non-specifically. (**c**) Cas12, the type V effectors, bound to its target DNA, induces a conformational change that traps Cas12 in an active state that can engage in non-specific cleavage of single-stranded DNAs from virus or host cells. (**d**) Cas13 binds the transcribed viral RNA, and induces a conformational change forming an accessible active site of Cas13, cleaving both phage and host RNAs

of CRISPR–Cas, which is to act as an immune system to protect the host cells from viral infection, and appear to serve to protect the population rather than the individual, preventing the generation of new phages at the expense of the infected host, thereby protecting neighboring cells. It is similar to an "abortive infection" process (Lopatina et al. 2020).

CRISPR–Cas participates in controlling gene expression. In many recent examples of CRISPR–Cas, a general trend seen in both class 1 and class 2 systems, the binding of guide RNA to target nucleic acids by base pairing allows CRISPR–Cas to exert its functions in controlling bacterial collective behavior. For example, in

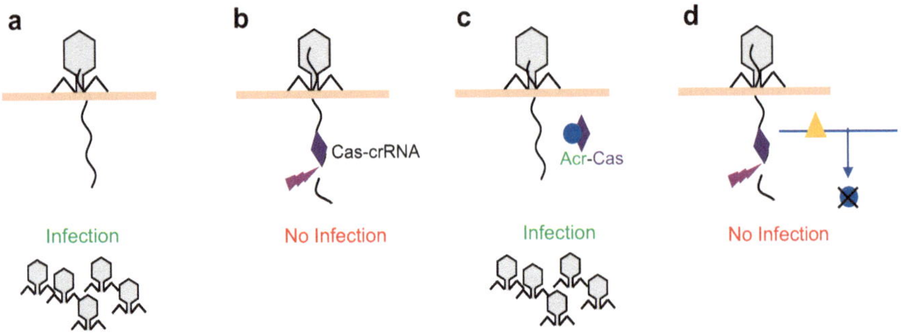

Fig. 10.7 Evolutionary arms race between phages and host cells. (**a**) Lytic phage infects host bacterial cells. Most bacterial cells will undergo lysis on completion of the phage lytic cycle. (**b**) Phage cannot infect the cells harbouring an active CRISPR–Cas system, which cleaves the invading phage genomes. (**c**) Phage encoding Acr proteins (anti-CRISPR proteins) prevents Cas proteins from binding to or cleaving phage DNA. (**d**) Some bacteria encode repressors that silence expression of the phage gene encoding the Acr proteins, and infection by Acr-encoding phages is prevented

pathogenic bacteria, CRISPR–Cas systems have been found to coordinate evasion of the host (human) immune response to infection or to directly cause severe DNA damage by invading the nucleus of infected host (human) cells (Medina-Aparicio et al. 2018). To expand host-enemy conflicts, various ways have been known for phages and MGEs to evade or thwart CRISPR-based immunity. Phages can passively counter CRISPR–Cas defenses by introducing point mutations or deleting CRISPR target regions (Deveau et al. 2010; Semenova et al. 2011). In addition, they can actively counter CRISPR defenses by using anti-CRISPR molecules (Acr proteins) (Bondy-Denomy et al. 2013; Bondy-Denomy et al. 2015; Trasanidou et al. 2019) (Fig. 10.7). More than 90 different Acr protein families (Bondy-Denomy et al. 2018) have been identified that inhibit type I, II, III, V, and VI CRISPR–Cas systems. Their actions are diverse and can include directly binding to Cas proteins to prevent phage DNA binding and cleavage; block guide RNA loading; induce nonspecific DNA binding, cause degradation of Cas proteins, or degrade second messenger molecules. Another interesting mechanism to prevent MGEs from inactivating the CRISPR–Cas system has been found in some archaeal type I-B systems (Li et al. 2021). This system not only provides adaptive immunity, but also acts as a transcriptional repressor of the toxin-antitoxin system. In this system, the Cascade complex uses the antitoxin RNA as a partially complementary guide to repress the expression of toxin genes. Therefore, if the transposon is integrated within an operon encoding Cascade to inhibit production of Cascade, the toxin RNA will induce dormancy and ultimately cell death (Li et al. 2021). Thus, counteracting the transposon activity that causes “anti-CRISPR” in this case can be considered an anti-anti-CRISPR function (Fig. 10.7).

10.5 Future Perspective

The process of the adaptive immunity in prokaryotes is driven by the CRISPR–Cas function, which has been gradually elucidated over the past decade. As mentioned in this chapter, CRISPR–Cas systems are confined to a relatively narrow range of MGEs or prokaryotes, and therefore, they evolved relatively recently and likely have specialized functions. These systems have arisen in the "exotic" bacteria or archaea as host cells, making their biological characterizations difficult. There is no doubt many unknown derivatives are still buried in nature. The complexity of the function of the CRISPR–Cas system represents the depth of the basic biology of this system and will continue to be one of the most interesting research areas in life science. In particular, the aspects of microbial biology that drive its evolution remain largely unknown, and studies of the diversity of CRISPR–Cas systems including eukaryotic OMEGA systems, will continue to provide fascinating discoveries in both basic science and applied technology for many years to come.

References

Abudayyeh OO, Gootenberg JS, Konermann S, Joung J, Slaymaker IM, Cox DB, Shmakov S, Makarova KS, Semenova E, Minakhin L et al (2016) C2c2 is a single-component programmable RNA-guided RNA-targeting CRISPR effector. Science 353:aaf5573

Altae-Tran H, Kannan S, Demircioglu FE, Oshiro R, Nety SP, McKay LJ, Dlakić M, Inskeep WP, Makarova KS, Macrae RK et al (2021) The widespread IS200/IS605 transposon family encodes diverse programmable RNA-guided endonucleases. Science 374:57–65

Amitai G, Sorek R (2016) CRISPR–Cas adaptation: insights into the mechanism of action. Nat Rev Microbiol 14:67–76

Anantharaman V, Makarova KS, Burroughs AM, Koonin EV, Aravind L (2013) Comprehensive analysis of the HEPN superfamily: identification of novel roles in intra-genomic conflicts, defense, pathogenesis and RNA processing. Biol Direct 8:15

Andersson AF, Banfield JF (2008) Virus population dynamics and acquired virus resistance in natural microbial communities. Science 320:1047–1050

Bao W, Jurka J (2013) Homologues of bacterial TnpB_IS605 are widespread in diverse eukaryotic transposable elements. Mob DNA 4:12

Barrangou R, Fremaux C, Deveau H, Richards M, Boyaval P, Moineau S, Romero DA, Horvath P (2007) CRISPR provides acquired resistance against viruses in prokaryotes. Science 315:1709–1712

Béguin P, Charpin N, Koonin EV, Forterre P, Krupovic M (2016) Casposon integration shows strong target site preference and recapitulates protospacer integration by CRISPR–Cas systems. Nucleic Acids Res 44:10367–10376

Bolotin A, Quinquis B, Sorokin A, Ehrlich SD (2005) Clustered regularly interspaced short palindrome repeats (CRISPRs) have spacers of extrachromosomal origin. Microbiology 151:2551–2561

Bondy-Denomy J, Davidson AR, Doudna JA, Fineran PC, Maxwell KL, Moineau S, Peng X, Sontheimer EJ, Wiedenheft B (2018) A unified resource for tracking anti-CRISPR names. CRISPR J 1:304–305

Bondy-Denomy J, Garcia B, Strum S, Du M, Rollins MF, Hidalgo-Reyes Y, Wiedenheft B, Maxwell KL, Davidson AR (2015) Multiple mechanisms for CRISPR-Cas inhibition by anti-CRISPR proteins. Nature 526:136–139

Bondy-Denomy J, Pawluk A, Maxwell KL, (2013) Bacteriophage genes that inactivate the CRISPR/Cas bacterial immune system. Nature 493:429–432

Brouns SJ, Jore MM, Lundgren M, Westra ER, Slijkhuis RJ, Snijders AP, Dickman MJ, Makarova KS, Koonin EV, van der Oost J (2008) Small CRISPR RNAs guide antiviral defense in prokaryotes. Science 321:960–964

Burstein D, Sun CL, Brown CT, Sharon I, Anantharaman K, Probst AJ, Thomas BC, Banfield JF (2016) Major bacterial lineages are essentially devoid of CRISPR-Cas viral defence systems. Nat Commun 7:10613

Charpentier E, Richter H, van der Oost J, White MF (2015) Biogenesis pathways of RNA guides in archaeal and bacterial CRISPR-Cas adaptive immunity. FEMS Microbiol Rev 39:428–441

Chylinski K, Makarova KS, Charpentier E, Koonin EV (2014) Classification and evolution of type II CRISPR-Cas systems. Nucleic Acids Res 42:6091–6105

Cong L, Ran FA, Cox D, Lin S, Barretto R, Habib N, Hsu PD, Wu X, Jiang W, Marraffini LA, Zhang F (2013) Multiplex genome engineering using CRISPR/Cas systems. Science 339:819–823

Deveau H, Garneau JE, Moineau S (2010) CRISPR-Cas system and its role in phage- bacteria interactions. Ann Rev Microbiol 64:475–493

Dong D, Ren K, Qiu X, Zheng J, Guo M, Guan X, Liu H, Li N, Zhang B, Yang D, Ma C et al (2016) The crystal structure of Cpf1 in complex with CRISPR RNA. Nature 532:522–526

East-Seletsky A, O'Connell MR, Knight SC, Burstein D, Cate JH, Tjian R, Doudna JA (2016) Two distinct RNase activities of CRISPR-C2c2 enable guide-RNA processing and RNA detection. Nature 538:270–273

Fonfara I, Richter H, Bratovic M, Le Rhun A, Charpentier E (2016) The CRISPR-associated DNA-cleaving enzyme Cpf1 also processes precursor CRISPR RNA. Nature 532:517–521

Gao P, Yang H, Rajashankar KR, Huang ZD, Patel J (2016) Type V CRISPR-Cas Cpf1 endonuclease employs a unique mechanism for crRNA-mediated target DNA recognition. Cell Res 26:901–913

Garrett RA, Shah SA, Erdmann S, Liu G, Mousaei M, León-Sobrino C, Peng W, Gudbergsdottir S, Deng L, Vestergaard G, Peng X, She Q (2011) CRISPR-Cas adaptive immune systems of the *Sulfolobales*: unravelling their complexity and diversity. Lifestyles 5:783–817

Gasiunas G, Barrangou R, Horvath P, Siksnys V (2012) Cas9-crRNA ribonucleoprotein complex mediates specific DNA cleavage for adaptive immunity in bacteria. Proc Natl Acad Sci USA 109:E2579–E2586

Groenen PM, Bunschoten AE, van Soolingen D, van Embden JD (1993) Nature of DNA polymorphism in the direct repeat cluster of *Mycobacterium tuberculosis*; application for strain differentiation by a novel typing method. Mol Microbiol 10:1057–1065

Hale CR, Cocozaki A, Li H, Terns RM, Terns MP (2014) Target RNA capture and cleavage by the Cmr type III-B CRISPR-Cas effector complex. Genes Dev 28:2432–2443

Hale CR, Majumdar S, Elmore J, Pfister N, Compton M, Olson S, Resch AM, Glover CV, Graveley BR 3rd, Terns RM, Terns MP (2012) Essential features and rational design of CRISPR RNAs that function with the Cas RAMP module complex to cleave RNAs. Mol Cell 45:292–302

Hickman AB, Dyda F (2015) The casposon-encoded Cas1 protein from Aciduliprofundum boonei is a DNA integrase that generates target site duplications. Nucleic Acids Res 43:10576–10587

Hirano H, Gootenberg JS, Horii T, Abudayyeh OO, Kimura M, Hsu PD, Nakane T, Ishitani R, Hatada I, Zhang F, Nishimasu H, Nureki O (2016) Structure and engineering of *Francisella novicida* Cas9. Cell 164:950–961

Ishino Y, Krupovic M, Forterre P (2018) History of CRISPR-Cas from encounter with a mysterious repeated sequence to genome editing technology. J Bacteriol 12:200. pii: JB.00580-17

Ishino Y, Shinagawa H, Makino K, Amemura M, Nakata A (1987) Nucleotide sequence of the *iap* gene, responsible for alkaline phosphatase isozyme conversion in *Escherichia coli,* and identification of the gene product. J Bacteriol 169:5429–5433
Jackson RN, Wiedenheft B (2015) A conserved structural chassis for mounting versatile CRISPR RNA-guided immune responses. Mol Cell 58:722–728
Jansen R, Embden JD, Gaastra W, Schouls LW (2002) Identification of genes that are associated with DNA repeats in prokaryotes. Mol Microbiol 43:1565–1575
Jiang K, Lim J, Sgrizzi S, Trinh M, Kayabolen A, Yutin N, Bao W, Kato K, Koonin EV, Gootenberg JS et al (2023) Programmable RNA-guided DNA endonucleases are widespread in eukaryotes and their viruses. Sci Adv 9(Issue 39):eadk0171
Jiang F, Taylor DW, Chen JS, Kornfeld JE, Zhou K, Thompson AJ, Nogales E, Doudna JA (2016) Structures of a CRISPR-Cas9 R-loop complex primed for DNA cleavage. Science 351:867–871
Jinek M, Chylinski K, Fonfara I, Hauer M, Doudna JA, Charpentier EA (2012) A programmable dual-RNA-guided DNA endonuclease in adaptive bacterial immunity. Science 337:816–821
Jinek M, Jiang F, Taylor DW, Sternberg SH, Kaya E, Ma E, Anders C, Hauer M, Zhou K, Lin S, Kaplan M et al (2014) Structures of Cas9 endonucleases reveal RNA-mediated conformational activation. Science 343:1247997
Kapitonov VV, Makarova KS, Koonin EV (2015) ISC, a novel group of bacterial and archaeal DNA transposons that encode Cas9 homologs. J Bacteriol 198:797–807
Koonin EV, Gootenberg JS, Abudayyeh OO (2023) Discovery of diverse CRISPR-Cas systems and expansion of the genome engineering toolbox. Biochemistry 62:3465–3487
Koonin EV, Krupovic M (2015) Evolution of adaptive immunity from transposable elements combined with innate immune systems. Nat Rev Genet 16:184–192
Koonin EV, Makarova KS (2019) Origins and evolution of CRISPR-Cas systems. Philos Trans R Soc B 374:20180087
Koonin EV, Makarova KS, Zhang F (2017) Diversity, classification and evolution of CRISPR-Cas systems. Curr Opin Microbiol 37:67–78
Koonin EV, Zhang F (2017) Coupling immunity and programmed cell suicide in prokaryotes: life-or-death choices. BioEssays 39:1–9
Krupovic M, Béguin P, Koonin EV (2017) Casposons: mobile genetic elements that gave rise to the CRISPR-Cas adaptation machinery. Curr Opin Microbiol 38:36–43
Krupovic M, Koonin EV (2016) Self-synthesizing transposons: unexpected key players in the evolution of viruses and defense systems. Curr Opin Microbiol 31:25–33
Krupovic M, Makarova KS, Forterre P, Prangishvili D, Koonin EV (2014) Casposons: a new superfamily of self-synthesizing DNA transposons at the origin of prokaryotic CRISPR-Cas immunity. BMC Biol 12:36
Krupovic M, Shmakov S, Makarova KS, Forterre P, Koonin EV (2016) Recent mobility of casposons, self-synthesizing transposons at the origin of the CRISPR-Cas immunity. Genome Biol Evol 8:375–386
Kuznedelov K, Mekler V, Lemak S, Tokmina-Lukaszewska M, Datsenko KA, Jain I, Savitskaya E, Mallon J, Shmakov S, Bothner B, Bailey S et al (2016) Altered stoichiometry *Escherichia coli* Cascade complexes with shortened CRISPR RNA spacers are capable of interference and primed adaptation. Nucleic Acids Res 44:10849–10861
Lewis KM, Ke A (2017) Building the class 2 CRISPR-Cas Arsenal. Mol Cell 65:377–379
Li M, Gong L, Cheng F, Yu H, Zhao D, Wang R, Wang T, Zhang S, Zhou J, Shmakov SA, Koonin EV, Xiang H (2021) Toxin- antitoxin RNA pairs safeguard CRISPR- Cas systems. Science 372:eabe5601
Li M, Wang R, Zhao D, Xiang H (2014) Adaptation of the *Haloarcula hispanica* CRISPR-Cas system to a purified virus strictly requires a priming process. Nucleic Acids Res 42:2483–2492
Liu L, Chen P, Wang M, Li X, Wang J, Yin M, Wang Y (2017) C2c1-sgRNA complex structure reveals RNA-guided DNA cleavage mechanism. Mol Cell 65:310–322
Lopatina A, Tal N, Sorek R (2020) Abortive infection: bacterial suicide as an antiviral immune strategy. Annu Rev Virol 7:371–384

Luo ML, Jackson RN, Denny SR, Tokmina-Lukaszewska M, Maksimchuk KR, Lin W, Bothner B, Wiedenheft B, Beisel CL (2016) The CRISPR RNA-guided surveillance complex in *Escherichia coli* accommodates extended RNA spacers. Nucleic Acids Res 44:7385–7394

Makarova KS, Anantharaman V, Aravind L, Koonin EV (2012) Live virus-free or die: coupling of antivirus immunity and programmed suicide or dormancy in prokaryotes. Biol Direct 7:40

Makarova KS, Aravind L, Grishin NV, Rogozin IB, Koonin EV (2002) A DNA repair system specific for thermophilic archaea and bacteria predicted by genomic context analysis. Nucleic Acids Res 30:482–496

Makarova KS, Aravind L, Wolf YI, Koonin EV (2011) Unification of Cas protein families and a simple scenario for the origin and evolution of CRISPR-Cas systems. Biol Direct 6:38

Makarova KS, Grishin NV, Shabalina SA, Wolf YI, Koonin EV (2006) A putative RNA-interference-based immune system in prokaryotes: computational analysis of the predicted enzymatic machinery, functional analogies with eukaryotic RNAi, and hypothetical mechanisms of action. Biol Direct 1:7

Makarova KS, Wolf YI, Alkhnbashi OS, Costa F, Shah SA, Saunders SJ, Barrangou R, Brouns SJ, Charpentier E, Haft DH, Horvath P et al (2015) An updated evolutionary classification of CRISPR-Cas systems. Nat Rev Microbiol 13:722–736

Makarova KS, Wolf YI, Forterre P, Prangishvili D, Krupovic M, Koonin EV (2014) Dark matter in archaeal genomes: a rich source of novel mobile elements, defense systems and secretory complexes. Extremophiles 18:877–893

Makarova KS, Wolf Y, Iranzo J, Shmakov SA, Alkhnbashi OS, Brouns SJJ, Charpentier E, Cheng D, Haft DH, Horvath P, Moineau S, Mojica FJM et al (2020) Evolutionary classification of CRISPR-Cas systems: a burst of class 2 and derived variants. Nat Rev Microbiol 18:67–83

Makarova KS, Wolf YI, Koonin EV (2013) The basic building blocks and evolution of CRISPR-Cas systems. Biochem Soc Trans 41:1392–1400

Mali P, Yang L, Esvelt KM, Aach J, Guell M, DiCarlo JE, Norville JE, Church GM (2013) RNA-guided human genome engineering via Cas9. Science 339:823–826

Marraffini LA, Sontheimer EJ (2008) CRISPR interference limits horizontal gene transfer in Staphylococci by targeting DNA. Science 322:1843–1845

Marraffini LA, Sontheimer EJ (2010) CRISPR interference: RNA-directed adaptive immunity in bacteria and archaea. Nat Rev Genet 11:181–190

Medina-Aparicio L, Davila S, Rebollar-Flores JE, Calva E, Hernandez-Lucas I (2018) The CRISPR-Cas system in Enterobacteriaceae. Pathog Dis 76:fty002

Mohanraju P, Makarova KS, Zetsche B, Zhang F, Koonin EV, van der Oost J (2016) Diverse evolutionary roots and mechanistic variations of the CRISPR-Cas systems. Science 353:aad5147

Mohanraju P, Saha C, van Baarlen P, Louwen R, Staals RHJ, van der Oost J (2022) Alternative functions of CRISPR-Cas systems in the evolutionary arms race. Nat Rev Microbiol 20:351–364

Mojica FJM, Díez-Villaseñor C, García-Martínez J, Soria E (2005) Intervening sequences of regularly spaced prokaryotic repeats derive from foreign genetic elements. J Mol Evol 60:174–182

Mojica MJ, Juez G, Rodríguez-Valera F (1993) Transcription at different salinities of *Haloferax mediterranei* sequences adjacent to partially modified PstI sites. Mol Microbiol 9:613–621

Murugan K, Babu K, Sundaresan R, Rajan R, Sashital DG (2017) The revolution continues: newly discovered systems expand the CRISPR-Cas toolkit. Mol Cell 68:15–25

Nakata A, Amemura M, Makino K (1989) Unusual nucleotide arrangement with repeated sequences in the *Escherichia coli* K-12 chromosome. J Bacteriol 171:3553–3556

Niewoehner O, Jinek M (2016) Structural basis for the endoribonuclease activity of the type III-A CRISPR-associated protein Csm6. RNA 22:318–329

Nishimasu H, Cong L, Yan WX, Ran FA, Zetsche B, Li Y, Kurabayashi A, Ishitani R, Zhang F, Nureki O (2015) Crystal structure of *Staphylococcus aureus* Cas9. Cell 162:1113–1126

Nishimasu H, Ran FA, Hsu PD, Konermann S, Shehata SI, Dohmae N, Ishitani R, Zhang F, Nureki O (2014) Crystal structure of Cas9 in complex with guide RNA and target DNA. Cell 156:935–949

Nunez JK, Kranzusch PJ, Noeske J, Wright AV, Davies CW, Doudna JA (2014) Cas1–Cas2 complex formation mediates spacer acquisition during CRISPR-Cas adaptive immunity. Nat Struct Mol Biol 21:528–534

Nunez JK, Lee AS, Engelman A, Doudna JA (2015) Integrase-mediated spacer acquisition during CRISPR-Cas adaptive immunity. Nature 527:535–538

Pacesa M, Pelea O, Jinek M (2024) Past, present, and future of CRISPR genome editing technologies. Cell 187:1076–1100

Pasternak C, Dulermo R, Ton-Hoang B, Debuchy R, Siguier P, Coste G, Chandler M, Sommer S (2013) ISDra2 transposition in *Deinococcus radiodurans* is downregulated by TnpB. Mol Microbiol 88:443–455

Pourcel C, Salvignol G, Vergnaud G (2005) CRISPR elements in *Yersinia pestis* acquire new repeats by preferential uptake of bacteriophage DNA, and provide additional tools for evolutionary studies. Microbiology 151:653–663

Pourcel C, Touchon M, Villeriot N, Vernadet J-P, Couvin D, Toffano-Nioche C, Vergnaud C (2020) CRISPRCasdb a successor of CRISPRdb containing CRISPR arrays and *cas* genes from complete genome sequences, and tools to download and query lists of repeats and spacers. Nucleic Acids Res 48(D1):08

Rollie C, Schneider S, Brinkmann AS, Bolt EL, White MF (2015) Intrinsic sequence specificity of the Cas1 integrase directs new spacer acquisition. Elife 4:e08716

Rouillon C, Zhou M, Zhang J, Politis A, Beilsten-Edmands V, Cannone G, Graham S, Robinson CV, Spagnolo L, White MF (2013) Structure of the CRISPR interference complex CSM reveals key similarities with cascade. Mol Cell 52:124–134

Samai P, Pyenson N, Jiang W, Goldberg GW, Hatoum-Aslan A, Marraffini LA (2015) Co-transcriptional DNA and RNA cleavage during Type III CRISPR-Cas immunity. Cell 161:1164–1174

Saito M, Xu P, Faure G, Maguire S, Kannan S, Altae-Tran H, Vo S, Desimone A, Macrae RK, Zhang F (2023) Fanzor is a eukaryotic programmable RNA-guided endonuclease. Nature 620:660–668

Sapranauskas R, Gasiunas G, Fremaux C, Barrangou R, Horvath P, Siksnys V (2011) The *Streptococcus thermophilus* CRISPR/Cas system provides immunity in *Escherichia coli*. Nucleic Acids Res 39:9275–9282

Semenova E, Jore MM, Datsenko KA, Semenova A, Westra ER, Wanner B, van der Oost J, Brouns SJJ, Severinovet K (2011) Interference by clustered regularly interspaced short palindromic repeat (CRISPR) RNA is governed by a seed sequence. Proc Natl Acad Sci USA 108:10098–10103

Shmakov S, Abudayyeh OO, Makarova KS, Wolf YI, Gootenberg JS, Semenova E, Minakhin L, Joung J, Konermann S, Severinov K, Zhang F et al (2015) Discovery and functional characterization of diverse class 2 CRISPR-Cas systems. Mol Cell 60:385–397

Shmakov S, Smargon A, Scott D, Cox D, Pyzocha N, Yan W, Abudayyeh OO, Gootenberg JS, Makarova KS, Wolf YI, Severinov K et al (2017) Diversity and evolution of class 2 CRISPR-Cas systems. Nat Rev Microbiol 15:169–182

Silas S, Makarova KS, Shmakov S, Páez-Espino D, Mohr G, Liu Y, Davison M, Roux S, Krishnamurthy SR, Fu BXH et al (2017) On the origin of reverse transcriptase-using CRISPR-Cas systems and their hyperdiverse, enigmatic spacer repertoires. MBio 8:e00897–e00817

Silas S, Mohr G, Sidote DJ, Markham LM, Sanchez-Amat A, Bhaya D, Lambowitz AM, Fire AZ (2016) Direct CRISPR spacer acquisition from RNA by a natural reverse transcriptase-Cas1 fusion protein. Science 351:aad4234

Smargon AA, Cox DB, Pyzocha NK, Zheng K, Slaymaker IM, Gootenberg JS, Abudayyeh OA, Essletzbichler P, Shmakov S, Makarova KS, Koonin EV, Zhang F (2017) Cab is a type VI-B CRISPR-associated RNA-guided RNase differentially regulated by accessory proteins Csx27 and Csx 28. Mol Cell 65:618–630

Swarts DC, Jore MM, Westra ER, Zhu Y, Janssen JH, Snijders AP, Wang Y, Patel DJ, Berenguer J, Brouns SJJ, van der Oost J (2014) DNA-guided DNA interference by a prokaryotic Argonaute. Nature 507:258–261

Swarts DC, van der Oost J, Jinek M (2017) Structural basis for guide RNA processing and seed-dependent DNA targeting by CRISPR-Cas12a. Mol Cell 66:221–233

Trasanidou D, Gerós AS, Mohanraju P, Nieuwenweg AC, Nobrega FL, Staals RHJ (2019) Keeping CRISPR in check: diverse mechanisms of phage-encoded anti-CRISPRS. FEMS Microbiol Lett 366:fnz098

Venclovas Č (2016) Structure of Csm2 elucidates the relationship between small subunits of CRISPR-Cas effector complexes. FEBS Lett 590:1521–1529

Wang J, Li J, Zhao H, Sheng G, Wang M, Yin M, Wang Y (2015) Structural and mechanistic basis of PAM-dependent spacer acquisition in CRISPR-Cas systems. Cell 163:840–853

Weinberg Z, Perreault J, Meyer MM, Breaker RR (2009) Exceptional structured noncoding RNAs revealed by bacterial metagenome analysis. Nature 462:656–659

Westra ER, Dowling AJ, Broniewski JM, van Houte S (2016) Evolution and ecology of CRISPR *Annu.* Rev Ecol Evol Syst 7:307–331

Yamano T, Nishimasu H, Zetsche B, Hirano H, Slaymaker IM, Li Y, Fedorova I, Nakane T, Makarova KS, Koonin EV, Ishitani R et al (2016) Crystal structure of Cpf1 in complex with guide RNA and target DNA. Cell 165:949–962

Yang H, Gao P, Rajashankar KR, Patel DJ (2016) PAM-dependent target DNA recognition and cleavage by C2c1 CRISPR-Cas endonuclease. Cell 167:1814–1828

Yoon PH, Skopintsev P, Shi H, Chen L, Adler BA, Al-Shimary M, Craig RJ, Loi KJ, DeTurk EC, Li Z et al (2023) Eukaryotic RNA-guided endonucleases evolved from a unique clade of bacterial enzymes. Nucleic Acids Res 51:12414–12427

Yosef I, Goren MG, Qimron U (2012) Proteins and DNA elements essential for the CRISPR adaptation process in *Escherichia coli*. Nucleic Acids Res 40:5569–5576

Zetsche B, Gootenberg JS, Abudayyeh OO, Slaymaker IM, Makarova KS, Essletzbichler KS, Volz SE, Joung J, van der Oost J, Regev A, Koonin EV, Zhang F (2015) Cpf1 is a single RNA-guided endonuclease of a class 2 CRISPR-Cas system. Cell 163:759–771

Zetsche B, Heidenreich M, Mohanraju P, Fedorova I, Kneppers J, DeGennaro EM, Winblad N, Choudhury SR, Abudayyeh OO, Gootenberg JS, Wu WY, Scott DA, Severinov K, van der Oost J, Zhang F (2017) Multiplex gene editing by CRISPR-Cpf1 using a single crRNA array. Nat Biotechnol 35:31–34

Chapter 11
Small Self-Cleaving Ribozymes in Eukaryotes

Marcos de la Peña, Olga Rueda, and Amelia Cervera

Abstract Small self-cleaving ribozymes are compact catalytic RNA motifs capable of mediating site-specific self-cleavage. Originally discovered in the minimal circular RNA genomes of plant viroids and viral satellites, they were long considered rare biological curiosities limited to these unusual infectious RNAs. However, subsequent discoveries revealed that these small ribozymes are widespread across all domains of life, including eukaryotic genomes, where they are often embedded within retrotransposons or even within host genes. Their structural diversity reflects a multitude of evolutionary origins and functional adaptations, ranging from genome processing during rolling-circle replication to potential regulatory roles in mRNA maturation, stability, and gene expression in mammals and other vertebrates. Notably, the discovery of retrozymes, a family of non-autonomous retrotransposons with embedded ribozymes that generate abundant circular RNAs, highlights a conserved and efficient natural pathway for circRNA biogenesis in both plants and animals. These findings collectively underscore that small self-cleaving ribozymes are not only ancient and pervasive but also highly adaptable, influencing genome evolution, the mobility of genetic elements, and the regulation of gene expression. As genomic technologies continue to advance, it is likely that additional classes and functions of small ribozymes will be uncovered, further illuminating the fundamental roles of RNA catalysis in biology.

Keywords Ribozyme · circRNA · Viroid · Retrotransposon · Retrozyme

---

M. de la Peña (✉) · O. Rueda · A. Cervera
Instituto de Biología Molecular y Celular de Plantas, UPV-CSIC, Valencia, Spain
e-mail: rivero@ibmcp.upv.es

A. Kanai, A. W. Nicholson (eds.), *Molecular Evolution of RNA Regulatory Enzymes and Their Systems*, Nucleic Acids and Molecular Biology 38,
https://doi.org/10.1007/978-3-032-03300-0_11

11.1 Introduction: The Small Family of Small Ribozymes

Small self-cleaving ribozymes are a group of enigmatic catalytic RNA motifs (~50–200 nucleotides) that mediate intramolecular cleavage of their phosphodiester backbone via an internal phosphoester transfer mechanism. This reaction of self-cleavage is carried out at rates only a few-fold slower than their protein counterparts do, producing RNA fragments with 2′,3′-cyclic phosphate and 5′-hydroxyl termini. As shown in Table 11.1, the major and well-known classes of small ribozymes include the Hammerhead (HHR) (Hutchins et al. 1986; Prody et al. 1986), Hairpin (HPR) (Buzayan et al. 1986), Deltavirus (DVR) (Kuo et al. 1988), Varkud satellite (VSR) (Saville and Collins 1990), GlmS (Winkler et al. 2004), Twister (TWR)(Roth et al. 2014), Twister sister (TSR), Hatchet, and Pistol (Weinberg et al. 2015) ribozymes. More recently, a few novel classes of small ribozymes have also been discovered in mammalian genomes (Chen et al. 2021; Hernandez et al. 2020; Zhang et al. 2024). However, these newly found motifs have not yet been characterized in detail at the biochemical and structural levels (see below). While they all share a common catalytic centre involving nucleophilic attack by a 2′-OH group, their structural diversity likely reflects distinct evolutionary pathways, origins and functional adaptations across time and genomes (Bevilacqua and Yajima 2006; Fedor and Williamson 2005).

Table 11.1 Small ribozymes and their abbreviations

Ribozyme	Abbreviation
Hammerhead	HHR
Discontinuous hammerhead	Dis-HHR
Hairpin	HPR
Deltavirus	DVR
Varkud satellite	VSR
Glucosamine-6-phosphate	glmS
Twister	TWR
Twister sister	TSR
Twister sister-like	TS-like
Hatchet	
Pistol	
B2 and ALU	
Hovlinc	

11.2 Small Ribozymes Were Discovered in Strange Entities Having Minimal Genomes of Circular RNA

The HHR, the first small ribozyme ever reported, was initially discovered in viroids and viral satellites of circular RNA (circRNA) (De la Peña et al. 2017). These agents are a family of minimal infectious molecules (~300 nt) affecting plants, and which are all characterized by a naked genome of circRNA. Since its discovery almost 40 years ago, the HHR has been extensively used as a model catalytic RNA for biochemical, structural, and biological studies. It is composed of ~50 nt that adopt a Y-shaped tertiary structure with three helical stems (I, II, III) radiating from a conserved catalytic core. Depending on the open-ended helix, there are three classical circularly permuted forms, named type I, II or III (Fig. 11.1a) (De la Peña et al. 2017), although other non-circularly permuted motifs are also possible (Eckert et al. 2023; Forgia et al. 2023). The key features of this ribozyme include (i) a 15-nucleotide

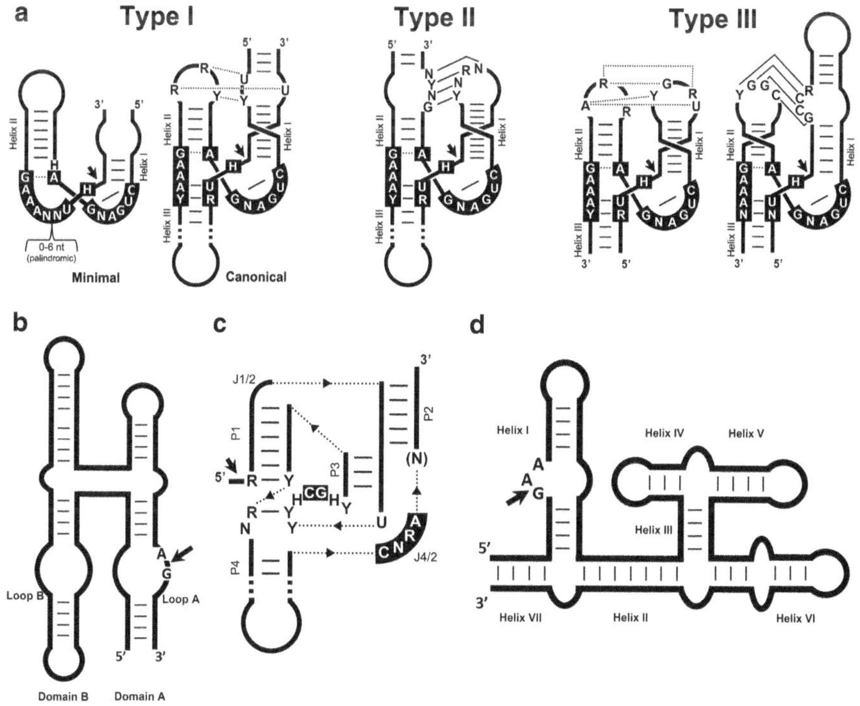

Fig. 11.1 Schematic representations of (**a**) the three major topologies, type I, II and III, described for the Hammerhead ribozyme. Depending on the tertiary interactions between loops, two main classes of HHRs are described for type I (minimal on the left and canonical on the rigth, detected in Penelope transposons and metazoan retrozymes, respectively) and for type III motifs (viroid-like motifs on the left and Delta-like motifs on the right); (**b**) the Hairpin ribozyme; (**c**) the Delta virus ribozyme; and (**d**) the Varkud satellite ribozyme. Sites of self-cleavage are indicated by an arrow. Highly conserved nucleotides are shown in black boxes

conserved region with a cytosine (C17 in type I) acting as a general base, which coordinates a Mg^{2+} ion to stabilize the transition state, and (ii) a tertiary interaction between loops in stems I and II that stabilizes the active conformation, critical for high-speed cleavage (De la Peña et al. 2003; Khvorova et al. 2003) (Fig. 11.1a).

Almost simultaneously with the HHR discovery, a second small ribozyme, the HPR, was also detected in the circRNA genome (minus polarity) of the Satellite of Tobacco Ringspot Virus or satTRSV (Buzayan et al. 1986). HPR folds into a secondary structure comprising two domains: Domain A (helix I—loop A—helix 2), which contains the substrate and the primary substrate-recognition region of the ribozyme, and Domain B (helix 3—loop B—helix 4), which houses the primary catalytic determinants of the ribozyme including a guanine (G8) that acts as a general base. The two domains are covalently joined via a phosphodiester linkage that connects helix 2 to helix 3. These domains must interact with one another for catalysis to occur (Fig. 11.1b).

A third small ribozyme was discovered in 1988 in another atypical infectious agent with a genome of circRNA: the Human Deltavirus. For this reason, the new ribozyme was called the Deltavirus ribozyme or DVR (Kuo et al. 1988). This RNA motif shows five helical segments connected by a nested double pseudoknot and self-cleaves at its 5′ end (Fig. 11.1c). Like the two previous ribozymes detected in circRNAs infecting plants, the main role for DVR is to process genome units of the Human Delta virus during its replication through a rolling circle mechanism.

A final example of the originally discovered ribozymes is the Varkud Satellite ribozyme or VSR, a slightly larger and more complex catalytic RNA (~150 nt) (Fig. 11.1d) found in a circRNA of the *Neurospora* mitochondria (VSRNA), which, however, is encoded by a low-copy DNA plasmid (Saville and Collins 1990). This self-cleaving reaction may be involved in the natural processing pathway of multimeric VSRNAs *in vivo*. Likely due to the discovery of these first four ribozymes in a rare family of molecules such as circular RNAs, small self-cleaving ribozymes were generally regarded for decades as biological oddities restricted to unusual RNAs.

11.3 The First Examples of DNA-Encoded Ribozymes in Eukaryotic Genomes

Small ribozymes present in small replicating circRNAs catalyse the specific RNA self-cleavage of these minimal genomes. The biological role of this activity is to process genome units during the replication of these circRNAs through a classical rolling-circle mechanism (Symons et al. 1987). However, soon after its discovery, examples of HHR motifs were also found encoded in the satellite DNA of newts and salamanders (Epstein and Gall 1987), an unexpected finding whose biological role remained unclear for many years (Cremisi et al. 1992). A few more unanticipated examples of genome-encoded HHRs were also described in other eukaryotes, such as plants — carnation (Daròs and Flores 1995) and, later, *Arabidopsis thaliana* (Przybilski et al. 2005) — and in the repetitive DNA of invertebrates — *Schistosoma mansoni* (Ferbeyre et al. 1998) and cave crickets (Rojas et al. 2000). Previous bioinformatic studies had already identified hundreds of HHR motifs in genomic databases (Ferbeyre et al. 2000). However, these analyses utilized methodologies with highly relaxed sequence and structural constraints, potentially generating a significant number of false positives. With the discovery in 2005 of two canonical type III HHRs in the genome of *A. thaliana* (Przybilski et al. 2005), it was suggested that the full landscape of ribozyme diversity was far from complete, and that deeper bioinformatic analyses could be key to solving this question (Hammann and Westhof 2007).

Among self-cleaving RNAs, a unique motif was reported in 2004: the glmS ribozyme (Winkler et al. 2004). This motif functions as a metabolite-sensing RNA or riboswitch found in the 5′ untranslated region (UTR) of bacterial mRNAs encoding glucosamine-6-phosphate (GlcN6P) synthetase (glmS). Structurally, the ribozyme forms a conserved catalytic core with a pseudoknot arrangement, where GlcN6P acts as a general acid-base catalyst, directly participating in the cleavage reaction. This riboswitch is widely conserved in bacteria but is absent in eukaryotes. However, the discovery of the bacterial thiamine pyrophosphate (TPP)-responsive thi-box riboswitch (Winkler et al. 2002) encoded in eukaryotes such as fungi, algae, or plants (Croft et al. 2007; Sudarsan et al. 2003; Wachter et al. 2007), fuelled the idea that other functional RNA motifs could be present in eukaryotes.

11.4 Examples of Small Ribozyme Motifs Conserved in Mammalian Genomes

An exceptional discovery in the field took place in 2006. Through *in vitro* selection experiments, the laboratory of Jack Szostak reported a genomic DVR encoded in the human genome (Salehi-Ashtiani et al. 2006). The DVR motif mapped to the third intron of the *CPEB3* gene, which encodes the cytoplasmic polyadenylation element-binding protein 3, and was conserved in the same intron of diverse marsupial and placental mammals (Fig. 11.2a). The biological function and mechanism of a self-cleaving motif within a gene-intron is still under research. However, a previous study described an intriguing relationship between the intronic *CPEB3* DVR and human memory performance (Vogler et al. 2009), and more recently, it has been found that the inhibition of the ribozyme increases protein expression and enhances object location memory in mice (Chen et al. 2024).

A second unexpected discovery was the so-called discontinuous hammerhead ribozyme or dis-HHR in a mammalian genome (Martick et al. 2008). Bioinformatic analyses revealed the presence of an atypical HHR in the 3′ untranslated regions of

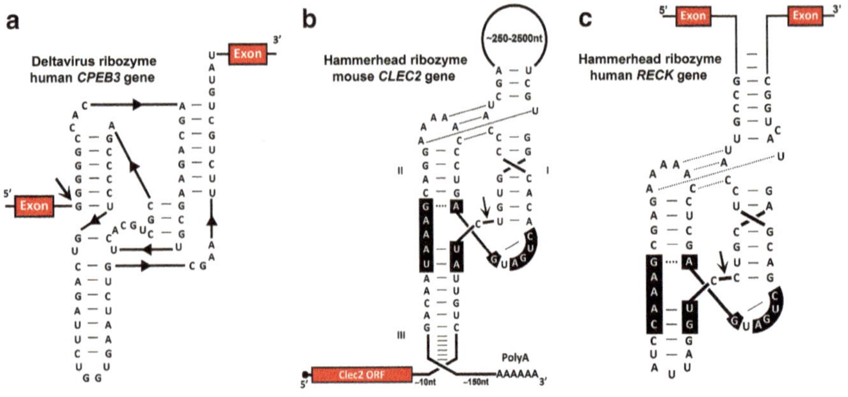

Fig. 11.2 Examples of small ribozymes conserved in mammalian genomes. (**a**) An intronic deltavirus ribozyme present in the human CPEB3 gene. (**b**) A discontinuous hammerhead ribozyme located at the 3' UTR of the mouse CLEC2 genes. (**c**) An intronic hammerhead ribozyme conserved in the human RECK gene and in all homeothermic vertebrates, including birds and mammals (monotremes, marsupials, and placentals). Conserved catalytic nucleotides of the hammerhead ribozyme are shown in black boxes

C-type lectin-like family 2 (CLEC2) genes of diverse mammals (Fig. 11.2b). In this case, the biological role of a self-cleaving ribozyme in the 3' UTR is unknown, but original work using a reporter gene demonstrated that the presence of this dis-HHR in the 3' UTR induced an 80% reduction of the gene expression compared with the same reporter gene carrying a mutated ribozyme (Martick et al. 2008). Most likely, 3′ UTR cleavage through dis-HHR may eventually remove the polyA tail, thus preventing proper nuclear export and, consequently, translation of the mRNA. However, more sophisticated roles in gene expression regulation can be envisaged for a self-cleaving motif in the 3' UTR, such as control of mRNA decay, microRNA (miRNA) binding, or alternative polyadenylation sites, among others (Scott et al. 2009; Scott et al. 2013).

Moreover, similar studies found conserved type I HHRs as intronic motifs in diverse genes of amniotes (reptiles, birds, and mammals), including humans (De la Peña and Garcia-Robles 2010) (Fig. 11.2c). These ribozymes mapped to critical genes such as the tumour suppressor RECK (in all mammals and birds) or DTNB (dystrobrevin beta, in lizards and birds). The analysis of RECK expressed sequence tags revealed fusion events between self-cleaved intronic RNAs and U5/U6 small nuclear RNAs, implicating HHRs in spliceosomal interactions or mRNA biogenesis (Garcia-Robles et al. 2012). The clear sequence and structural similarities of these type I HHRs with those found in repetitive DNA of lower metazoans suggested an exaptation of HHRs from mobile genetic elements (see below) into regulatory roles, with strong selective pressure maintaining their sequences over 300 million years of amniote evolution.

11.5 The Explosion of DNA-Encoded Ribozymes

Our understanding of the distribution of genomic small ribozymes underwent a paradigm shift in 2009 and 2010 with landmark discoveries concerning Deltavirus and hammerhead ribozymes, respectively. First, using structure-based bioinformatics to systematically search genomes and transcriptomes across the tree of life, the team of Andrej Lupták revealed a previously unrecognized diversity of self-cleaving ribozymes with the characteristic double-pseudoknot fold of the DVR motifs (Webb et al. 2009). These ribozymes were present in a wide variety of cellular transcripts, including those of animals, plants, fungi, and protists. Another study demonstrated that many of these ribozymes are located at the 5′-ends of non-long terminal repeat (non-LTR) retrotransposons, and particularly in the R2 family (Eickbush and Eickbush 2010). The study showed that the 5′-untranslated region of the R2 element in *Drosophila* species folds into a DVR motif that catalyses rapid and efficient self-cleavage at the junction between the 28S rRNA and the R2 element. Further expanding the functional significance of genomic DVRs, a new analysis demonstrated their role not only in transcript processing but also in translation initiation of non-LTR retrotransposons, including other Long Interspersed Elements (LINEs), through an unknown mechanism (Ruminski et al. 2011). Together, these findings pointed to a dual role for DVRs in both the processing and translation of non-LTR retrotransposon RNAs, highlighting their evolutionary importance in shaping the expression and mobility of these pervasive genomic elements.

In 2010, new studies extended these results with the discovery of the ubiquitous presence of another small ribozyme, the HHR motif, across all kingdoms of life, including bacteria, plants, and animals (De la Peña and García-Robles 2010; Hammann et al. 2012; Perreault et al. 2011). Using similar sequence and structure-based bioinformatic approaches, researchers identified HHRs in intergenic regions of bacterial genomes and metagenomic data from the Galapagos Islands, but also in the repetitive DNA of eukaryotes like water moulds, corals, molluscs, and insects. Most notably, HHRs were detected in the genomes of 20 plant species, from unicellular algae to vascular plants, often embedded in tandem repeats or telomeric regions. These findings placed HHRs as evolutionarily ancient motifs linked to satellite DNA and transposable elements.

A direct link between HHRs and retroelements deepened with the discovery of minimalist HHR variants in Penelope-like elements (PLEs), an ancient family of eukaryotic retrotransposons present in fungal, protist, plant and metazoan genomes (Arkhipova 2006; Evgen'ev et al. 1997). PLEs, which encode rare telomerase-like reverse transcriptases and even spliceosomal introns, were found to harbour HHRs in their flanking long terminal repeats (PLTRs) (Cervera and de la Peña 2014). These HHRs likely facilitate RNA processing during retrotransposition, in a similar way as DVRs in non-LTR retrotransposons.

Finally, the discovery of the twister ribozyme (TWR) class in 2013 revealed a novel and widespread self-cleaving RNA motif across diverse prokaryotic and eukaryotic genomes, including plants and animals. In eukaryotes, TWR variants are often found embedded in untranslated or repetitive regions, and, as described for DVRs and HHRs, are also related to diverse families of retrotransposons in plants (Martín-Guindal 2018) and animals (Liu et al. 2021). Altogether, these discoveries underscore that self-cleaving ribozymes are far more abundant than previously recognized, with at least three mayor actors (HHR, DVR and TWR) widespread in eukaryotic genomes, and intimately linked to the retrotransposition of DNA.

11.6 Retrozymes: Connecting Genomic Ribozymes, Retroelements and Circular RNAs

In 2016 it was reported that most of the HHR motifs previously detected in plant genomes were part of a novel family of retrotransposons. These new plant Retroelements with ribozymes were named Retrozymes, and found to be a unique and widespread group of small (~1 kb), non-autonomous (non-protein coding) retrotransposons with Long Terminal Repeats (LTR) (Cervera et al. 2016). Their main feature was the presence of a type III HHR embedded within each retrozyme LTR (Fig. 11.3a). These elements, found in a broad range of plant species including eudicots, monocots, ferns, and algae, typically consist of two LTRs (about 350 bp each) flanking an internal region of 300–600 bp. The HHRs catalyse self-cleavage reactions that process the primary transcript into linear and, notably, covalently closed circular RNAs (circRNAs) of ~600–1000 nt. These circRNAs accumulate

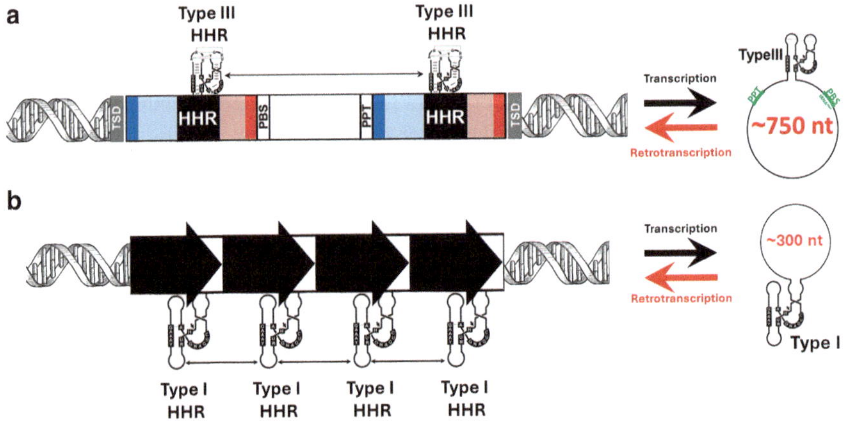

Fig. 11.3 (**a**) Schematic representation of an LTR-retrozyme from plants. The 5′ and 3′ regions of the LTRs are shown in blue and red colors, respectively. Darker blue and red regions correspond to the conserved domains also present in Ty3-gypsy retrotransposons. (**b**) A non-LTR retrozyme from metazoans. Transcription of the plant and metazoan retrozymes results in a circular RNA (right) that, eventually, can be retrotranscribed and integrated in a new locus of the genome. TSD refers to target site duplications, HHR refers to hammerhead ribozyme, PBS refers to primer binding site, and PPT refers to polypurine tract

abundantly, with their expression levels and sequence heterogeneity varying with the tissue and developmental stage. Retrozyme circRNAs are highly structured, with substantial self-complementarity, and their sequences are highly heterogeneous except for the conserved HHR motif and two short domains reminiscent of Ty3-gypsy LTR-retrotransposons (Cervera et al. 2016). Evidence suggests that retrozyme circRNAs may serve as templates for reverse transcription and genomic reintegration, relying on the enzymatic machinery of autonomous LTR retrotransposons of the plant for mobility, such as Ty3-gypsy retrotransposons (Cervera et al. 2016; De la Peña and Cervera 2017). Intriguingly, retrozyme circRNAs may potentially undergo rolling-circle replication, a mechanism reminiscent of viroids and viral satellite RNAs, further linking these genomic elements to the evolutionary origins of infectious plant circRNAs.

There also exist retrozymes in metazoan genomes, which are widespread among invertebrates and, to a lower extent, vertebrates (Cervera and de la Peña 2020). Animal retrozymes represent a different class of non-autonomous, non-LTR retrotransposons characterised by the presence of type I HHRs within short tandem DNA repeats of 170 to 400 bp in length (Fig. 11.3b). These elements have been identified in diverse metazoan phyla, including cnidarians (e.g., corals), molluscs (e.g., mussels), and chordates (e.g., salamanders and axolotls), often existing in thousands to over a hundred thousand genomic copies. Animal retrozymes are found to be actively transcribed in both somatic and, notably, germline tissues, producing abundant linear and circular RNAs. The type I HHRs within these repeats catalyse self-cleavage efficiently, generating monomeric RNAs with 5′-hydroxyl and 2′,3′-cyclic phosphate ends. Circularization of these RNAs is probably mediated by host tRNA ligases rather than by the ribozymes themselves, although the actual mechanism is still unknown. The resulting circRNAs are stable, highly structured, and can constitute a significant fraction of total cellular RNA. Unlike plant retrozymes, animal retrozymes lack LTRs, primer binding sites, or polypurine tracts, suggesting they are mobilized by the machinery of non-LTR retrotransposons such as LINEs or Penelope-like elements. The discovery of retrozymes in animals reveals a conserved and efficient natural pathway for circRNA biosynthesis via self-cleaving ribozymes, suggesting that this mechanism is ancient and pervasive across eukaryotes, with potential implications for genome evolution and the regulation of gene expression.

11.7 How Many more Small Ribozymes Are There?

In 2020, two additional catalytic RNAs encoded in mammalian genomes were reported: the B2 and ALU ribozymes (Hernandez et al. 2020). These two self-cleaving RNA elements derive from two of the most abundant short interspersed nuclear elements (SINEs) in mammals: B2 in mice and ALU in humans. Both RNAs can self-cleave *in vitro*, with ribozyme activity depending on magnesium and monovalent cations. The self-cleavage is intrinsic to the RNA, though they exhibit much lower activity (~100 fold less) compared to classic small motifs such as HHRs or DVRs. However, the Polycomb protein EZH2, known for its role in chromatin modification, binds to B2 and ALU RNAs and accelerates their self-cleavage by more than 100-fold (Hernandez et al. 2020). EZH2 is not a nuclease but promotes a conformation of the RNA that is prone to self-cleavage. Under normal conditions, B2 and ALU RNAs bind to chromatin at stress-responsive genes by RNA-protein interactions, acting as "speed bumps" that retard RNA Polymerase II elongation and suppress transcription. Upon stress (such as heat shock or ER stress), EZH2 is recruited to these genes and induces rapid cleavage of B2/ALU RNAs. This cleavage releases the transcriptional block, enabling a swift upregulation of stress response genes (Hernandez et al. 2020). Although the cleavage mechanism of these RNAs is unclear, it is noteworthy to highlight that the ALU domain adopts a hammerhead-like fold (De la Peña et al. 2003; De la Peña et al. 2009; Weichenrieder et al. 2000). These similarities suggests that all these RNAs may share some catalytic strategies centred around their three-way Y-junction architecture.

Using a human genomic library of small (200-1000 bp) DNA fragments flanked by the T7 promoter, a novel self-cleaving RNA of about 168 nucleotides, named Hovlinc ribozyme, was recently reported (Chen et al. 2021). This catalytic RNA mapped within a very long noncoding RNA (vlincRNA) in humans, chimpanzees, bonobos and even gorillas. However, the hovlinc motif exhibited catalytic inactivity specifically in the gorilla homolog. This ribozyme does not seem to fall into any known class, forming a novel structure with four stem loops and two pseudoknots. Its evolutionary emergence is estimated at 10–13 million years ago, coinciding with the last common ancestor of humans, chimpanzees, and gorillas. The biological function of Hovlinc remains unclear, but its presence in hominids and absence of activity in gorillas raises intriguing evolutionary questions.

More recently, minimal twister sister-like (TS-like) self-cleaving ribozymes encoded in the human genome have been reported in a study using deep mutational scanning and covariance analysis (Zhang et al. 2024). Structurally, these ribozymes are similar to the catalytic core of the twister sister ribozyme or TSR family, featuring two nearly identical catalytic internal loops and a lantern-like shape. However, they are circularly permuted and lack some peripheral elements found in canonical TSRs, which affect their stability and catalytic efficiency. Despite these differences, their sequence and secondary structure homology strongly support their classification as TS-like ribozymes. The exact physiological roles of these minimal ribozymes have not yet been established. However, their conservation, abundance (especially in LINE-1 elements), and robust *in vitro* self-cleaving activity suggest that they may play a role in RNA processing or regulation, possibly influencing retrotransposon mobility.

11.8 Conclusions

Over the past several decades, extensive biochemical, structural, and bioinformatic analyses have uncovered multiple classes of small self-cleaving ribozymes from their original discovery in minimal genomes of circRNA to all domains of life. These studies are moving the family of the simplest catalytic RNAs to a complex landscape of motifs encoded in DNA genomes, ranging from simple bacteriophages to humans. In eukaryotes, the characterized ribozymes show a remarkable structural and functional diversity, far beyond their initial discovery in the processing of minimal genomes during rolling circle replication. They have been notably found to be closely linked to diverse retrotransposons where they have multiple functions, from RNA processing to RNA translation initiation through unknown mechanisms. Moreover, in diverse cases they seem to have been exapted into regulatory roles within host genes and genomes, influencing processes such as mRNA maturation, stability, and gene expression. The identification of diverse mammalian ribozymes highlights the ongoing expansion of this field and suggests that more functional ribozymes remain to be discovered. These findings underscore the evolutionary adaptability of catalytic RNAs and their influence on genome dynamics and gene regulation. As new technologies enable deeper exploration of genomic and transcriptomic data, the landscape of small self-cleaving ribozymes will likely continue to grow, offering new insights into the fundamental roles of RNA catalysis in biology.

References

Arkhipova IR (2006) Distribution and phylogeny of Penelope-like elements in eukaryotes. Syst Biol 55:875–885

Bevilacqua PC, Yajima R (2006) Nucleobase catalysis in ribozyme mechanism. Curr Opin Chem Biol 10:455–464. https://doi.org/10.1016/J.CBPA.2006.08.014

Buzayan JM, Gerlach WL, Bruening G (1986) Satellite tobacco ringspot virus RNA: a subset of the RNA sequence is sufficient for autolytic processing. Proc Natl Acad Sci USA 83:8859–8862

Cervera A, de la Peña M (2014) Eukaryotic penelope-like retroelements encode hammerhead ribozyme motifs. Mol Biol Evol 31:2941–2947. https://doi.org/10.1093/molbev/msu232

Cervera A, de la Peña M (2020) Small circRNAs with self-cleaving ribozymes are highly expressed in diverse metazoan transcriptomes. Nucleic Acids Res 48:5054–5064. https://doi.org/10.1093/nar/gkaa187

Cervera A, Urbina D, de la Peña M (2016) Retrozymes are a unique family of non-autonomous retrotransposons with hammerhead ribozymes that propagate in plants through circular RNAs. Genome Biol 17:135. https://doi.org/10.1186/s13059-016-1002-4

Chen Y, Qi F, Gao F, Cao H, Xu D, Salehi-Ashtiani K, Kapranov P (2021) Hovlinc is a recently evolved class of ribozyme found in human lncRNA. Nat Chem Biol 17(5):601–607. https://doi.org/10.1038/s41589-021-00763-0

Chen CC, Han J, Chinn CA, Rounds JS, Li X, Nikan M, Myszka M, Tong L, Passalacqua LFM, Bredy T, Wood MA, Luptak A (2024) Inhibition of Cpeb3 ribozyme elevates CPEB3 protein expression and polyadenylation of its target mRNAs and enhances object location memory. elife 13. https://doi.org/10.7554/ELIFE.90116

Cremisi F, Scarabino D, Carluccio MA, Salvadori P, Barsacchi G (1992) A newt ribozyme: a catalytic activity in search of a function. Proc Natl Acad Sci USA 89:1651–1655
Croft MT, Moulin M, Webb ME, Smith AG (2007) Thiamine biosynthesis in algae is regulated by riboswitches. Proc Natl Acad Sci USA 104:20770–20775. https://doi.org/10.1073/pnas.0705786105
Daròs JA, Flores R (1995) Identification of a retroviroid-like element from plants. Proc Natl Acad Sci USA 92:6856–6860. https://doi.org/10.1073/pnas.92.15.6856
De la Peña M, Cervera A (2017) Circular RNAs with hammerhead ribozymes encoded in eukaryotic genomes: the enemy at home. RNA Biol 14:985–991. https://doi.org/10.1080/15476286.2017.1321730
De la Peña M, Garcia-Robles I (2010) Intronic hammerhead ribozymes are ultraconserved in the human genome. EMBO Rep 11:711–716. https://doi.org/10.1038/embor.2010.100
De la Peña M, García-Robles I (2010) Ubiquitous presence of the hammerhead ribozyme motif along the tree of life. RNA 16:1943–1950. https://doi.org/10.1261/rna.2130310
De la Peña M, Gago S, Flores R (2003) Peripheral regions of natural hammerhead ribozymes greatly increase their self-cleavage activity. EMBO J 22:5561–5570. https://doi.org/10.1093/emboj/cdg530
De la Peña M, Dufour D, Gallego J (2009) Three-way RNA junctions with remote tertiary contacts: a recurrent and highly versatile fold. RNA 15:1949–1964. https://doi.org/10.1261/rna.1889509
De la Peña M, Garcia-Robles I, Cervera A (2017) The hammerhead ribozyme: a long history for a short RNA. Molecules 22:1–11. https://doi.org/10.3390/molecules22010078
Eckert I, Friedrich R, Weinberg CE, Weinberg Z (2023) Discovery of natural non-circular permutations in non-coding RNAs. Nucleic Acids Res 51:2850–2861. https://doi.org/10.1093/NAR/GKAD137
Eickbush DG, Eickbush TH (2010) R2 retrotransposons encode a self-cleaving ribozyme for processing from an rRNA cotranscript. Mol Cell Biol 30:3142–3150. https://doi.org/10.1128/MCB.00300-10
Epstein LM, Gall JG (1987) Self-cleaving transcripts of satellite DNA from the newt. Cell 48:535–543
Evgen'ev MB, Zelentsova H, Shostak N, Kozitsina M, Barskyi V, Lankenau DH, Corces VG (1997) Penelope, a new family of transposable elements and its possible role in hybrid dysgenesis in Drosophila virilis. Proc Natl Acad Sci USA 94:196–201
Fedor MJ, Williamson JR (2005) The catalytic diversity of RNAs. Nat Rev Mol Cell Biol 6:399–412. https://doi.org/10.1038/NRM1647
Ferbeyre G, Smith JM, Cedergren R (1998) Schistosome satellite DNA encodes active hammerhead ribozymes. Mol Cell Biol 18:3880–3888
Ferbeyre G, Bourdeau V, Pageau M, Miramontes P, Cedergren R (2000) Distribution of hammerhead and hammerhead-like RNA motifs through the GenBank. Genome Res 10:1011–1019
Forgia M, Navarro B, Daghino S, Cervera A, Gisel A, Perotto S, Aghayeva DN, Akinyuwa MF, Gobbi E, Zheludev IN, Edgar RC, Chikhi R, Turina M, Babaian A, Di Serio F, de la Peña M (2023) Hybrids of RNA viruses and viroid-like elements replicate in fungi. Nat Commun 14(1):1–11. https://doi.org/10.1038/s41467-023-38301-2
Garcia-Robles I, Sanchez-Navarro J, de la Peña M (2012) Intronic hammerhead ribozymes in mRNA biogenesis. Biol Chem 393:1317–1326. https://doi.org/10.1515/hsz-2012-0223
Hammann C, Westhof E (2007) Searching genomes for ribozymes and riboswitches. Genome Biol 8:1–11. https://doi.org/10.1186/GB-2007-8-4-210/FIGURES/5
Hammann C, Luptak A, Perreault J, de la Peña M (2012) The ubiquitous hammerhead ribozyme. RNA 18:871–885. https://doi.org/10.1261/rna.031401.111
Hernandez AJ, Zovoilis A, Cifuentes-Rojas C, Han L, Bujisic B, Lee JT (2020) B2 and ALU retrotransposons are self-cleaving ribozymes whose activity is enhanced by EZH2. Proc Natl Acad Sci USA 117:415–425

Hutchins CJ, Rathjen PD, Forster AC, Symons RH (1986) Self-cleavage of plus and minus RNA transcripts of avocado sunblotch viroid. Nucleic Acids Res 14:3627–3640. https://doi.org/10.1093/nar/14.9.3627

Khvorova A, Lescoute A, Westhof E, Jayasena SD (2003) Sequence elements outside the hammerhead ribozyme catalytic core enable intracellular activity. Nat Struct Biol 10:708–712. https://doi.org/10.1038/nsb959. nsb959[pii]

Kuo MY, Sharmeen L, Dinter-Gottlieb G, Taylor J (1988) Characterization of self-cleaving RNA sequences on the genome and antigenome of human hepatitis delta virus. J Virol 62:4439–4444

Liu G, Jiang H, Sun W, Zhang J, Chen D, Murchie AIH (2021) The function of twister ribozyme variants in non-LTR retrotransposition in Schistosoma mansoni. Nucleic Acids Res 49:10573–10588. https://doi.org/10.1093/NAR/GKAB818

Martick M, Horan LH, Noller HF, Scott WG (2008) A discontinuous hammerhead ribozyme embedded in a mammalian messenger RNA. Nature 454:899–902. https://doi.org/10.1038/nature07117

Martín-Guindal, A. 2018. Análisis de RNAs catalíticos en retroelementos de plantas. TFM UPV. https://riunet.upv.es/handle/10251/97643

Perreault J, Weinberg Z, Roth A, Popescu O, Chartrand P, Ferbeyre G, Breaker RR (2011) Identification of hammerhead ribozymes in all domains of life reveals novel structural variations. PLoS Comput Biol 7:e1002031. https://doi.org/10.1371/journal.pcbi.1002031

Prody GA, Bakos JT, Buzayan JM, Schneider IR, Bruening G (1986) Autolytic processing of dimeric plant virus satellite RNA. Science 1979(231):1577–1580. https://doi.org/10.1126/science.231.4745.1577

Przybilski R, Graf S, Lescoute A, Nellen W, Westhof E, Steger G, Hammann C (2005) Functional hammerhead ribozymes naturally encoded in the genome of Arabidopsis thaliana. Plant Cell 17:1877–1885. https://doi.org/10.1105/tpc.105.032730

Rojas AA, Vazquez-Tello A, Ferbeyre G, Venanzetti F, Bachmann L, Paquin B, Sbordoni V, Cedergren R (2000) Hammerhead-mediated processing of satellite pDo500 family transcripts from Dolichopoda cave crickets. Nucleic Acids Res 28:4037–4043

Roth A, Weinberg Z, Chen AG, Kim PB, Ames TD, Breaker RR (2014) A widespread self-cleaving ribozyme class is revealed by bioinformatics. Nat Chem Biol 10:56–60. https://doi.org/10.1038/nchembio.1386

Ruminski DJ, Webb CHT, Riccitelli NJ, Lupta A, Luptak A (2011) Processing and translation initiation of non-long terminal repeat retrotransposons by hepatitis delta virus (HDV)-like self-cleaving ribozymes. J Biol Chem 286:41286–41295. https://doi.org/10.1074/jbc.M111.297283

Salehi-Ashtiani K, Luptak A, Litovchick A, Szostak JW (2006) A genomewide search for ribozymes reveals an HDV-like sequence in the human CPEB3 gene. Science 1979(313):1788–1792. https://doi.org/10.1126/science.1129308

Saville BJ, Collins RA (1990) A site-specific self-cleavage reaction performed by a novel RNA in Neurospora mitochondria. Cell 61:685–696

Scott WG, Martick M, Chi YI (2009) Structure and function of regulatory RNA elements: ribozymes that regulate gene expression. Biochim Biophys Acta 1789:634–641. https://doi.org/10.1016/j.bbagrm.2009.09.006

Scott WG, Horan LH, Martick M (2013) The hammerhead ribozyme: structure, catalysis, and gene regulation. Prog Mol Biol Transl Sci 120:1–23. https://doi.org/10.1016/B978-0-12-381286-5.00001-9

Sudarsan N, Barrick JE, Breaker RR (2003) Metabolite-binding RNA domains are present in the genes of eukaryotes. RNA 9:644–647. https://doi.org/10.1261/RNA.5090103

Symons RH, Hutchins CJ, Forster AC, Rathjen PD, Keese P, Visvader JE (1987) Self-cleavage of RNA in the replication of viroids and virusoids. J Cell Sci Suppl 7:303–318

Vogler C, Spalek K, Aerni A, Demougin P, Muller A, Huynh KD, Papassotiropoulos A, de Quervain DJ (2009) CPEB3 is associated with human episodic memory. Front Behav Neurosci 3:4. https://doi.org/10.3389/neuro.08.004.2009

Wachter A, Tunc-Ozdemir M, Grove BC, Green PJ, Shintani DK, Breaker RR (2007) Riboswitch control of gene expression in plants by splicing and alternative 3′ end processing of mRNAs. Plant Cell 19:3437. https://doi.org/10.1105/TPC.107.053645

Webb CHT, Riccitelli NJ, Ruminski DJ, Lupták A (2009) Widespread occurrence of self-cleaving ribozymes. Science 326:953. https://doi.org/10.1126/science.1178084

Weichenrieder O, Wild K, Strub K, Cusack S (2000) Structure and assembly of the Alu domain of the mammalian signal recognition particle. Nature 408:167–173. https://doi.org/10.1038/35041507

Weinberg Z, Kim PB, Chen TH, Li S, Harris KA, Lünse CE, Breaker RR (2015) New classes of self-cleaving ribozymes revealed by comparative genomics analysis. Nat Chem Biol 11:606–610. https://doi.org/10.1038/nchembio.1846

Winkler W, Nahvi A, Breaker RR (2002) Thiamine derivatives bind messenger RNAs directly to regulate bacterial gene expression. Nature 419:952–956. https://doi.org/10.1038/nature01145

Winkler WC, Nahvi A, Roth A, Collins JA, Breaker RR (2004) Control of gene expression by a natural metabolite-responsive ribozyme. Nature 428:281–286. https://doi.org/10.1038/nature02362

Zhang Z, Hong X, Xiong P, Wang J, Zhou Y, Zhan J (2024) Minimal twister sister-like self-cleaving ribozymes in the human genome revealed by deep mutational scanning. elife 12. https://doi.org/10.7554/ELIFE.90254

GPSR Compliance

The European Union's (EU) General Product Safety Regulation (GPSR) is a set of rules that requires consumer products to be safe and our obligations to ensure this.

If you have any concerns about our products, you can contact us on ProductSafety@springernature.com

In case Publisher is established outside the EU, the EU authorized representative is:

Springer Nature Customer Service Center GmbH
Europaplatz 3
69115 Heidelberg, Germany

Batch number: 09544921

Printed by Printforce, the Netherlands